Project Management for Engineering and Construction

Garold D. Oberlender, Ph.D., P.E.
Professor Emeritus of Civil Engineering
Oklahoma State University

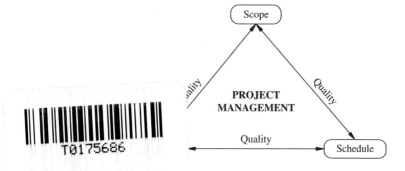

Third Edition

New York   Chicago   San Francisco   Athens
London   Madrid   Mexico City   Milan
New Delhi   Singapore   Sydney   Toronto

Contents

Chapter 11. Construction Phase   305

ABOUT THE AUTHOR

GAROLD D. OBERLENDER is professor emeritus of civil engineering at Oklahoma State University, where he served for 35 years as coordinator of the graduate program in construction engineering and project management. Prior to joining the faculty at Oklahoma State University, he lived nine years in the Dallas area and worked in the engineering and construction industry. He holds B.S., M.S., and Ph.D. degrees in civil engineering.

In addition to authoring this book, Dr. Oberlender is the coauthor with Robert L. Peurifoy of *Estimating Construction Costs*, Sixth Edition, and *Formwork for Concrete Structures*, Fourth Edition. His books are adopted by universities around the world and widely used by practicing civil engineers.

Dr. Oberlender is a fellow member of the American Society of Civil Engineers (ASCE) and has been inducted into the National Academy of Construction (NAC). He is also a fellow member of the National Society of Professional Engineers (NSPE). He has served as chairman of the Construction Engineering Division of the American Society for Engineering Education (ASEE). Currently he is a writer of construction engineering PE exam questions for the National Council of Examiners for Engineering and Surveying (NCEES).

Dr. Oberlender is frequently an invited speaker on the subject of project engineering and management by companies in industry and professional and technical organizations in the United States and other countries. He served on the Academic Council of the Construction Industry Institute (CII) and was principal investigator for several CII research teams. He was selected for the CII Outstanding Researcher Award for his research on improving early estimates.

In addition to his teaching and research, he maintained a consulting engineering practice for engineering and construction projects in the petrochemical and electrical power industries. Dr. Oberlender is a registered professional engineer in Oklahoma and Texas and a member of numerous honorary societies, including Chi Epsilon, Tau Beta Pi, Sigma Xi, and Phi Kappa Phi.

Preface

This book presents the principles and techniques of managing engineering and construction projects from the initial conceptual phase, through design and construction, to completion. It emphasizes project management during the early stages of project development because the ability to influence the quality, cost, and schedule of a project can best be achieved during the early stage of development. Most books discuss project management during construction, after the scope of work is fully defined, the budget is fixed, and the completion date is firm. It is then too late to make any significant adjustments to improve the quality, cost, or schedule of the project.

Although each project is unique, there is certain information that must be identified and organized at the beginning of a project, before any work is started. Numerous tables and graphs are presented and discussed throughout this book to provide guidelines for management of the three basic components of a project: scope, budget, and schedule. Throughout this book, achieving project quality to meet the owner's satisfaction is emphasized as an integral part of project management.

This third edition has a new chapter that addresses risk management. This topic is extremely important because owners, designers, and contractors are all exposed to risk from the start of a project through its completion. Risk assessment, analysis, and mitigation are key factors in project management of engineering and construction projects.

In preparing this third edition, the author has updated example problems in all chapters and added examples in many chapters. New sections have been added, including: ensuring quality in a project, the owner's team, the importance of the estimator, formats for work breakdown structures, design work packages, benefits of planning, and build-operate-transfer delivery methods.

The intended audience of this book is engineers in industry who aid the owner in the feasibility study, coordinate the design effort, and witness construction in the field. It is also intended for students of university programs in engineering, architecture, and construction because graduates of these programs usually are involved in project management as they advance in their careers.

This book is based on the author's experience in working with hundreds of project managers in the engineering and construction industry. Much of the material in this book is based on formal and informal discussions with these project managers, who are actively involved in the practice of project management. Although the author has observed that no two project managers operate exactly the same, there are common elements that apply to all projects and all project managers. The author presents these common elements of effective project management that have been successfully applied in practice.

The author would like to thank Glenn Barin and Rock Spencer for their careful review, helpful comments, and advice in the development of the new risk management chapter in this third edition. The author would also like to thank the many project managers in industry who have shared their successes, and problems, and who have influenced the author's thoughts in the development of this book.

Finally, the author greatly appreciates the patience and tolerance of his wife, Jana, and her support and encouragement during the writing and editing phases of the third edition of this book.

Garold D. Oberlender, Ph.D., P.E.

Introduction

Purpose of This Book

The purpose of this book is to present the principles and techniques of project management beginning in the conceptual phase by the owner, through coordination of design and construction, to project completion. Emphasis is placed on managing the project in its early stage of development, during the owner's study and design. It is presented from this perspective because the ability to influence the overall quality, cost, and schedule of a project can best be achieved early in the life of a project. Most books and articles discuss project management during the construction phase, after design is completed. At this time in the life of a project the scope of work is fully defined, the budget is fixed, and the completion date is firm. It is then too late to make any significant adjustments to improve the quality, cost, or schedule of the project.

Experienced project managers agree that the procedures used for project management vary from company to company and even among individuals within a company. Although each manager develops his or her own style of management, and each project is unique, there are basic principles that apply to all project managers and projects. This book presents these principles and illustrates the basic steps, and sequencing of steps, to develop a work plan to manage a project through each phase from conceptual development to completion.

Project management requires teamwork among the three principal contracting parties: the owner, designer, and contractor. The coordination of the design and construction of a project requires planning and organizing a team of people who are dedicated to a common goal of completing the project for the owner. Even a small project involves a large number of people who work for different organizations. The key to a successful project is the selection and coordination of people who have the ability to detect and solve problems to complete the project.

Throughout this book the importance of management skills is emphasized to enable the user to develop his or her own style of project management. The focus is to apply project management at the beginning of the project, when it is first approved. Too often the formal organization to manage a project is not developed until the beginning of the construction phase. This book presents the information that must be assembled and managed during the development and engineering design phase to bring a project to successful completion for use by the owner.

The intended audience of this book is students enrolled in university programs in engineering and construction. It is also intended for the design firms which aid the owner in the feasibility study, coordinate the design effort, and witness construction in the field. This book is also for persons in the owner's organization who are involved in the design and construction process.

Arrangement of This Book

A discussion of project management is difficult because there are many ways a project can be handled. The design and/or construction of a project can be performed by one or more parties. Regardless of the method that is used to handle a project, the management of a project generally follows these steps:

Step 1: Project Definition (to meet the needs of the end user)
Intended use by the owner upon completion of construction
Conceptual configurations and components to meet the intended use

Step 2: Project Scope (to meet the project definition)
Define the work that must be accomplished
Identify the quantity, quality, and tasks that must be performed

Step 3: Project Budgeting (to match the project definition and scope)
Define the owner's permissible budget
Determine direct and indirect costs plus contingencies

Step 4: Project Planning (the strategy to accomplish the work)
Select and assign project staffing
Identify the tasks required to accomplish the work

Step 5: Project Scheduling (the product of scope, budgeting, and planning)
Arrange and schedule activities in a logical sequence
Link the costs and resources to the scheduled activities

Step 6: Project Tracking (to ensure the project is progressing as planned)
Measure work, time, and costs that are expended
Compare "actual" to "planned" work, time, and cost
Step 7: Project Close Out (final completion to ensure owner satisfaction)
Perform final testing and inspection, archive documents, and confirm payments
Turn over the project to the owner

These steps describe project management in its simplest form. In reality there is considerable overlap between the steps, because any one step may affect one or more other steps. For example, budget preparation overlaps project definition and scope development. Similarly, project scheduling relates project scope and budget to project tracking and control.

The topic of project management is further complicated because the responsibility for these steps usually involves many parties. Thus, the above steps must all be integrated together to successfully manage a project. Subsequent chapters of this book describe each of these steps.

Chapter 1 defines general principles related to project management. These basic principles must be fully understood because they apply to all the remaining chapters. Many of the problems associated with project management are caused by failure to apply the basic management principles that are presented in Chapter 1.

Chapter 2, Working with Project Teams, presents the human aspects of project management. The project team is a group of diverse individuals, each with a special expertise, that performs the work necessary to complete the project. As leader of the project team, the project manager acts as a coach to answer questions and to make sure the team understands what is expected of them and the desired outcome of the project.

Chapter 3, Project Initiation, presents material that is generally performed by the owner. However, the owner may contract the services of a design organization to assist with the feasibility study of a project. The project manager should be involved at the project development or marketing phase to establish the scope. This requires input from experienced technical people that represent every aspect of the proposed project.

Chapter 4, Early Estimates, presents the techniques and processes of preparing estimates in the early phase of a project. Preparation of early estimates is a prerequisite to project budgeting. For engineering and construction projects, the early cost estimate is used by the owner in making economic decisions to approve the project. The early cost estimate is a key project parameter for cost control during the design process.

Chapter 5, Project Budgeting, applies to all parties in a project: the owner, designer, and contractor. The budget must be linked to the quantity, quality, and schedule of the work to be accomplished. A change in scope or schedule almost always affects the budget, so the project manager must continually be alert to changes in a project and to relate any changes to the budget.

Chapter 6, Development of Work Plan, applies to the project manager who is responsible for management of the design effort. Generally, he or she is employed by the professional design organization, which may be an agency of the owner or under contract by the owner to perform design services. The material presented in this chapter is important because it establishes the work plan which is the framework for guiding the entire project effort. The information in this chapter relates to all the project management steps and chapters of this book.

Chapter 7, Design Proposals, presents the process of preparing proposals from the design organization to the owner. After the owner has defined the goals, objectives, intended use, and desired outcome of the project, a request for proposals is solicited from the design organization. The design organization must convert the owner's expectations of the project into an engineering scope of work, budget, and schedule.

Chapter 8, Project Scheduling, provides the base against which all activities are measured. It relates the work to be accomplished to the people who will perform the work as well as to the budget and schedule. Project scheduling cannot be accomplished without a well-defined work plan, as described in Chapter 6, and it forms the basis for project tracking, as described in Chapter 9.

Chapter 9, Tracking Work, cannot be accomplished without a well-defined work plan, as described in Chapter 6, and a detailed schedule, as described in Chapter 8. This chapter is important because there is always a tendency for scope growth, cost overrun, or schedule delays. A control system must simultaneously monitor the three basic components of a project: the work accomplished, the budget, and the schedule. These three components must be collectively monitored, not as individual components, because a change in any one component usually will affect the other two components.

Chapter 10, Design Coordination, applies to the project manager of the design organization. The quality, cost, and schedule of a project is highly dependent on the effectiveness of the design effort. The end result of the design process is to produce plans and specifications in a timely manner that meet the intended use of the project by the owner. The product of design must be within the owner's approved budget and schedule and must be constructable by the construction contractor.

Chapter 11, Construction Phase, is important because most of the cost of a project is expended in the construction phase, and the quality of the final project is highly dependent upon the quality of work that is performed by the construction contractors. Most of the books that have been written on project management have been directed toward a project in the construction phase. This book emphasizes project management from the initial conception of the project by the owner, through coordination of design and development of the construction documents, and into the construction phase until project close out.

Chapter 12, Project Close Out, discusses the steps required to complete a project and turn it over to the owner. This is an important phase of a project because the owner will have expended most of the budget for the project, but will not receive any benefits from the expenditures until it is completed and ready for use. Also it is sometimes difficult to close a project because there are always many small items that must be finished.

Chapter 13, Personal Management Skills, addresses the human aspects of project management. Although the primary emphasis of this book is on the techniques of project management, it is the project manager working with his or her people who ensures the successful completion of a project.

Chapter 14, Risk Management, presents the process of assessing potential risks and methods of analyzing risks in engineering and construction projects. The design and construction is a risk endeavor through all phases of a project, beginning with the owner's feasibility and economic study, through preliminary engineering and final design, procurement of materials and equipment, and the many field operations during construction.

Definition of a Project

A project is an endeavor that is undertaken to produce the results that are expected from the requesting party. For this book a project may be design only, construction only, or a combination of design and construction. A project consists of three components: scope, budget, and schedule. When a project is first assigned to a project manager, it is important that all three of these components be clearly defined. Throughout this book, the term *Scope* represents the work to be accomplished, that is, the quantity and quality of work. *Budget* refers to costs, measured in dollars and/or labor-hours of work. *Schedule* refers to the logical sequencing and timing of the work to be performed. The quality of a project must meet the owner's satisfaction and is an integral part of project management as illustrated in Figure 1-1.

Figure 1-1 is shown as an equilateral triangle to represent an important principle of project management: a balance is necessary between the scope, budget, and schedule. For any given project, there is a certain amount of work that must be performed and an associated cost and schedule for producing the work. Any increase in the scope of work requires a corresponding increase in budget and schedule. Conversely, any decrease in scope of work results in a corresponding decrease in budget and schedule. This principle applies between any and all of the three components of a project. For example, any adjustment in budget

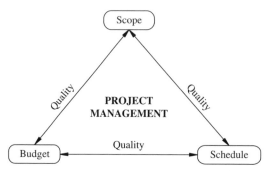

Figure 1-1 Quality is an integral part of scope, budget, and schedule.

and/or schedule requires a corresponding adjustment in scope. This simple concept of a balance between scope, budget, and schedule is sometimes not fully recognized during early project development as well as during design and construction.

The source of many problems associated with a project is failure to properly define the project scope. Too often the focus is just on budget or schedule. Not only should the scope, budget, and schedule be well defined, but each must be linked together since one affects the other, both individually and collectively.

Since the project scope defines the work to be accomplished, it should be the first task in the development of a project, prior to the development of either the budget or the schedule. Experienced project managers agree that the budget and schedule are derived from the scope. Too often, top management specifies a project budget or schedule and then asks the project team to define a scope to match the budget. This is the reverse order of defining a project and is not a good project management practice. It is the duty of a project manager to ensure that the project scope, budget, and schedule are linked together.

Budgeting is important because it establishes the amount of money the owner will spend to obtain the project and the amount of money that the design and construction organizations will be compensated for performing the work. Each party is concerned about project cost overrun because it adversely affects profitability and creates adverse relationships between the parties.

Scheduling is important because it brings together project definition, people, cost, resources, timing, and methods of performing work to define the logical sequencing of activities for the project. The schedule is the final product of scope definition, budgeting, and planning and forms the base against which all activities are measured. Project tracking and control cannot be accomplished without a good plan and schedule.

Quality is an element that is integrated into and between all parts of a project: scope, budget, and schedule. It should not be construed as merely creating drawings with a minimum number of errors, furnishing equipment that meets specifications, or building a project to fulfill the requirements of a contract. Certainly these factors are a part of quality, but it involves much more. Quality is meeting the needs and satisfaction of the ultimate end user of the project, the owner.

Quality is the responsibility of all participants in a project, including all levels of management and workers in each of the principal parties. An attitude of achieving quality must be instilled in everyone and perpetuate throughout the work environment. The attitude should not be "what can we do to pass quality control or final inspection?" Instead, it should be "what can we do to improve our work and what is the best way we can furnish a project that meets the needs and satisfaction of the owner?"

Ensuring Quality in a Project

Ensuring quality is the responsibility of all parties in a project: the owner, designer, and contractor. It is the owner's duty to define and communicate the desired level of quality in a project. The designer is responsible for producing contract documents that ensures the owner's level of quality will be achieved during construction. The contractor is responsible for providing the materials, equipment, and labor skills to ensure the work will be performed in accordance with the design drawings and specifications.

Early in the project, before contracts are signed and commencing design, the owner's organization needs to hold a meeting to discuss, and agree upon, the end product and the level of quality the project is to provide. Although upper management in the owner's organization is responsible for final approval of the level of quality, it is the end user of the project in the owner's organization that should be involved in defining the level of quality. The end user is the party that will use and maintain the project after it is completed. Members of the owner's staff that should be involved in defining quality should include financing, engineering, operations, and maintenance personnel. This group must separate "what the owner needs" from "what the owner wants," while ensuring the completed project can be used as intended by the owner. If the owner does not have staff that is able to identify the desired quality, it may be necessary to involve the design team or outside consultants to assist the owner in defining the level of quality that can be achieved within the constraints of budget and schedule. It is important to maintain continuous communications with the owner throughout the project, so the owner understands the total affect quality can have on the cost and schedule of a project.

The design organization plays a major role in quality. The owner's level of quality must be communicated to the design team, so the proper level of quality is translated into the design drawings and specifications. The project manager of the design team must ensure the level of quality is clearly defined in the designer's project proposal, which requires the designer to explain "design quality" in terms the owner can understand. The designer must develop a complete, and as error free as possible, set of contract documents that ensure the level of quality that is expected from the owner. Regularly scheduled meetings should be held to ensure the design team is on track regarding the expected level of quality, because it is often too easy for project participants to get distracted by costs and schedule, and forget that quality is a "must" in a project. There needs to be a thorough review of plans and specifications by experienced field construction personnel, before release to the contractor, to ensure constructability of the design. The design team should remain involved in the project during construction, to ensure the plans and specifications are being turned into a quality project during construction.

To ensure quality the contractor must have a quality control system that is workable and understood by all parties. To work effectively, any quality control system requires a buy-in by all parties involved in a project. Quality control should concentrate on those parts of a project where quality is important and not get consumed in expending large amounts of effort inspecting work that has only a small impact on the performance of a project when it is completed and in use by the owner. Contractors, subcontractors, and suppliers should be carefully selected based on their record of quality work, rather than selecting them solely on the lowest cost they quote. A work environment should be established that promotes quality and pride in workmanship among all parties in the project. There must be sincere interest in the importance of quality and it should be expected in all aspects of the project. Management needs to instill an attitude of achieving quality work by everyone in the project by recognizing and rewarding good quality work. Regularly held meetings need to be scheduled to address quality, just like meetings that are held to review budgets, schedules, and safety. Quality should be unique, tied to each specific project, rather than a standardization.

Responsibilities of Parties

Each of the three principal parties in a project has a role to fulfill in the various phases of design development and construction. A team approach between the owner, designer, and contractor must be created with a cooperative relationship to complete the project in the most efficient manner. Too often an adverse relationship develops that does not serve the best interest of anyone.

The owner is responsible for setting the operational criteria and level of quality for the completed project. Examples are usage of a building, barrels per day of crude oil to be refined, millions of cubic feet per hour of gas to be transported in a pipeline, and so on. Any special equipment, material, or company standards that are to apply to the project must also be defined. Owners also need to identify their level of involvement in the project, such as, the review process, required reports, and the levels of approval. The owner is also responsible for setting parameters on total cost, payment of costs, major milestones, and the project completion date.

Many problems can be created when owners fail to fulfill their responsibility of clearly defining the operational criteria of a project. If the operational criteria is not clearly defined, it will have to be defined as the project progresses through design and/or into construction, which causes confusion for all parties. The owner may end up paying for things in a project that are not needed, because participants in the project do not clearly understand the operational criteria of the project. The owner runs the risk of receiving a project that is within approved budget and on schedule, but may not be useable as intended. The designer may over-design the project, trying to second-guess what the owner wants, which may cause the project to be more expensive than necessary. Or, the designer may under-design the project, not realizing what the owner expects in the project, which can lead to expensive additions during construction. Changes of the owner's operational criteria during design cause rework for the designer, which will increase the cost of the design effort and may adversely affect the budget and schedule of the project.

The probability of rework during construction is greatly increased when the owner has not adequately defined the operational criteria of the project. Timely responses to contractors questions may be impaired because the owner is not sure of what they want in the project. The number of change orders are likely to increase, which may result in cost overruns, schedule delays, and claims against the owner by the contractor. The productivity and morale of construction workers may be impaired because of continuous changes by the owner to adjust the project during construction so it will meet the operational needs of the owner.

The designer is responsible for producing design alternatives, computations, drawings, and specifications that meet the needs of the owner. In addition there may be other duties that are delegated to the designer by the owner, such as, on-site or periodic inspection, review of shop drawings, and in some instances the acquisition of land and/or permits. It is the duty of the designer to produce a project design that meets all federal, state, and local codes; standards; and environmental and safety regulations. In addition a budget for the design should be prepared, along with a design schedule that matches the owner's schedule. The design schedule should be directly correlated to the construction schedule so the project can be completed by the construction contractor when the owner needs it.

Many problems may arise when a designer fails to give adequate attention to the impact of design alternatives on the cost or schedule during construction. Sometimes a design alternative is selected that may cause restrictions in the method of construction that a contractor may want to use during construction, which can adversely affect the contractor's operations. The selection of design alternatives also can impact the procurement of major equipment required in a project, which can influence the scheduling of construction work. Also, lack of attention to details in the design drawings may cause erection problems at the job-site, which can lead to delays in work, increased costs, and claims from the construction contractor. Inadequate attention to details during design can cause confusion during construction, including delays in work in the field and unnecessary rework.

Designers also need to make a special effort to eliminate poor wording in the written specifications that can cause misinterpretation of the requirements of the work to be performed, which can lead to poor quality work or costly changes orders. Overly restrictive wording in the specifications may place restraints on the contractor's operations and/or sequencing of construction work, which can affect costs and schedules. Designs that are poorly assembled can impair productivity of crafts, morale of workers on the job-site, and lead to legal claims against the designer and/or owner from the contractors. The quality of design documents has a significant impact on the quality of construction and the long-term maintenance and operations after the project is completed.

Generally the designers are not obligated under standard-form contracts to guarantee the construction cost of a project, although there have been some cases where the designer has been held legally responsible for the construction price. As part of their design responsibility, designers usually prepare an estimate of the probable construction cost for the design they have prepared. Major decisions by the owner to proceed with the project are made from the designer's cost estimate.

The cost and operational characteristics of a project are influenced most, and are more easy to change, during the design phase. Because of this, the designer plays a key role during the early phase of a project by working with the owner to keep the project on track so the owner/contractor relationship will be in the best possible form.

The construction contractor is responsible for the performance of all work in accordance with the contract documents that have been prepared by the designer. This includes furnishing all labor, equipment, material, and know-how necessary to build the project. The construction phase is important because most of the project budget is expended during construction. Also, the operation and maintenance of the completed project is highly dependent on the quality of work that is performed during construction. The contractor must prepare an accurate estimate of the project, develop a realistic construction schedule, and establish an effective project control system for cost, schedule, and quality.

Serious problems can arise when contractors fail to perform their work in accordance with the contract documents. If the contractor does not perform work as specified in the contract, the owner has the option to call the performance bond, which brings a new contractor to the job to complete the project. The quality of the project may be impaired when another construction contractor must be brought to the job-site to properly redo the defaulting contractor's work. When the owner calls the performance bond on a defaulting contractor, the ability of the defaulting contractor to acquire future contracts is greatly impaired. The contractor may be disqualified from bidding future work of the owner, and the designer may discourage other owners from allowing the contractor to bid on the designer's work. Also, the bonding capacity of the defaulting contractor may be reduced, or completely eliminated, by the bonding company. Adverse publicity may damage the reputation of the construction contractor, which can adversely affect the success of the contractor in securing future work.

Who Does the Project Manager Work For?

The project manager works for the project, although he or she may be employed by the owner, designer, or contractor. Therefore, for any project there are at least three project managers. Although these three individuals work for a different organization, they must develop methods to ensure good working relationships. At the beginning of a project, all project managers should meet together to define and agree upon the authority and responsibility of each party. They need to agree on procedures for review and approval of documents and the distribution of documents between the parties of each organization. There needs to be regularly scheduled joint project managers' meetings, with a set agenda to share information and coordinate the work in an organized manner. Sharing information in a timely manner is necessary for a successful project. Each project manager needs to make a special effort to work together as a team to help each other. They need to show trust, respect, and confidence in the work of others with an attitude of finding solutions instead of finding blame.

The Construction Industry Institute (CII) has sponsored research and published numerous papers on a variety of topics related to project management. *Organizing for Project Success,* a CII publication, provides a good description of the interface between project managers for the owner, designer, and contractor. The following paragraphs are a summary of the project management teams that are discussed in the publication.

After commitment has been made by an owner to invest in a project, an Investment Management Team is formed within the owner's organization to provide overall project control. The major functions, such as marketing, engineering, finance, and manufacturing, are usually represented. A Project Executive usually leads the team and reports to the head of the business unit which made the decision to proceed with the project. A member of this team is the Owner's Project Manager.

The Owner's Project Manager leads a Project Management Team which consists of each Design Project Manager and Construction Project Manager that is assigned a contract from the owner. Their mission is to accomplish the work, including coordinating the engineering, procurement, and construction phases. The Owner's Project Manager leads this team, which is one of the most important management functions of the project. The Owner's Project Manager is responsible for the accomplishment of all work, even though he or she has limited resources under his or her direct control because the work has been contracted to various organizations.

Reporting to each Design Project Manager and Construction Project Manager are the Work Managers who fulfill the requirements of their contracts. Each Design and Contractor Project Manager reports to the Owner's Project Manager for contractual matters and to his or her parent organization for business matters.

The Work Managers are the design leaders and supervisors who lead the teams actually accomplishing the work. They are directly responsible for the part of the contract assigned to them by their Project Manager. They must also communicate and coordinate their efforts with Work Managers from other organizations. Usually this communication does not flow vertically through a chain of command, but instead flows horizontally between people actually involved in the work. It is their responsibility to also work with their Project Manager and keep them informed. This is further discussed in Chapters 2 and 11.

Purpose of Project Management

For the purpose of this book, project management may be defined as:

> The art and science of coordinating people, equipment, materials, money, and schedules to complete a specified project on time and within approved cost.

Much of the work of a project manager is organizing and working with people to identify problems and determine solutions to problems. In addition to being organized and a problem solver, a manager must also work well with people. It is people who have the ability to create ideas, identify and solve problems, communicate, and get the work done. Because of this, people are the most important resource of the project manager. Thus, the project manager must develop a good working relationship with people in order to benefit from the best of their abilities.

It is the duty of a project manager to organize a project team of people and coordinate their efforts in a common direction to bring a project to successful completion. Throughout the project management process there are five questions that must be addressed:

1. Who will do the work?
2. What work will be performed?
3. When will the work be done?
4. How much will the work cost?
5. What can go wrong?

The work required often involves people outside of the project manager's organization. Although these individuals do not report directly to the project manager, it is necessary that effective working relationships be developed.

A manager must be a motivated achiever with a "can do" attitude. Throughout a project there are numerous obstacles that must be overcome. The manager must have perspective with the ability to forecast methods of achieving results. The drive to achieve results must always be present. This attitude must also be instilled in everyone involved in the project.

Good communication skills are a must for a manager. The management of a project requires coordination of people and information. Coordination is achieved through effective communication. Most problems associated with project management can be traced to poor communications. Too often the "other person" receives information that is incorrect, inadequate, or too late. In some instances the information is simply never received. It is the responsibility of the project manager to be a good communicator and to ensure that people involved in a project communicate with each other.

Types of Management

Management may be divided into at least two different types: functional management (sometimes called discipline management) and project management. Functional management involves the coordination of repeated work of a similar nature by the same people. Examples are management of a department of design engineering, surveying, estimating, or purchasing. Project management involves the coordination of one time work by a team of people who often have never previously worked together. Examples are management of the design and/or construction of a substation, shopping center, refinery unit, or water treatment plant. Although the basic principles of management apply to both of these types of management, there are distinct differences between the two.

Most individuals begin their career in the discipline environment of management. Upon graduation from college, a person generally accepts a position in a discipline closely related to his or her formal education. Typical examples are design engineers, estimators, schedulers, or surveyors. The work environment focuses on how and who will perform the work, with an emphasis on providing technical expertise for a single discipline. Career goals are directed toward becoming a specialist in a particular technical area.

Project management requires a multi-discipline focus to coordinate the overall needs of a project with reliance on others to provide the technical expertise. The project manager must be able to delegate authority and responsibility to others and still retain focus on the linking process between disciplines. Project managers cannot become overly involved in detailed tasks or take over the discipline they are educated in, but should focus on the project objectives.

A fundamental principle of project management is to organize the project around the work to be accomplished. The work environment focuses on what must be performed, when it must be accomplished, and how much it will cost. Career development for project managers must be directed toward the goal of becoming a generalist with a broad administrative viewpoint.

The successful completion of a project depends upon the ability of a project manager to coordinate the work of a team of specialists who have the technical ability to perform the work. Table 1-1 illustrates the relationship between project management and discipline management.

TABLE 1-1   Distinguishing between Project Management and Discipline Management

Project management is concerned with	Discipline management is concerned with
What must be done	How it will be done
When it must be done	Who will do it
How much it will cost	How well it will be done
Coordinating overall needs	Coordinating specific needs
Multi-discipline focus	Single-discipline focus
Reliance on others	Providing technical expertise
Project quality	Technical quality
Administrative viewpoint	Technical viewpoint
A generalist's approach	A specialist's approach

Functions of Management

Management is often summarized into five basic functions: planning, organizing, staffing, directing, and controlling. Although these basic management functions have been developed and used by managers of businesses, they apply equally to the management of a project.

Planning is the formulation of a course of action to guide a project to completion. It starts at the beginning of a project, with the scope of work, and continues throughout the life of a project. The establishment of milestones and consideration of possible constraints are major parts of planning. Successful project planning is best accomplished by the participation of all parties involved in a project. There must be an explicit operational plan to guide the entire project throughout its life.

Organizing is the arrangement of resources in a systematic manner to fit the project plan. A project must be organized around the work to be performed. There must be a breakdown of the work to be performed into manageable units, which can be defined and measured. The work breakdown structure of a project is a multi-level system that consists of tasks, subtasks, and work packages.

Staffing is the selection of individuals who have the expertise to produce the work. The persons that are assigned to the project team influence every part of a project. Most managers will readily agree that people are the most important resource on a project. People provide the knowledge to design, coordinate, and construct the project. The numerous problems that arise throughout the life of a project are solved by people.

Directing is the guidance of the work required to complete a project. The people on the project staff that provide diverse technical expertise must be developed into an effective team. Although each person provides work in his or her area of expertise, the work that is provided by each must be collectively directed in a common effort and in a common direction.

Controlling is the establishment of a system to measure, report, and forecast deviations in the project scope, budget, and schedule. The purpose of project control is to determine and predict deviations in a project so corrective actions can be taken. Project control requires the continual reporting of information in a timely manner so management can respond during the project rather than afterward. Control is often the most difficult function of project management.

Key Concepts of Project Management

Although each project is unique, there are key concepts that a project manager can use to coordinate and guide a project to completion. A list of the key concepts is provided in Table 1-2.

TABLE 1-2   Key Concepts of Project Management

1. Ensure that one person, and only one person, is responsible for the project scope, budget, and schedule
2. Don't begin work without a signed contract, regardless of the pressure to start
3. Confirm that there is an approved scope, budget, and schedule for the project
4. Lock in the project scope at the beginning and ensure there is no scope growth without approval
5. Make certain that scope is understood by all parties, including the owner
6. Determine who developed the budget and schedule, and when they were prepared
7. Verify that the budget and schedule are linked to the scope
8. Organize the project around the work to be performed, rather than trying to keep people busy
9. Ensure there is an explicit operational work plan to guide the entire project
10. Establish a work breakdown structure that divides the project into definable and measurable units of work
11. Establish a project organizational chart that shows authority and responsibilities for all team members
12. Build the project staff into an effective team that works together as a unit
13. Emphasize that quality is a must, because if it doesn't work it is worthless, regardless of cost or how fast it is completed
14. Budget all tasks; any work worth doing should have compensation
15. Develop a project schedule that provides logical sequencing of the work required to complete the job
16. Establish a control system that will anticipate and report deviations on a timely basis so corrective actions can be taken
17. Get problems out in the open with all persons involved so they can be resolved
18. Document all work, because what may seem irrelevant at one point in time may later be very significant
19. Prepare a formal agreement with appropriate parties whenever there is a change in the project
20. Keep the client informed; they pay for everything and will use the project upon completion

Each of the key concepts shown in Table 1-2 is discussed in detail in subsequent chapters of this book. It is the responsibility of the project manager to address each of these concepts from the beginning of a project and through each phase until completion.

Role of the Project Manager

The role of a project manager is to lead the project team to ensure a quality project within time, budget, and scope constraints. A project is a single, non-repetitive enterprise, and because each project is unique, its outcome can never be predicted with absolute confidence. A project manager must achieve the end results despite all the risks and problems that are encountered. Success depends on carrying out the required tasks in a logical sequence, utilizing the available resources to the best advantage. The project manager must perform the five basic functions of management: planning, organizing, staffing, directing, and controlling.

Project planning is the heart of good project management. It is important for the project manager to realize that he or she is responsible for

TABLE 1-3   Project Manager's Role in Planning

1. Develop planning focused on the work to be performed
2. Establish project objectives and performance requirements early so everyone involved knows what is required
3. Involve all discipline managers and key staff members in the process of planning and estimating
4. Establish clear and well-defined milestones in the project so all concerned will know what is to be accomplished, and when it is to be completed
5. Build contingencies into the plan to provide a reserve in the schedule for unforeseen future problems
6. Avoid reprogramming or replanning the project unless absolutely necessary
7. Prepare formal agreements with appropriate parties whenever there is a change in the project and establish methods to control changes
8. Communicate the project plan to clearly define individual responsibilities, schedules, and budgets
9. Remember that the best-prepared plans are worthless unless they are implemented

project planning, and it must be started early in the project (before starting any work). Planning is a continuous process throughout the life of the project, and to be effective it must be done with input from the people involved in the project. The techniques and tools of planning are well established. Table 1-3 provides guidelines for planning.

A project organizational chart should be developed by the project manager for each project. The chart should clearly show the appropriate communication channels between the people working on the project. Project team members must know the authority of every other team member in order to reduce miscommunications and rework. Organized work leads to accomplishments and a sense of pride in the work accomplished. Unorganized work leads to rework. Rework leads to errors, low productivity, and frustrated team members. Table 1-4 provides guidelines for organizing.

Project staffing is important because people make things happen. Most individuals will readily agree that people are the most important resource on a project. They create ideas, solve problems, produce designs, operate equipment, and install materials to produce the final product. Because each project is unique, the project manager must understand the work to be accomplished by each discipline. The project manager should then work with his or her supervisor and appropriate discipline managers to identify the persons who are best qualified to work on the project. Table 1-5 provides guidelines for project staffing.

TABLE 1-4   Project Manager's Role in Organizing

1. Organize the project around the work to be accomplished
2. Develop a work breakdown structure that divides the project into definable and measurable units of work
3. Establish a project organization chart for each project to show who does what
4. Define clearly the authority and responsibility for all project team members

TABLE 1-5   Project Manager's Role in Staffing

1. Define clearly the work to be performed, and work with appropriate department managers in selecting team members
2. Provide an effective orientation (project goals and objectives) for team members at the beginning of the project
3. Explain clearly to team members what is expected of them and how their work fits into the total project
4. Solicit each team member's input to clearly define and agree upon scope, budget, and schedule

The project manager must direct the overall project and serve as an effective leader in coordinating all aspects of the project. This requires a close working relationship between the project manager and the project staff to build an effective working team. Because most project team members are assigned (loaned) to the project from their discipline (home) departments, the project manager must foster the development of staff loyalty to the project while they maintain loyalty to their home departments. The project manager must be a good communicator and have the ability to work with people at all levels of authority. The project manager must be able to delegate authority and responsibility to others and concentrate on the linking process between disciplines. He or she cannot become overly involved in detailed tasks, but should be the leader of the team to meet project objectives. Table 1-6 provides guidelines for directing the project.

Project control is a high priority of management and involves a cooperative effort of the entire project team. It is important for the project manager to establish a control system that will anticipate and report deviations on a timely basis, so corrective action can be initiated before more serious problems actually occur. Many team members resist being controlled; therefore the term *monitoring a project* may also be used as a description for anticipating and reporting deviations in the project. An effective project control system must address all parts of the project: quality, work accomplished, budget, schedule, and scope changes. Table 1-7 provides guidelines for project control.

TABLE 1-6   Project Manager's Role in Directing

1. Serve as an effective leader in coordinating all important aspects of the project
2. Show interest and enthusiasm in the project with a "can do" attitude
3. Be available to the project staff, get problems out in the open, and work out problems in a cooperative manner
4. Analyze and investigate problems early so solutions can be found at the earliest possible date
5. Obtain the resources needed by the project team to accomplish their work to complete the project
6. Recognize the importance of team members, compliment them for good work, guide them in correcting mistakes, and build an effective team

TABLE 1-7   Project Manager's Role in Controlling

1. Maintain a record of planned and actual work accomplished to measure project performance
2. Maintain a current milestone chart that displays planned and achieved milestones
3. Maintain a monthly project cost chart that displays planned expenditures and actual expenditures
4. Keep records of meetings, telephone conversations, and agreements
5. Keep everyone informed, ensuring that no one gets any "surprises," and have solutions or proposed solutions to problems

Professional and Technical Organizations

Due to the increased cost and complexity of projects, the interest in developing and applying good project management principles has gained considerable attention by owners, designers, and contractors. Numerous organizations have made significant contributions related to project management by conducting research, sponsoring workshops and seminars, and publishing technical papers. The following paragraphs describe some of these organizations.

The American Society of Civil Engineers (ASCE), founded in 1852, is the oldest national engineering society in the United States. Membership consists of civil engineers working in government, education, research, construction, and private consulting. The construction division of ASCE has many councils and technical committees that have published technical papers related to project management in its *Journal of Construction Engineering and Management.*

The National Society of Professional Engineers (NSPE), founded in 1936, is the national engineering society of registered professional engineers from all disciplines of engineering. NSPE membership includes engineers who are organized in five practice divisions: construction, education, government, industry, and private practice. The construction practice division has numerous committees that have contributed to contract documents and legislation related to engineers in the construction industry.

The Project Management Institute (PMI), founded in 1969, consists of members from all disciplines and is dedicated to advancing the state-of-the-art in the profession of project management. PMI has a certification program for project management professionals and publishes a *Project Management Book of Knowledge* (PMBOK).

The Association for Advancement of Cost Engineering-International (AACE-I), founded in 1956, is an organization of worldwide members. It serves total cost management professionals in disciplines such as cost engineering, cost estimating, planning & scheduling, decision and risk management, project management, project control, cost/schedule control, earned value management, claims, and more. Since 1976 AACE has administered a certification program for individuals as Certified Cost

Consultant, Certified Cost Engineer, Certified Estimating Professional, Earned Value Professional, Planning & Scheduling Professional, and others.

The Construction Management Association of America (CMAA), founded in 1981, is an organization of corporate companies, public agencies, and individual members who promote the growth and development of construction management (CM) as a professional service. CMAA publishes documents related to CM, including the *Standard CM Services and Practice*.

The Construction Industry Institute (CII), founded in 1983, is a national research organization consisting of an equal number of owner and contractor member companies, and research universities from across the United States. CII is organized into committees, councils, and research teams which are comprised of owners, contractors, and academic members who work together to conduct research and produce publications on a variety of topics related to project management.

The following list of organizations is provided to the reader as sources for information related to project management:

American Institute of Architects

American Society of Civil Engineers

American Society of Military Engineers

Association for Advancement of Cost Engineering-International

Construction Industry Institute

Construction Management Association of America

Design Build Institute of America

National Society of Professional Engineers

Project Management Institute

Society of American Value Engineers

References

1. Adams, J. R. and Campbell, B., *Roles and Responsibilities of the Project Manager,* Project Management Institute, Newtown Square, PA, 1982.
2. Adams, J. R., Bilbro, C. R., and Stockert, T. C., *An Organization Development Approach to Project Management,* Project Management Institute, Newtown Square, PA, 1986.
3. Knutson, K., Schexnayder, C. J., Flori, C. M., and Mayo, R., *Construction Management Fundamentals.* 2nd ed., McGraw-Hill, New York, NY, 2009.
4. Levy, Sidney M., *Project Management in Construction, 6th ed.*, McGraw-Hill, New York, NY, 2012.
5. "Organizing for Project Success," Publication No. 12-2, Construction Industry Institute, Austin, TX, February, 1991.
6. Stuckenbruck, L. C., *The Implementation of Project Management: The Professional's Handbook,* Project Management Institute, Newtown Square, PA, 1981.

Working with Project Teams

Project Teams

Project teams must be assembled to accomplish the work necessary to complete engineering and construction projects. Team members are vital to the success of the project. The project manager depends on the team because he or she typically does not have the expertise to do all the work required to complete the project. For any team, there must be a leader to guide the overall efforts. In many respects the project manager acts as a coach, answering questions, making sure the team understands the desired outcome of the project, and ensuring that team members know what is expected of them and the importance of sharing information. The project manager must make sure that his or her team understands and is focused on the desired outcome of the project. The project manager also acts as a facilitator in project communications for conflict resolution and team performance.

Project teams are made up of all the participants in the project, including in-house personnel and outside consultants. Team members report either part-time or full-time to the project manager and are responsible for some aspect of the project's activities. Teamwork must be well coordinated with effective interaction to achieve the shared objective of completing the scope, budget, and schedule constraints of the project. Managing project teams is a fundamental skill within the area of human resources management. The Project Management Institute (PMI) defines human resources management as the art and science of directing and coordinating human resources throughout the life of a project, by using administrative and behavioral knowledge to achieve predetermined project objectives of scope, cost, time, quality, and participant satisfaction.

For a successful project, the project manager must build and lead an effective project team. Team building is the process of influencing a group of diverse individuals, each with his or her own goals, needs, and perspectives, to work together effectively for the good of the project. The team effort should accomplish more than the sum of the individual efforts. Every team needs motivation. Team motivation is the process by which project managers influence the team members to do what it takes to get the job done. The key problem is "How do you motivate team members when they are borrowed resources?" Usually, members of the project team are individuals who are assigned from other departments to the project manager's project. Because these individuals are borrowed from other departments or hired from outside organizations, the project manager must devise a method to motivate them to be dedicated to the project while remaining loyal to their home departments and organizations. This presents a real challenge to the project manager.

Teamwork

Teamwork is not a new concept; it has been used for centuries. Essentially a *team* is a group of people that work together to achieve a common goal. An effective team can accomplish more and better work than any one person acting alone. Projects require people with diverse expertise and different experience. Each member of the team brings an expertise that contributes to successful completion of the project. There are four types of people:

- Those who make things happen
- Those who watch things happen
- Those who don't know what is happening
- Those who don't want to know what is happening

Every project manager wants people who are proactive to make things happen. However, there are people who are passive and tend to not take actions until they are given precise directions, or until they feel safe in doing work. Unfortunately there are people who do not appear to know or understand what is happening or what needs to be done. Sometimes there are people in upper management that do not want to know what is happening in a project, but instead they do not want to be bothered and only want the project team to do their assigned tasks and resolve problems. Often the project manager does not have the opportunity to select every member of the project team. Instead, assignment of team members is made by upper management. Therefore, the project manager has to work with all types of people.

Reorganizing, downsizing, and merging of businesses have created new emphasis on teamwork. Many companies outsource work to finish projects. Everyone agrees that teamwork is important; the real task is organizing a successful team to achieve a successful project.

Teamwork starts with the sponsor of the project who defines goals, objectives, needs, and priorities. For successful projects, teamwork starts with the team's formation at the beginning of the project and continues throughout the life of a project. A well-organized team resolves disputes, solves problems, and communicates effectively. Effective teamwork discourages fault finding and accusations and promotes unity and a common focus on the same set of project goals and priorities. Although everyone is a key player on a successful team, every team must have a leader. The project manager is the leader of the team.

Teams for Small Projects

A team is two or more people working together to accomplish a common goal. When managing multiple small projects, the project manager is usually required to share team members with other project managers. Generally the project duration is short with minimal contact between the project manager and team members. Sometimes the team members are specialists who are hired by contracts from outside sources to perform a specific task or function.

Since the project manager of a project is generally responsible for managing multiple projects at the same time, it is often difficult for him or her to give the needed attention to each project, which complicates scheduling and resource control. Only minimal staffs can be afforded on small projects. This means that the few individuals assigned must take responsibility for multiple functions. In this type of work environment, the skill of the project manager to maneuver through the various departments within his or her organization to get people to do the work on the project is crucial in completing the projects on time and within budget. There is less potential for comprehensive look-ahead planning and attention to those functions not currently experiencing problems. The ability to meet project deadlines is highly dependent on the schedules of others.

For engineering work, it is difficult to have a core discipline team assigned to each project. As a consequence, time is wasted while team members wait for information. Since small projects have short durations, there is often insufficient time for detailed planning and in-process correction of problems. The learning curve for personnel is still climbing when the project is over.

Although managing multiple small projects may not have the formality of managing a single large project, the principles of working with people in the spirit of cooperation and teamwork still applies. Typically the project manager relies on frequent phone calls or e-mail in lieu of formal face-to-face team meetings.

Working with Multiple Teams

As a project progresses through design into construction, the work of the owner's, designer's, and contractor's teams must merge into a collective effort. Although each of these teams have their own objectives, the diverse expertise that each possesses must converge into an overlapping environment as illustrated in Figure 2-1.

Each triangle in Figure 2-1 represents a team. Although each team performs a different function, each team must develop an attitude of shared ownership in the project. The project manager from each organization

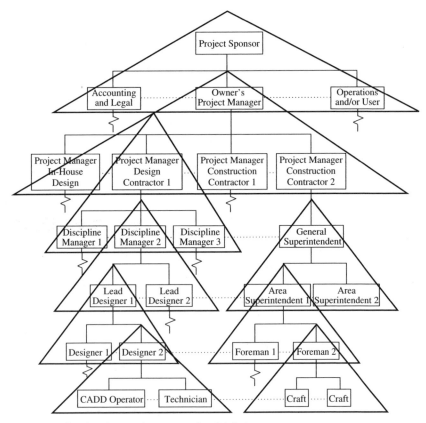

Figure 2-1 Overlapping environment of multiple teams.

must create and foster an environment where team members contribute to solving problems and doing their jobs well, rather than trying to do just what they feel is necessary to get by. The team building and teamwork that was started at the beginning of the project must be continued into the construction.

Regardless of the size and number of teams, there must be a single head project manager to make final decisions and keep focus on the project. The owner's project manager has overall responsibility and final authority for the total project. Serving on the owner's project management team are project managers who are responsible for leading lower levels of teams that are responsible for engineering design and construction of the project. As illustrated in Figure 2-1 there is a project manager for in-house design, a project manager for each design contractor, and a project manager for each construction contractor. Each of these project managers leads the team for his or her organization. Below these managers are lower tiers of teams who are led by work managers. As shown in Figure 2-1, the manager of the lower-level team serves as a member of the higher-level team. There must be one head of each team.

Owner's Team

As shown in Figure 2-1, the project manager of the owner's team is the highest level in the project team. The owner's project manager generally has limited staff support from his or her company. For a small project the owner's project manager may be the only person in the owner's organization that works on the project. For larger projects there may be several assigned support personnel from the owner's organization, such as clerical staff, project engineers, contracts, procurement, and safety personnel.

The owner's project manager leads a team that consists of each design project manager and construction project manager that is assigned a contract from the owner. The owner's project manager is responsible for the accomplishment of all work, even though he or she has limited resources under his or her direct control because most of the work has been contracted to various organizations. Therefore, much of the work of the owner's project manager is administering design and construction contracts and updating the project sponsor in the owner's organization.

The role of the owner's project manager is important to the successful completion of a project. Generally he or she wants a competent designer that has experience in developing plans and specifications for the type of project needed by the owner. It is important to have a designer that can relate the design and construction requirements in terms that can be understood by the owner, in particular the operations people in the owner's organization that will use the project when it is completed. The owner's project manager also needs an experienced construction contractor that does quality work, is reasonable in costs, and will finish the project within the time required by the owner. A good working relationship among all parties in the project requires good communications regarding the status and progress of work.

Design Teams

Members of the design team are selected based on the specific expertise needed for a particular project. The team is composed of individuals with diverse backgrounds, including design disciplines [architectural, civil, mechanical, electrical, structural, computer-aided design and drafting (CADD), etc.], project control individuals (cost control, estimating, quality control, safety, etc.), non-technical people (purchasing, legal, financial, permitting, regulatory, etc.), and the sponsor's representative.

The work of the design team occurs in the early stage of a project, when there are many unknowns. Often the scope of work is not clearly or completely defined. Sometimes the budget is simply the amount of money the owner has available to spend, and there is no tie of the cost budget to the scope of work. Occasionally the only known information about the schedule is the finish date of the project. All these issues create challenges and difficulties for project managers responsible for engineering design of a project. Table 2-1 gives common problems of engineering project managers.

Every design team must have a sponsor's representative, who may be the owner's project manager or appointed by the owner's project manager. This individual must communicate to the project team the corporate policies and the funding limit of the sponsoring organization. He or she must have the responsibility and authority to act on behalf of the sponsoring organization.

The sponsor's representative plays a key role in resolving issues related to the project requirements and costs that will impact the

TABLE 2-1   Common Problems of Engineering Project Managers

1. Project scopes are not well defined
2. Too much growth from the beginning to the end of projects
3. Project budgets are not tied to scope
4. Inconsistency of workload, either too busy or nothing to do
5. Trying to handle too many projects at the same time
6. Shifting priorities from one project to another project
7. Poor communications, misunderstandings, and too much rework
8. The company is not organized to handle projects
9. No system exists to handle projects
10. No overall, understandable work plan exists

sponsor's organization after the project is completed, when it will be used by the sponsor. He or she reviews and approves the evolving scope, budget, and schedule. As a team member, the sponsor's representative approves any changes in scope, budget, and schedule before commitments are made. This individual should be an active participant on the design team, answering questions and providing information needed by the team to accomplish the work. Unfortunately, the sponsoring organization's representative is sometimes not involved until the project gets into trouble. Early involvement of the sponsor's representative can prevent future problems. The design project manager must make this individual feel that he or she is part of the project team.

Selecting design team members is an important step in project management because it begins the team building process. How team members are selected varies, depending on the policies of the design project manager's organization and the persuasiveness of the project manager to get the people on the project that he or she wants. Chapter 6 presents various organization structures for design firms. To illustrate selection of design team members, consider a design firm that is organized as shown in Figure 6-5. The project manager and his or her supervising manager will review the project's needs to identify required discipline expertise and required personnel. Then, a meeting is arranged with the manager of appropriate discipline managers to request team members. The people assigned to the project are selected by the department manager of engineers for each respective discipline.

Obviously, the project manager always wants the best and most qualified workers assigned to his or her project. However, the assignment of team members is often based on who is available at the time assignments are made. If the project manager feels a person who is assigned to the team lacks the required skills, then the project manager must act as a coach to assist the team member and/or make arrangements for additional training to ensure the work can be completed.

Construction Teams

The work environment and culture of a construction project is unique compared to most working conditions. A typical construction project consists of groups of people, normally from several organizations, that are hired and assigned to a project to build the facility. Due to the relatively short life of a construction project, these people may view the construction project as accomplishing short-term tasks. However, the project manager of the construction team must instill in the team that building long-term relationships is more important in career advancement than trying to accomplish short-term tasks.

Even small-sized construction projects involve a large number of people. Organizing their efforts is complex, even if they all work for the same organization. Sources of information, location, timing, and problem complexity change as people enter the project, perform their assigned duties, and depart.

With all the diversity involved in a construction project, people must be managed so they work together efficiently to accomplish the goal. This requires skilled people who are willing to sacrifice short-term gratification for the long-term satisfaction of achieving a larger goal. Common sense and flexibility are essential to working with construction teams. The key to a successful construction project is properly skilled construction managers. These individuals possess the ability to recognize the degree of uncertainty at any point in the execution of the project and to manage the efforts of others to achieve clearly defined objectives that result in successful completion of the final product.

The organizational chart for a construction project consists of lines and boxes that show the division of work and the relationship of the workers to formal authority. The boxes in the chart depict the tasks to be performed and the lines depict the coordination required.

The number of construction teams for a project depends upon the number of contracts awarded by the owner. For each construction contractor, and subsequent tier of subcontractors, a construction team is formed to perform the work in accordance with the contract documents issued by the owner.

Team Management

Successful team management requires the team to be an integral unit of the organization. A team must have a well-defined mission with common goals, objectives, and strategies. The role of each team member must be clearly defined. The project manager must learn the needs of team members and encourage team participation. Team members will put extra effort into accomplishing work when they know the project manager cares about them and their careers. This can be accomplished only by effective communications and feedback throughout the project. Trust is instilled among team members and the project manager by creating an environment of understanding and teamwork. Open and honest communications are necessary to instill integrity and support for each other. Trust is essential to effective and successful teams.

It is the project manager's responsibility to ensure that individuals are assigned primary responsibility for discrete work. Most workers want to do what is expected of them and will do the work, provided there are clear instructions and understandings. This requires a collective culture of mutual agreements between the project manager and team members. All individuals must have the common goal of creating a team that plans and executes the work with a clear knowledge of what they are going to do, who is going to do it, and when it will be done. Sometimes it is necessary to know where it will be done or how it will be done. For example, for some instances it may be necessary to know what method of analysis will be used in a design.

Teams and the Project Manager's Responsibilities

When working with team members, the project manager must cross many boundaries in the organizational structure to develop the project team into a cohesive group. This must be done quickly in spite of constraints imposed by others. The project manager must combine administrative and behavioral knowledge to work well with people. People skills are vital for effective management of team members. The project manager must create a cordial environment that enables the team to work together so members will motivate themselves to peak performance.

Attributes of the project manager include the following:

- Organizes thoughts and actions
- Identifies problems and finds solutions
- Works well under pressure
- Accepts responsibility
- Motivates people with a "can do" attitude
- Anticipates and forecasts future events
- Works well with others, inside and outside of the company
- Is honest, fair, and has respect for others
- Communicates well—writing, speaking, and listening
- Leads and encourages the team
- Forward leading, not backward reporting

The project manager is responsible for resolving conflicts between team members in addition to organizing, coordinating, and directing the project. The project manager is the team leader who is responsible for developing the project requirements. This is accomplished by effective communications.

When working with teams, the project manager acts as the leader in acquiring resources, selecting team members, developing the sponsor's requirements, defining scope and quality, defining budgets, and determining schedules. The project manager must establish a control system to complete the project in accordance with the expected requirements. The project manager is expected to control project activities within a defined scope, budget, and schedule. Situations will arise when design differences must be resolved. Trade-offs will have to be made to comply with the budget and schedule.

An important responsibility of the project manager is decision making. During team meetings, numerous decisions must be made. The process used in making decisions can have a direct impact on team performance. In some situations the decision can be made solely by the project manager, possibly with input from one or more team members. However, there are other situations when the decision making should involve the entire team. The project manager must establish a process for decision making that matches the decision to be made. For example, the decision may be to resolve the best way to perform a design or produce drawings while another may involve generating ideas, solving a problem with one correct answer, or deciding issues with multiple correct answers. The project manager must develop a leadership style that is respected and accepted by the project team for decision making.

It is the project manager's responsibility to ensure that each team member is assigned primary responsibility for specific work. Recognizing there are numerous work items in a project, the project manager must devise methods to ensure there will be no overlap or gaps in the work of the project team. Each team member should be asked to clearly define the work they expect to perform for the project. Team members should also identify potential gaps or overlaps in their work that may be performed by others. For example, will the anchor bolts be designed by the structural engineer who will be designing the superstructure or by the foundation engineer who will design the substructure? There is a risk that anchor bolts may be designed by both engineers, or neither of them. After review by all team members the project manager should develop an overall work plan that integrates the work of all team members. Then regularly scheduled meetings should be held to ensure no overlaps or gaps in the work of the team.

Sometimes when a new team is formed, the project manager may feel that one of the team members is lacking in the required skills to perform the work. To address this issue the project manager should meet with the team member to discover his or her work experience relative to the project. This meeting should ascertain the confidence level of the team member in his or her ability to perform the work. It may be necessary to identify internal training or mentoring that may be available to assist the team member. The team member can be asked to investigate seminars that may be available to assist in performing the work. The project manager should show support in assisting the team member in their career growth and professional development. Periodic checks with the team member may be necessary to ensure work is progressing as expected.

The composition of a design team represents individuals with diverse backgrounds, including civil, mechanical, electrical, structures, etc. The project manager must devise methods to ensure mutual respect among team members and to reinforce the understanding that everyone on the team is an important player in the success of the project. The project manager needs to ensure that each discipline is given adequate recognition during team meetings. Meetings should be conducted in a manner that shows the importance of integrating design disciplines. The project manager should show a genuine interest and appreciation for diversity of design disciplines. A team environment should be created that fosters mutual respect of design disciplines. The project manager needs to instill a team culture that everyone is a key contributor in a successful project. It is helpful for the project manager to identify the impact of each team member's work and how it affects other members on the team.

Key Factors in Team Leadership

Developing a culture where each team member feels that he or she is a part of the team and wanted by the team is essential to a successful team. Individuals who feel they are an important part of the team will develop a sense of pride because they are a part of the team and will become enthusiastic and motivated to assist others to ensure the successful performance of the entire team.

To be an effective leader, the project manager must possess hard and soft skills. Hard skills include technical expertise and knowledge about the work performed by the project team. These skills are generally attained by formal education and further enhanced by on-the-job experience. Soft skills involve people issues that are not normally attained in formal education, but acquired by experience in working with people. As the project manager rises in the ranks, soft skills become more important to effective leadership. Most people are hired on their hard skills, but promoted on their soft skills.

Project managers typically begin their career in a technical area where hard skills are important. They are promoted based on performance of their technical skills and seldom ask for help, especially if they are good problem solvers. They may become overly independent and fail to cultivate relationships with people who could become useful to them in the future. They rely on their technical skills rather than learning how to influence people. Much of the work of project manager is working with people with diverse backgrounds and expertise to solve problems and get the work done.

The behavior and leadership style of the project manager has a significant influence on the team. The project manager must have high ethics and a sense of fairness and honesty while dealing with members of the team as well as others who are not on the team. In many respects the project manager is the role model for the team. It is difficult for people to be highly motivated and productive when they do not have the respect of their leader. The project manager must communicate the desired goals, objectives, values, and outcomes of the project. The team can then translate these issues into producing quality work. The project manager must also keep members informed of the status of the project.

Each project manager develops a leadership style as he or she progresses in his or her career. However, the common end result of any project manager is to lead the team that maximizes performance of individual team members and for the team as a group. For any project manager the key factors shown in Table 2-2 are important for successful team leadership.

Team communications are vital to a successful team because highly motivated and dedicated workers want, and need, to be informed. Regularly scheduled team meetings are essential. There may be numerous meetings between key team members to exchange detailed information, but a regularly scheduled weekly team meeting should be held for sharing status, making decisions, and documenting information. Because team members are frequently located at different physical locations, regular face-to-face meetings are necessary to keep the sense of team unity.

A well-defined scope for the team guides the progress of work and provides clear goals that can be used as guidelines for decision making. The project manager must ensure the scope is defined and understood by each team member before work is started. A firm scope that is clearly understood provides empowerment to team members. It also allows

TABLE 2-2   Key Factors in Team Leadership

1. Show interest and enthusiasm in the project and the project team
2. Be available to answer questions and assist team members in their work
3. Provide the resources needed by the team members to perform their work
4. Show trust and confidence in the team's ability to accomplish the work
5. Ensure that everyone on the team knows what is expected of him or her
6. Give credit to a team member when he or she has performed outstanding work
7. Hold each team member responsible for their work
8. Conduct regularly scheduled team meetings to maintain open communications
9. Exercise caution when dealing with topics that may be sensitive to team members
10. Be ethical in dealing with team members
11. Show interest and expectations in producing quality work
12. Show honesty and integrity in all relationships with team members

more independence and autonomy of individual team members to perform their work in the most efficient and expeditious manner because their assigned work and desired outcome is known. Individuals who know their responsibilities and the required outcome of their work are free to be innovative and creative, thereby producing high-quality work with performance. The result is a successful project.

Team Building

Effective teamwork is a key element in any successful project. Teamwork must be started early in the process, and it must be continuous throughout the life of a project. Experienced engineers and project managers all agree that teamwork is necessary, but the real question is "How does one organize a successful team?"

Effective communications is essential to team building. In simple terms, effective communications means the other person has received and understood the information that is being given to him or her. The giver of information must obtain feedback from the receiver to ensure effective communications. The project team cannot function when there are breakdowns in communications. Misinformation or incomplete information is a major deterrent to team building. Effective communications ensures that everyone knows what is expected and when it is expected.

All participants on the team have a common customer, the sponsor or user of the finished project. Team building starts with the sponsoring organization, with its project charter and mission statement. The project sponsor must be informed on established objectives, and he or she needs to be clear in their commitments. The project sponsor must have a good prequalification process for selecting designers, contractors, and other third-party participants. In addition, the sponsor must know and communicate his or her goals and aspirations for the project and must set priorities related to cost, schedule, safety, and the expected level of quality. Everyone on the project must realize that the project sponsor pays for everything and is therefore the common customer of all parties.

Designers want an educated sponsor who is knowledgeable in the process of designing and constructing the project, but sometimes that is not the case. Sometimes the project manager must help the project sponsor to understand the importance of sequencing work and the impact of decisions that must be made in designing and constructing the project.

From the first day of the project, there must be continuity in the project team. High turnover of team members creates wasted time in educating new people and lost knowledge of previous developments in the project. The contractor should be brought into the project at the earliest possible date. Construction contractors have excellent knowledge of costs and methods of construction that are extremely helpful during the design phase of a project. People experienced in construction can provide valuable input in the constructability of a project.

Many private-sector projects are financed by lending institutions. Too often, for these types of projects the lender is not an active participant in the team and everyone suffers from it. Unfortunately, the lender stays too far from the project and does not become involved until problems arise. There are other parties that are sometimes not brought into the team from the beginning of the project. For example, the purchasing agent is an important participant in teams that must procure large amounts of material or major equipment. Early involvement of the person who will be involved in issuing purchase orders, tracking vendors' shipping dates and delivery dates, and receipt of procured material and equipment has a significant impact on meeting installation deadlines in a project.

Key words for team building are pressure, responsibility, honesty, kindness, respect, and communications. Engineering and construction projects usually require tight schedules which create pressure on the project team to complete the work at the earliest possible date. This requires cooperation among team members who must assume responsibility and perform their work in the most expeditious manner. The ground rule should be "Everyone is a contributor and winner on a successful team." The team must stop worrying about the 1% that is wrong and focus on the 99% that is right. The project team must have open communications and avoid hiding mistakes and pointing blame. A successful team can achieve a successful project and have fun doing it.

Team building is very important for teams that consist of people who have never worked together. For example, a team that consists of multiple engineering disciplines combined with people from purchasing, quality, and cost control. To facilitate team building, the project manager needs to hold a preproject meeting with all team members to allow them to get acquainted informally. The project manager needs to emphasize the importance of each team member and how they are an important part of the project. Adequate time should be allowed during team meetings to emphasize the importance of integrating the team's work. The project manager should create a work environment of mutual respect by giving credit and recognition to each member of the project team and encouraging cooperation.

Some companies have begun the team building process by holding a weekend retreat for team members, including their family members. The retreat is usually held in a resort setting to allow interaction of team members related to their personal interests. This allows everyone to realize that team members have similar aspirations and common interests.

For example, understanding that others on the team have children with special talents or must care for elderly parents can provide a sense of bonding and mutual respect, which is the first step in effective team building.

Motivating Teams

For years managers and supervisors have struggled with methods of motivating workers. In the early 1950s A. H. Maslow developed a theory of motivation called the *hierarchy of needs*. Maslow's theory has been used by managers, as well as educators, to try to understand why people behave the way they do, how to motivate them, and how to enlist their commitment.

Maslow proposed that humans are a wanting animal and their wants become needs. It is the quest to satisfy their needs that drives or motivates people. The needs include basic physical and psychological needs, which are perceived in the human mind and satisfied by material things. Maslow proposed that the needs of humans follow a hierarchy, beginning with comfort and basic survival, such as food, clothing, and shelter. Once this first need is satisfied, humans seek to satisfy the next higher level of need: safety and security. After this need is satisfied, individuals seek progressively higher levels of need, including belonging, ego, and finally self-fulfillment. Self-fulfillment is the highest level of need and may include an urge to have increasing influence or give back to the world, such as establishing an endowment. Figure 2-2 is a graphical representation of the five levels in Maslow's hierarchy of human needs.

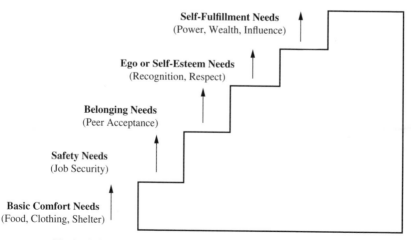

Figure 2-2 Maslow's hierarchy of needs.

As the lower levels of needs are satisfied, it is only natural for a person to become motivated to achieve the next higher level in Maslow's hierarchy of needs. A project manager must use his or her skills to determine the specific needs of team members in order to motivate them to their full potential.

Determining the needs of people is not easy to accomplish in the day-to-day work environment. Sometimes it may be necessary to spend time in an informal setting to build a relationship that provides understanding of what is important to a team member. For example, a team member may be concerned about having adequate time to attend a child's school function or care for an ill family member. A team member in this situation may be motivated by being allowed flexible work hours with the understanding that the time will be made up at nights or on weekends.

Effective methods should be used by the project manager to identify the needs of team members so he or she can be responsive to those needs in motivating the team. It may be necessary to converse with each team member outside the formal work environment to better know them. The project manager needs to get to know team members in a personal, nonintrusive, manner to show special interest in them. It is important to reinforce confidence in team members and let them know that the project manager wants to assist them in their work. It is desirable to determine what is important to team members as related to their work. Sometimes it may be advantageous to discretely ask team members what is important to them outside their work. When working with people, it is necessary to be sensitive to each team member and show genuine interest in supporting their work.

The project manager must also realize that the needs of people can change. For example, personal finances or problems within the family can change the needs of a person rapidly. Each project manager must devise a way to ascertain the needs of their team members. Once the needs are identified, then an appropriate method of motivation can be determined. Attempting to motivate without sincerity or integrity can cause teams to fragment or disintegrate. Thus, the project manager must use good people skills to motivate team members.

Conflict Management

Due to the dynamic nature of project environments, it is inevitable that conflicts among team members will arise. Conflicts can arise over the distribution of resources, access to information, disagreement about decisions, or the perception by an individual that he or she is not respected or fully a part of the team. Conflict can have a negative influence by fostering interpersonal hostility, reduced performance, and dissension in the team.

Conflicts can arise due to different agendas among team members since some of the team members may work for different companies. For example, on the owner's project management team there will be members from one or more design firms as well as one or more construction companies. Likewise, the design team's project manager may have team members who are in-house as well as design consultants who are under contract to perform special parts of the design. The construction contractor's project manager usually has many diverse personnel from both in-house and subcontractors.

It should be recognized that conflicts are often a result of changes, for example, modification of drawings, reassignment of a team member, or changes in meeting dates that are not communicated to other team members. The project manager and team should evaluate each conflict resolution to capture lessons learned, both positive and negative, for the benefit of future project work. A good project manager must and can manage conflicts.

Table 2-3 gives examples of potential conflicts that may arise during the design of a project and strategies to resolve or prevent the conflict.

To manage conflict the project manager must use techniques to deal with disagreements, both technical and personal in nature, that inevitably develop among team members. Project managers and team members may perceive that conflicts are bad, shouldn't exist, are caused by troublemakers, or should be avoided. However, all project participants

TABLE 2-3   **Examples of Conflicts and Resolutions of Conflicts during Design**

Potential conflict	Strategy to resolve the conflict
Changes in the design parameters after partial completion of design	Hold routine team meetings to discuss impact of changes on other members of the project team
Changes in dates or time of design team meetings	Maintain consistent times and agendas of design team meetings
Confusion about revisions of drawings or specifications during design	Establish and maintain a master directory of drawings and specification revisions
Changes by one team member that adversely affect other designers	Develop a procedure to discuss changes before they are approved
Recurring complaints about changes related to design of the project	Create a log of potential changes for review before submitting a change in design

should recognize that conflicts are inevitable and actually can be beneficial if resolved in an appropriate manner. Resolution of conflicts can lead to innovation and to ideas about how to improve work efficiency.

Withdrawal or giving up is a poor way of managing conflicts. It is a stopgap attempt to resolve a conflict and does not solve the problem. Withdrawal is a passive approach to solving a problem and only temporarily delays the inevitable future reoccurrence of the problem. Another method, smoothing, is a more active technique to managing conflicts. However, smoothing only temporarily avoids the conflict by appeasing one or more of the parties involved in the conflict. Smoothing does not provide long-lasting solutions.

Compromising is another approach to settling conflicts. This approach involves bargaining between the disputing parties to reach an acceptable agreement. The disputing parties make trade-offs that often fall short of ideal solutions. Too often, compromising does not result in a definitive resolution and leaves opportunities for a reoccurrence of the conflicts. Depending on the circumstances, however, compromising may be the best method of resolving conflicts.

Confronting and problem solving is a method of resolving conflicts that requires participation by all parties involved in the conflict. It requires an open dialogue to identify the root problems and a joint effort to use problem-solving techniques to objectively resolve the conflict. The potential to find final, mutually acceptable solutions is usually higher when the project manager and team use this method.

For some situations it may be necessary for the project manager to exercise authority to force a resolution of the conflict. Forcing a resolution can only be used when the project manager has the authority and is willing to use his or her power to settle the conflict without producing hurt feelings. The project manager must understand that forcing a solution may create resentment or other adverse reactions that can affect the team's future performance.

Developing a Consensus

Sometimes multiple solutions to a problem exist, each of which may yield the same result. Team members may not agree on any one particular solution. Some members may be indifferent regarding a particular solution, while others may have strong feelings about a solution. The project manager must work with the team to develop a consensus regarding selection of the best solution. For these situations, the project manager must help the team members to focus on solving the problem.

Voting, trading, or averaging can be helpful in reaching a consensus. The project manager must lead the team in seeking facts to avoid dilemmas and indecisions. During discussions, conflict should be accepted as helpful and every effort should be made to prevent threats, offensive comments, or defensive actions. Team members should avoid personal interests and behaviors that exclude the opinions and positions of other team members. Instead, they should exhibit mutual respect for each other with a special effort to focus on what is best overall for the project.

Team Conduct

A team is simply defined as two or more people who, by working together, accomplish more than if they worked separately. Projects typically consist of multiple design and construction packages, each with specific assignments and responsibilities. For a successful team, specific rules are required.

Goals for a particular team explicitly direct the project requirements of each specific group. For example, the safety team should know that it is supposed to keep health and accident incidents to a quantified level, or the scheduling team should know the number of days allowed to develop the CPM schedule. Teams need to know the required time to do their work and the desired outcome of their work.

In addition to knowing the goals, the team must also know how it is expected to operate. The design and construction teams should know from the beginning the extent of their power and authority to act. For example, some teams may be expected only to solve problems, while others are only required for routine reviews, and others exist to make decisions.

Every project requires a set of rules to be obeyed. Without rules, disorder and frustration will likely occur. Rules should not be viewed as restrictive. Instead, a good set of rules provides freedom to the team members to perform their work because they know what is expected of them. Generally, when the rules are unclear or absent, teams limit themselves from reaching their full potential.

Working relationships must be clearly defined and understood. To be effective, each team should know where it fits in the overall scheme of other operating teams. The leader of each team must answer questions regarding internal team relationships. Team members must show trust and respect for other team members, but close personal friendship is not always necessary. In engineering and construction projects, teams often function on the basis of professional relationships. Interrelationships between teams is the responsibility of upper management.

Personal values of individual team members often play an important role in teams. Each team member brings his or her own set of principles and values to the team. The extent and determination to which they hold these values influences how well the team will be able to work together. Team members should not have to compromise their personal values and principles. The team leader must be committed to respecting team members, promoting openness and flexibility. Situations will arise when compromises are necessary to resolve issues.

References

1. Culp, G, and Smith, R. Anne, *Managing People (Including Yourself) for Project Success,* Van Nostrand Reinhold, New York, NY, 1992.
2. Hensey, Mel, "Making Teamwork Work," *Civil Engineering,* American Society of Civil Engineers, Reston, VA, February 1992.
3. Maslow, Abraham, *Motivation and Personality,* Harper & Row, New York, NY, 1954.
4. Lewis, Chat T., Garcia, Joseph E., and Jobs, Sarah M., *Managerial Skills in Organizations,* Allyn and Bacon, Boston, MA, 1990.
5. Ruskin, Arnold M. and Estes, W. Eugene, "Organizational Factors in Project Management," *Journal of Management in Engineering,* American Society of Civil Engineers, Reston, VA, January 1986.
6. Sanvido, Victor E. and Riggs, Leland S., "Managing Successful Retrofit Projects," *Cost Engineering,* Association for Advancement of Cost Engineering–International, Morgantown, WV, December 1993.
7. Tatum, Clyde B., "Designing Project Organizations: An Expanded Process," *Journal of Construction Engineering and Management,* American Society of Civil Engineers, Reston, VA, June 1986.
8. Whetton, David A. and Cameron, Kim S., *Developing Management Skills,* 2nd ed., Harper-Collins, New York, NY, 1991.

Project Initiation

Design and Construction Process

Early in a project, the owner must select a process for design and construction. There are many choices of processes, each with advantages and disadvantages. The process selected affects financing; selection of team members; and the project cost, quality, and schedule. Although the process selected is important, selecting good-quality people is more important. A successful project is achieved by people working together with clear responsibilities.

Design and construction projects progress through three phases: project definition, design, and construction. It should be mentioned that for a total project there are business planning steps that precede design and there is an operations and maintenance phase that follows construction. This book focuses on the design and construction of projects. Project definition sets the stage for design work, and design work sets the stage for construction work. The project definition phase involves discovery to identify and analyze project requirements and constraints. Although the initial focus is on the owner's requirements and constraints, it must be recognized that the owner's requirements and constraints carry over to both the designer and contractor. Integration of the owner's requirements and constraints provides a description of the project and helps identify a plan for the time and cost of delivering the project.

Projects can generally be classified into three sectors: buildings, infrastructure, and process. Examples of building-sector projects include commercial buildings, schools, office buildings, and hospitals. For building-sector projects, where the architect is the prime designer, the design follows three stages: schematic design, design development, and contract documents. The schematic design produces the basic appearance of the project, building elevations, layout of floors, room arrangements within the building, and overall features of the project. At the conclusion of schematic design the owner can review the design configuration and the estimated cost before giving approval to proceed into design development. Design development defines the functional use and systems in the project in order to produce the contract documents, the plans and specifications for constructing the project.

Infrastructure-sector projects include transportation systems, such as city streets, county roads, state and federal highways, airports, or navigational waterways. The infrastructure sector also includes utility projects, such as water and sewer line systems, gas distribution lines, electrical transmission and distribution, telephone, and cable lines. For these types of projects the owner may be a private company or an agency of the government. The prime designer is the engineer, who generally prepares a complete design before construction contracts are created.

Process-sector projects include chemical plants, oil refining, pharmaceuticals, pulp and paper, and electrical generating. Engineers are the prime designers of process-sector projects. The stages of design include preliminary engineering, detailed engineering, and development of the contract documents. For example, preliminary engineering may involve designing the process flow sheets and mechanical flow sheets for a chemical processing plant. The preliminary engineering produces the major processes and major equipment required in the project. Detailed engineering involves the actual sizing of pipes that will connect to the equipment and control systems to operate the facility, such as piping and instrumentation drawings. The contract documents are the final drawings and specifications for constructing the project.

Depending on the project delivery method, procurement may start during the design phase. For example, as soon as the specification is completed for a major piece of equipment, a purchase order may be issued to procure the equipment if it is a long lead-time item that must be ordered in advance of construction to ensure that it can be installed without delaying the project. Procurement is not restricted only to equipment. Procurement may also apply to long lead-time acquisition of bulk material or procurement of construction contractors.

In the current practice of competitive-bid projects, contractors bid the project after the contract documents are completed. After accepting the bid, the contractor must develop shop drawings to build the project. Shop drawings are prepared by the contractor and submitted to the designer for approval. The shop drawings show the detailed fabrication and installation that will be used during construction. Thus, the contractor is also involved in design. The production of shop drawings impacts the quality of fabrication of manufactured items that will be installed at the job-site. Site construction involves labor, material, and construction equipment to physically build the project.

For non-competitive–bid projects, the owner negotiates a contract with a firm to provide engineering and/or construction services. Typically, the cost of the project is negotiated on some type of cost-reimbursable basis. The agreement also specifies how the engineering design will be integrated with the construction process.

Advances in the Engineering and Construction Process

The construction industry has matured and continued to enhance the integration of activities in the design, fabrication, construction, and operation of constructed facilities. Major advancements in computer hardware and software have produced two-dimensional (2-D) and three-dimensional (3-D) computer-aided design (CAD) systems. The CAD technology has progressed to versatile modeling systems that can be used throughout the design, engineering, and construction phases to greatly improve the capability to detect and prevent interference during field construction. The result is more efficient construction operations and less rework.

The biggest improvement using CAD is better coordination of activities within an integrated process, rather than automating individual activities within the existing fragmented design/construction process. The design intent may not be fully realized in the field using traditional information flow to the field through the use of drawings and other hardcopy documents. Traditional paper-based construction documents do not permit field personnel to interact with the 3-D model to extract information that meets their needs. Communication that uses 3-D modeling coupled with improved representation of design intent and other supplemental information can help alleviate many typical construction problems associated with material availability, work packaging, construction sequencing, and field changes.

Private versus Public Projects

Projects may also be classified as private-sector or public-sector projects. The owner of a private-sector project is typically a business that provides goods and services for a profit. Examples include commercial retail stores, manufacturing facilities, industrial process plants, and entertainment facilities. Since the owner is a private business, the business administrators have the flexibility to choose any engineering and construction services that suit their specific needs. For example, they can competitive bid the project or select a sole source firm to provide engineering and construction services. They are not restricted to accepting the lowest bid for the work and can choose any form of payment for services.

The owner of public-sector projects is typically a government agency, such as city, county, state, or federal. Examples include local school boards, state highway departments, or the federal department of energy or defense. For public-sector projects the owner typically uses the competitive-bid method based on the lowest bid price for securing engineering and construction services. However, in recent years there has been an increase in qualification-based selection (QBS) for securing engineering and construction services. Using the QBS process, the owner selects engineering and construction services based on specific qualifications and other factors, rather than only price.

Contractual Arrangements

Project management requires teamwork among the three principal contracting parties. Members of the owner's team must provide the project's needs, the level of quality expected, a permissible budget, and the required schedule. They must also provide the overall direction of the project. The designer's team must develop a set of contract documents that meets the owner's needs, budget, required level of quality, and schedule. In addition, the work specified in the contract documents must be constructable by the contractor. The contractor's team must efficiently manage the physical work required to build the project in accordance with the contract documents.

There are numerous combinations of contract arrangements for handling a project. Figure 3-1 illustrates the fundamental arrangements in their simplest form. Each of these arrangements is briefly described in the following paragraphs.

A design/bid/build contract is commonly used for projects that have no unusual features and a well-defined scope. It is a three-party arrangement involving the owner, designer, and contractor. This method involves three steps: a complete design is prepared, followed by solicitation of competitive bids from contractors, and the award of a contract to a construction contractor to build the project. Two separate contracts are awarded, one to the designer and one to the contractor. Since a complete design is prepared before construction, the owner knows the project's configuration and approximate cost before commencing construction. Considerable time can be required because each step must be completed before starting the next step. Also changes during construction can be expensive because the award of the construction contract is usually based upon a lump-sum, fixed-price bid before construction, rather than during construction.

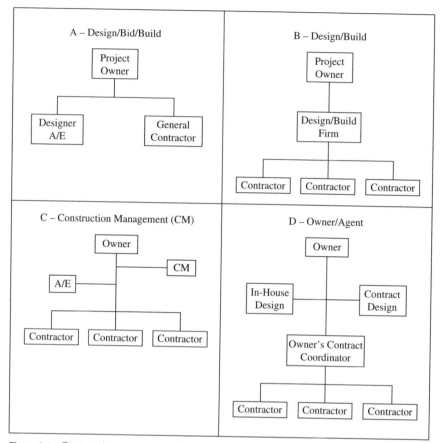

Figure 3-1 Contracting arrangements.

Table 3-1 gives several advantages and disadvantages of using the design/bid/build method of contracting construction work.

A design/build contract is often used to shorten the time required to complete a project or to provide flexibility for the owner to make changes in the project during construction. It is a two-party arrangement between the owner and the design/build firm. Since the contract with the design/build firm is awarded before starting any design or construction, a cost-reimbursable arrangement is normally used instead of a lump-sum, fixed-cost arrangement. This method requires extensive involvement of the owner for decisions that are made during the selection of design alternatives and the monitoring of costs and schedules during construction.

Table 3-2 gives several advantages and disadvantages of using the design/build method of contracting construction work.

TABLE 3-1   Advantages and Disadvantages of Design/Bid/Build Contracts

Advantages	Disadvantages
Cost of project is known before owner commits to signing construction contract	Changes by owner after the contract is signed can be costly and cause delays
Responsibilities and risks of parties are defined in contract documents	Major changes, outside of the original contract bid documents, may be difficult and costly
Owner has a high level of involvement and control during design development	Combined time required to design, bid, and build the project can be extensive
Selected for projects when cost is primary and schedule is secondary	Owner has little control of the project during construction

A construction management (CM) contract can be assigned to a CM firm to coordinate the project for the owner. The CM contract is a four-party arrangement involving the owner, designer, CM firm, and contractor. During the past twenty years there has been considerable debate regarding the CM process and the amount of responsibility assigned to the CM firm by the owner. The basic CM concept is that the owner assigns a contract to a firm that is knowledgeable and capable of coordinating all aspects of the project to meet the intended use of the project by the owner. The CM method of contracting is discussed further in Chapter 11.

An owner/agent arrangement is sometimes used for handling a project. Some owners perform part of the design with in-house personnel and contract the balance of design to one or more outside design consultants. Construction contracts may be assigned to one contractor or to multiple contractors. Although uncommon, an owner may perform all design and construction activities with in-house personnel. When a

TABLE 3-2   Advantages and Disadvantages of Design/Build Contracts

Advantages	Disadvantages
Less time to complete the project because construction can overlap design	Difficulty in evaluating design/build contractor with respect to total cost of project
Reduced adversarial relationship between design and construction personnel	Total cost of project is not known at the beginning of project because of no design
Provides flexibility to the owner to make changes during design and construction	Handling of inspection is an issue that must be addressed because the designer is the builder
Having one contracting firm reduces the owner's management responsibility	Considerable time may be required of owner to make major decisions and monitor costs

project is handled in this manner, it is sometimes referred to as a force-account method.

There are two general types of owners: single-builder owners and multiple-builder owners. Single-builder owners are organizations that do not have a need for projects on a repetitive basis, normally have a limited project staff, and contract all design and construction activities to outside organizations. They usually handle projects with a design/bid/build or construction management contract.

Multiple-builder owners are generally large organizations that have a continual need for projects, and generally have a staff assigned to project work. They typically will handle small-sized, short-duration projects by design/bid/build. For a project in which they desire extensive involvement, a design/build, construction management, or owner/agent contract arrangement is often used.

An owner can select a variety of ways to handle a project. The contract arrangement that is selected depends on the resources available to the owner, the amount of project control the owner wishes to retain, the amount of involvement desired by the owner, the amount of risk that is shared between the owner and contractor, and the importance of cost and schedule.

Phases of a Project

A project is in a continual state of change as it progresses from its start, as a need by the owner, through design development and, finally, construction. Figure 3-2 shows the various phases during the life of a project. As the project moves from one phase to another, additional parties become involved and more information is obtained to better identify scope, budget, and schedule. There are times when a project recycles through a phase before gaining management approval to proceed to the subsequent phase. During each phase, it is the responsibility of the project manager to keep all work within the approved scope, budget, and schedule.

In the early phases of design development, there may not be sufficient information to define the scope accurately enough to know the work to be performed. A characteristic of most project managers is "I can do it." This characteristic often leads to assignment of work to the project manager before the work is completely defined or officially approved. This applies to the project manager in either the owner, designer, or contractor organization. The people who work around the project manager include clients, subordinates, project team members, upper management, and colleagues who are themselves project managers. A project manager cannot efficiently utilize his or her time or effectively manage when special requests are made for work that is not well defined. If

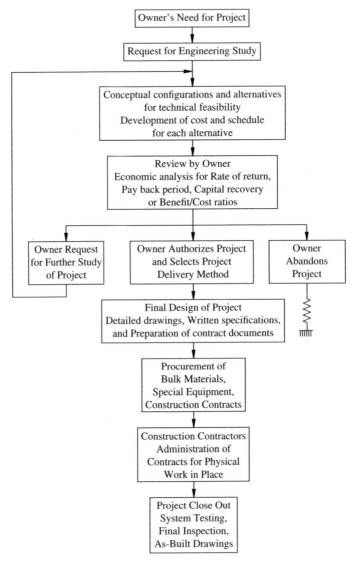

Figure 3-2 Phases of a project.

these conditions exist, the work should be performed on a time and material basis for actual work accomplished, until an adequate scope, budget, and schedule can be determined. Another option is to define a scope, with a matching budget and schedule. Then when there is a deviation from the defined scope, the project manager can advise the owner of the readjusted budget and schedule caused by the change in scope and obtain the owner's approval before proceeding with the work.

During the development of conceptual configurations and alternatives, the quality and total cost of the project must be considered. This can only be achieved through extensive input from the owner who will ultimately use the project, since the cost to operate and maintain the facility after completion is a major factor in project design. Sometimes the budget is a controlling factor, which causes the owner's contemplated scope to be reduced, or expanded. If this condition exists, care must be exercised to ensure the project meets the minimum needs of the owner and there is a clear understanding of the level of quality that is expected by the owner. It is the duty of the project manager to ensure that project development meets the owner's expectations.

The owner's authorization to proceed with final design places pressure on the designer to complete the contract documents at the earliest possible date. However, the quality and completeness of the bid documents have a great influence on the cost of the project. Adequate time should be allocated to the designer to produce a design for the project that is constructable and will perform for the owner with the least amount of maintenance and operating costs.

For large projects the procurement of bulk material and special equipment has a large impact on the construction schedule. The project manager must ensure that long lead-time purchase items are procured. This must be coordinated with the owner's representative on the project team.

The type of contract chosen and the contractors selected to bid the project influence cost, schedule, and quality. The project manager plays an important role in the process of qualifying of contractors, the evaluation of bids, and recommendations of the award of construction contracts.

Owner's Study

A project starts as a need by the owner for the design and construction of a facility to produce a product or service. The need for a facility may be recognized by an operating division of the owner, a corporate planning group, a top executive, a board of directors, or an outside consulting firm. Generally one or more persons within the owner's organization are assigned to perform a needs assessment to study the merits of pursuing the project.

The first requirement of the owner is objective setting. This is important because it provides a focus for scope definition, guides the design process, and influences the motivation of the project team. The process of setting objectives involves an optimization of quality, cost, and schedule. The owner's objectives must be clearly communicated and understood by all parties and serve as a benchmark for the numerous decisions that are made throughout the duration of the project.

The magnitude of the owner's study varies widely, depending on the complexity of a project and the importance of the project to the owner. It is an important study because the goals, objectives, concepts, ideas, budgets, and schedule that are developed will greatly influence the design and construction phases.

A part of the owner's study is defining the requirements of the proposed project. However, some owners do not have the experience or expertise to conduct a study to effectively define, in engineering and construction terms, the operational needs of their company. The owner may be very successful in running a manufacturing operation, but not be familiar with defining scope, budget, and schedule for engineering and construction projects. Therefore, the project manager of the design team may need to assist the owner with the important owner's study to define the requirements of a proposed project. Table 3-3 gives methods to assist the owner in defining project requirement.

The owner's study must conclude with a well-defined set of project objectives and needs, the minimum requirements of quality and performance, an approved maximum budget, and a required project completion date. Failure to provide any of the above items starts a project in the wrong direction and leads to future problems. Sometimes an owner will contract parts of the study to an outside consulting firm. If an outside firm is utilized, the owner must still be involved to be certain his or her needs are represented.

The thoroughness and completeness of the owner's study has a significant impact on total project cost. An inadequately defined project scope leads to changes during design and/or construction. An incomplete scope

---

TABLE 3-3   Methods of Assisting the Owner in Defining Project Requirements

Meet with the owner's staff, in particular the operations and maintenance people who will use the project when it is completed

Spend adequate time with the operations and maintenance people, to become acquainted with the work they do, how they do it, and what they are trying to accomplish

Conduct a series of meetings to separate "what they need" from "what they want," and relate those needs to the requirements that must be designed and built into the proposed project

Develop preliminary scope, budget, and schedule requirements of the project and translate those requirements into terms that can be understood and agreed upon by the owner's operating group and management

Obtain feedback from the owner's staff to verify they understand the project requirements that must be accomplished in order to satisfy the needs of the owner's operating group

Document the results of the meetings, discussions, alternatives, and the decisions that were made and ensure the results are distributed to the design group

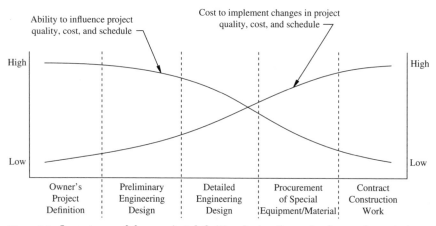

Figure 3-3 Importance of clear project definition during the early phases of a project.

leads to costly change orders and, frequently, to claims and disputes which lead to major cost overruns, delays, and other problems. Experienced managers agree the time to achieve savings and reduce changes is in the early life of the project, not at the start of construction. This concept is illustrated in Figure 3-3.

Owner's Needs and Project Objectives

An owner must know his or her needs and objectives before any productive project work can be started. If the owner doesn't know what the project requires, then no one knows what to do. Defining owner needs is the first step in a broad range of preproject activities that lead to scope definition. A project manager cannot form the project team to execute the project without a clear scope definition.

The process of identifying owner needs and objectives requires the involvement of a wide range of people within the owner's organization. This includes top managers and investors, financial personnel, and in particular the people who will use and/or operate the project after it is constructed. The process of identifying owner needs and objectives usually involves numerous activities and discussions. It is important that "what is needed" be separated from "what is wanted." Without constraints of cost and schedule, focus easily shifts from what is needed to what is wanted. This makes a project unaffordable and non-feasible. Because there are always constraints of cost and schedule, the owner must develop a project definition based upon need. This process involves an optimization of quantity, quality, cost, and schedule.

Members of the owner's organization must realize that it is their responsibility to resolve all issues related to project needs and objectives before assigning the project to the project manager. It is not the duty of the project manager or the project team to define the owner needs. Vague owner needs lead to project changes, scope growth, cost overruns, rework, and misunderstandings among team members. The best way to determine needs, and information related to needs, is to talk to the people who will use the facility after it is constructed.

The owner sets the project objective, which is the purpose, aim, goal, intent, or reason the project is to be designed and constructed. The project manager in charge of the project defines the project scope, which is work that must be performed to achieve the owner's project objectives. Setting project objectives and establishing project scope requires an interactive role between the owner and designer. Table 3-4 shows the interactive role of the owner and designer in the process to ensure a well-defined understanding between the two parties.

The following paragraphs present a hypothetical example of the development of an owner's needs. An owner may define a company goal of centralization of its operations to streamline operating efficiency. To achieve this goal, company management may set the objective of consolidating the service facility of each of its five operating districts into a single location. Thus, there is a need to design and build a service facility that will serve the five operating districts. Key people, from each district, must meet and agree on what is needed in a facility that satisfies the intended usage by each operating district. Negotiations between the

TABLE 3-4   Owner's Project Objectives and Designer's Project Scope

Project Objectives
The owner has the primary responsibility for setting project objectives.
Project objectives define the needs of the owner that must be met by the project.
A detailed definition of project objectives is a prerequisite to defining project scope.
There needs to be a separation between "what is needed" from "what is wanted."
People involved should include operations, financial, and upper management.

Project Scope
The designer has the primary responsibility of defining the scope of the project.
The project scope identifies work items, "deliverable," required to achieve the project.
Each work item in the scope should be budgeted and scheduled before starting work.
There needs to be a separation between "what should be done" and "what should not be done."
People involved should include designers, lead designers, and upper management.

Interactive Roles of Owner and Designer
A joint meeting of the owner and designer is necessary to clarify objectives and scope.
The owner should relate objectives in terms that can be understood by the designer.
The designer should relate scope in terms that can be understood by the owner.
Both the owner and designer should agree that the objectives and scope are compatible.
The preliminary project budget and schedule can be developed after scope is defined.

people should focus on what is best overall for the company in order to achieve efficiency of operations, which is the company's goal. Compromise is often necessary to separate "what is needed" from "what is wanted." The end result should be a facility that meets the needs of all five districts and can be operated more efficiently than five separate service facilities. For example, agreement may be reached that the owner needs a facility consisting of three buildings: an employee's office building, a warehouse, and a maintenance shop. Additionally, an outside heavy equipment and bulk materials storage area may be needed. These minimum requirements of the facility then initiate the process of project definition and scope.

A part of the owner's needs and objectives study is assessment of the total project budget because management generally will not approve starting the design of a project unless the probable total cost is known. The project budget at this stage of development is based on parameter costs, such as cost per square foot of building or cost per acre of site development. If the anticipated project cost exceeds the amount that management is willing to approve, then it is necessary to reduce the scope of work. For example, the employee's office building and maintenance shop may be retained in the project and the warehouse eliminated. This decision would be made if the warehouse is the lowest priority of the three buildings. Consideration would be given to adding the warehouse at a future date when funds are available, after completion of the site-work, the employee's office building, and the maintenance shop. The project management of this type of project is further discussed in Chapters 6, 8, and 9, and the Appendix.

Project Scope Definition

Project scope identifies those items and activities that are required to meet the needs of the owner. For example, a project may need three buildings consisting of an employee's office building, a warehouse, and a maintenance shop. In addition, the project may need a crushed aggregate area for storage of heavy equipment and bulk materials. Each of the above items should be defined in further detail, such as number of employees in each building, type and amount of storage needed in the warehouse, type of maintenance required, and size and weight of equipment. This type of information is needed by the project manager and team to define the work required to meet the owner's needs and objectives.

The purpose of project scope definition is to provide sufficient information to identify the work to be performed, to allow the design to proceed without significant changes that may adversely affect the project budget and schedule. Just to state that a project consists of three buildings and an outside storage area is not enough information to start the design phase.

TABLE 3-5 Abbreviated Checklist for Project Scope Definition of a Petrochemical Project

1. General
- 1.1 Size of plant capacity
- 1.2 Process units to be included
- 1.3 Type of plant feedstock
- 1.4 Products to be made, initial and future
- 1.5 Should plant be designed for minimum investment
- 1.6 Horizontal vs. stacked arrangement of equipment
- 1.7 Layout and provisions for future expansion
- 1.8 Any special relationships (e.g., involvements of other companies)

2. Site information
- 2.1 Access to transportation: air, waterway, highway, railway
- 2.2 Access to utilities: water, sewer, electrical, fire protection
- 2.3 Climate conditions: moisture, temperature, wind
- 2.4 Soil conditions: surface, subsurface, bearing capacity
- 2.5 Terrain: special precautions for adjacent property
- 2.6 Acquisition of land: purchase, lease, expansion potential
- 2.7 Space available for construction

3. Buildings
- 3.1 Number, types, and size of each
- 3.2 Occupancy: number of people, offices, laboratories
- 3.3 Intended usage: offices, conferences, storage, equipment
- 3.4 Special heating and cooling requirements
- 3.5 Quality of finish work and furnishings
- 3.6 Landscaping requirements
- 3.7 Parking requirements

4. Regulatory requirements
- 4.1 Permits: construction, operation, environmental, municipal
- 4.2 Regulations and codes: local, state, federal
- 4.3 Safety: detection systems, fires, emergency power
- 4.4 Environmental: air, liquids, solids, wetlands
- 4.5 Preservations restrictions

To assist the owner in this effort, a comprehensive checklist of items should be prepared. Table 3-5 is an abbreviated checklist for project scope definition of a petrochemical project. The table is provided for illustrative purposes only and does not include all the items that should be considered. A similar checklist should be prepared for other types of projects. Experienced design and construction personnel can provide valuable input to assist an owner in the development of a checklist for project scope.

Before design is started, scope must adequately define deliverables, that is, what will be furnished. Examples of deliverables are design drawings, specifications, assistance during bidding, construction inspection, record drawings, and reimbursable expenses. All this information must be known before starting design because it affects the project budget and schedule. To accomplish this, the project manager from the design organization must be involved early in the project; and he or she will require input from experienced technical people to represent every aspect of the proposed project.

A realistic budget and schedule cannot be determined for a project without a well-defined scope of work. Thus, the project scope should be developed first, then a project budget and schedule developed that matches the scope. It is the responsibility of all project managers to keep all work within the approved scope, and all costs and schedule within approved limits.

There are times when an owner may become excited about the merits of a project and anxious to begin work as soon as possible. This usually occurs when a new product is developed or a government official decides a facility should be built at a particular time or location. The project manager must thoroughly review the project scope and be certain that it is sufficiently well defined before starting work on the project. If this is not done, the project team is forced into defining scope while work is being performed, which leads to frustration and adverse relationships. The simple solution to this problem is to lock in the scope at the beginning of the project, before starting work, to make certain all parties know the full extent of the required work.

Project Strategy

In the early stages of project development the owner must develop the project strategy, a plan to carry out tasks in a timely manner. Project strategy forms the framework for handling the project. It includes the contracting strategy, the roles and responsibilities of the project team, and the schedule for design, procurement, and construction.

Contract strategy identifies the overall organizational structure and the allocation of risk among the contracting parties. In the early stages of project development the owner must decide the work that can be performed by in-house personnel and the work that must be contracted to outside organizations. The owner may have a large engineering staff that can handle the entire project: design, procurement, and construction. In other cases the owner may only have a limited staff for projects, which requires the assignment of contracts to outside organizations that have the capability to perform the necessary work.

Although a large organization may have the in-house capability, it may not be able to schedule the work when it is needed due to prior commitments. The owner's organization must make a realistic assessment of the work that can be accomplished in-house and an outside firm's capability to perform the work, and then evaluate the cost and schedule trade-offs of purchasing outside services.

The type of contract chosen defines the allocation of responsibilities and risks for each party and influences the project schedule. If a fast-track schedule is important in order to obtain an early return on the project investment, then a cost-plus-fee contracting strategy may be desirable. Government projects of an emergency nature are sometimes handled in this manner. If there is ample time to complete the entire design, a traditional design/bid/build approach with a lump-sum contract may be desirable. The owner must evaluate all possibilities, identify the advantages and disadvantages, and consider what best meets his or her needs, objectives, budget constraints, and schedule requirements.

The project strategy includes a schedule for the timing of design, procurement, and construction tasks. The purpose of the owner's schedule is to identify and interface overall project activities: design, procurement, and construction. A workable schedule must be developed that integrates the activities of all parties involved in the project. Any change in the project schedule should be approved by all parties.

Selection of Design Firms and Construction Contractors

Selection of the designer and constructor varies depending on many factors including the type, size, and complexity of the project; the owner's knowledge in handling engineering and construction projects, and how soon the owner wants the project completed. The method of selection depends on the owner's project strategy and the contract arrangement chosen by the owner.

When the owner plans to complete all the design before selecting a construction contractor, then a procedure must be initiated for selecting the designer. Typically, an owner selects a designer that he or she has used before and with which he or she has had a satisfactory experience. For private-sector projects, owners can simply choose their preferred designer or they may desire to obtain proposals from several design organizations that they have used in the past. A request for proposal (RFP) is issued to the prospective designers who then each prepare a design proposal as discussed in Chapter 7. After the design organizations have submitted their proposals, the owner can review and evaluate the proposals and make a decision for award of the design contract. For public-sector projects, selection of the designer depends on the policies and restrictions of the owner's organization. Generally, designers are selected from a list of prequalified firms. Chapter 5 presents methods of compensation for professional design services.

If the owner has no prior experience in working with designers, a procedure must be established to select the designer. After the owner has studied the proposed project and its need for design services, a list of prospective design organizations is identified. Often the list is compiled based on recommendations of other owners or those acquainted with design firms who are known to have the expertise required to design the project. Generally the list consists of at least three design firms that appear to be best qualified for the particular project. Each design firm is sent a letter that briefly describes the proposed project and inquires about its interest in the project. Upon receipt of confirmation that the design firm is interested in the project, the owner then conducts a separate interview of each design firm. At the interview the owner reviews the qualifications and records of the firm to assess its capability to complete the work within the allotted time and to review specific key personnel that would be assigned to the project. It is important for the owner to meet the specific people who will be performing the design work to ensure compatibility of personalities.

Typically after all interviews are conducted, the owner lists the design firms in the order of their desirability, taking into account their location, reputation, size, experience, financial stability, available personnel, quality of references, work load, and other factors related to the proposed project. Based on the evaluation, one or more additional interviews of the top design firms may be conducted before a decision is made on the final selection.

If the design is 100% complete, the owner may issue requests for bids (RFB) to construction contractors. For most private-sector projects the contract documents generally state that selection of the construction contractor will be based on the lowest and best bid. Typically for public-sector projects the contract documents state the selection of the construction contractor will be based on the lowest qualified bidder. However, the lowest bid is generally the criteria for selection of a construction contractor when the design is 100% complete.

Sometimes the owner may desire to start construction before design is completed. For example, the construction contractor may be chosen after 70% design completion, or the construction contractor may be selected at the same time the designer is selected in order to take advantage of the contractor's knowledge of building the project. When the owner desires to start construction before design is complete, selection of the construction contractor cannot be made on price alone because the design documents are not completed. When the construction contractor is selected before design is completed, a procedure is established to review and evaluate prospective construction contractors similar to the procedures presented in preceding paragraphs for selection of the designer. A more detailed discussion of project delivery methods for construction is presented in Chapter 11.

Partnering

The competitive environment and the rigid requirements of contracts have, at times, caused adverse relationships in the construction industry. Traditionally, contractors and vendors have been selected on a competitive-bid basis to provide construction services, under formal contracts, to meet the requirements specified in the drawings and specifications. A short-time commitment is made for the duration of the project. Thus, contractors and vendors work themselves out of a job.

A relatively new concept called *partnering* is an approach that focuses on making long-term commitments, with mutual goals for all parties involved, to achieve mutual success. The Construction Industry Institute (CII) established a task force on partnering to evaluate the feasibility of this method of doing business in the construction industry. CII Publication 17-1, entitled *In Search of Partnering Excellence,* is a report that discusses the research findings on partnering practices. The following paragraphs are excerpts from the report.

Partnering is a business strategy that offers many advantages to the parties involved; however, its success depends on the conduct of the parties and their ability to overcome barriers related to doing business differently than in the past. Companies agree to share resources in a long-term commitment of trust and shared vision, with an agreement to cooperate to an unusually high degree to achieve separate yet complementary objectives. Partnering is not to be construed as a legal "partnership" with the associated joint liabilities.

The first known partnering relationship in the construction industry was between an oil company and a contractor. The owner approached the contractor and proposed that some of the existing engineering blanket work be accomplished using a new set of relationships and accountabilities. Hence, both agreed to enter into a "partnering relationship" to perform multiple projects in different locations. The services provided by the contractor included project-execution related services, while the owner provided technical assurance and approved only primary funding documents and scoping documents developed by the contractor. Twenty-five different projects were performed with this relationship.

From a contractual point of view, this first partnering relationship differed from traditional contracts because the bureaucratic procedures were removed and all issues were open for negotiations. In this relationship the owner agreed to carry the financial burden of any risks that might occur during the duration of the relationship. The parties agreed to set performance evaluation criteria for major areas that were important to the projects. An incentive system based on the performance criteria was utilized, including monetary awards given by the owner to the contractor for doing a good job. Contractor incentives to employees included both monetary and non-monetary incentives.

A cultural change is required by all parties in a partnering relationship. The three key elements of any successful partnering relationship are trust, long-term commitment, and shared vision. As these three elements are developed, other subelements are achieved and the benefits to all parties are maximized. Both customer and supplier can profit from reduced overhead and work load stability. Competitive advantage is enhanced through improved cost, quality, and schedule. Growth and balance are important to the continual improvement of the partnering agreement. For example, in developing long-term commitment, a partnering agreement may grow from single to multiple projects. Likewise, trust may evolve from competitive bidding through complete disclosure of project costs in a cost-plus relationship. Shared vision can expand to open sharing and mutual development of business objectives.

The CII publication discusses applications of partnering to small businesses and projects, guidelines on selecting partners, and guidelines for implementing a partnering relationship.

References

1. Barrie, D. S. and Paulson, B. C., *Professional Construction Management,* 3rd ed., McGraw-Hill, Inc., New York, NY, 1992.
2. Barrie, D. S., *Directions in Managing Construction,* Wiley, New York, NY, 1981.
3. Haltenhoff, C. E., *Construction Management: A State-of-the-Art Update,* Proceedings of the Construction Division, American Society of Civil Engineers, Reston, VA, 1986.
4. Hinze, Jimmie, *Construction Contracts 3rd ed.,* McGraw-Hill, New York, NY, 2010.
5. Haltenhoff, C. E., *The CM Contracting System—Fundamentals and Practices,* Prentice Hall, Upper Saddle River, NJ, 1999.
6. Hancher, D. E., *In Search of Partnering Excellence,* Publication No. 17-1, Construction Industry Institute, Austin, TX, July 1991.
7. Laufer, A., *Owner's Project Planning: The Process Approach,* Source Document No. 45, Construction Industry Institute, Austin, TX, March 1989.
8. *Organizing for Project Success,* Publication No. 12-2, Construction Industry Institute, Austin, TX, February 1991.
9. Rowings, J. E., *Project Objective Setting,* Publication 12-1, Source Document No. 31, Construction Industry Institute, Austin, TX, April 1989.
10. *Scope Definition and Control,* Publication No. 6-2, Construction Industry Institute, Austin, TX, July 1986.
11. *Standard CM Services and Practice,* Construction Management Association of America, Washington, D.C., 1988.

4

Early Estimates

Importance of Early Estimates

For engineering and construction projects, accurate early cost estimates are extremely important to the sponsoring organization and the engineering team. For the sponsoring organization, early cost estimates are often a basis for business unit decisions, including asset development strategies, screening of potential projects, and committing resources for further project development. Inaccurate early estimates can lead to lost opportunities, wasted development effort, and lower than expected returns.

An early estimate is also important to the project team because it becomes one of the key project parameters. It helps formulate execution strategies and provides a basis to plan engineering and construction. The early estimate often serves as a baseline for identifying changes as the project progresses from design to construction. In addition, the performance of the project team and overall project success is often measured by how well the final cost compares to the early cost estimate.

Importance of Estimator

The skills, knowledge, and experience of the estimator are crucial for preparing accurate estimates for capital projects. Tools can aid the estimating process, but cannot replace the judgment and experience of a competent estimator. Skills can be learned in the classroom, but knowledge is gained by experience.

Most efforts to promote the practice of good estimating have focused on developing cost information, performing quantity takeoffs, adjusting costs for time/size/location, and creating estimating tools. Often, there is heavy emphasis on automating the estimating process because of the many calculations that are involved in preparing estimates.

Spreadsheets and commercial software can greatly assist in preparing estimates, but there are other factors that are equally important, including knowledge of materials, methods of construction, common sense, and good judgment.

Automated estimating procedures can not replace experience. A competent estimator has the skills to run spreadsheets and computer software, but also knows when to do periodic "do these numbers look right" checks of an estimate. There are times when the estimator uses his or her experience to see the big picture and realize what is important and apply judgment and do reality checks of numbers in the estimate.

Classification of Early Estimates

There are many estimates and reestimates for a project, based on the stage of project development. Estimates are performed throughout the life of a project, beginning with the first estimate through the various phases of design and into construction as shown in Figure 4-1. Initial cost estimates form the basis to which all future estimates are compared. Future estimates are often expected to agree with (i.e., be equal to or less than) the initial estimates. However, too often the final project costs exceed the initial estimates.

Various names have been given to estimates by several organizations. However, there is no industry standard that has been established for defining estimates. In general, an early estimate is defined as an estimate that has been prepared after the business unit study but prior to completion of detailed design.

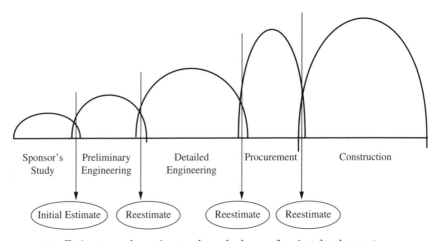

Figure 4-1 Estimates and reestimates through phases of project development.

Estimate class	Level of project definition	End usage (Typical purpose of estimate)	Expected accuracy range
Class 5	0% to 2%	Concept Screening	−50% to 100%
Class 4	1% to 5%	Study or Feasibility	−30% to +50%
Class 3	10% to 40%	Budget, Authorization, or Control	−20% to +30%
Class 2	30% to 70%	Control or Bid/Tender	−15% to +20%
Class 1	50% to 100%	Check Estimate or Bid/Tender	−10% to +15%

Figure 4-2   AACE International cost estimation classifications (18R-97).

Individual companies define estimate names and percent variations that they use. Various organizations have also defined classifications of cost estimates. Two examples are the cost estimate classifications by the Association for Advancement of Cost Engineering (AACE) International shown in Figure 4-2 and by the Construction Industry Institute (CII) shown in Figure 4-3.

In general, an early estimate is defined as an estimate that has been prepared before completion of detailed engineering. This definition applies to class 5, class 4, and early class 3 estimates of AACE International. This definition also applies to order-of-magnitude and factored estimates described in CII publications.

Estimating Work Process

Estimating is a process, just like any endeavor that requires an end product. Information must be assembled, evaluated, documented, and managed in an organized manner. For a process to work effectively, key information must be defined and accumulated at critical times. The primary factors in preparing estimates are shown in Table 4-1.

These concepts are illustrated in the estimating work process of Figure 4-4. The first step in the estimating work process is alignment. Alignment between the customer and the estimating team must be established before starting an estimate. As discussed on the following pages, alignment is accomplished by early communications to ensure a clear understanding of the customer's expectations and the estimating team's ability to meet those expectations. Close alignment helps mitigate estimate inaccuracies that can result from misunderstandings and

Estimate class	Percent range	Description/methodology
Order-of-Magnitude	+/−30 to 50%	Feasibility study—cost/capacity curves
Factored Estimate	+/−25 to 30%	Major equipment—factors applied for costs
Control Estimate	+/−10 to 15%	Quantities from mech/elect/civil drawings
Detailed or Definitive	+/−<10%	Based on detailed drawings

Figure 4-3   Construction industry institute cost estimate definitions (CII SD-6).

TABLE 4-1   **Primary Factors in Preparing Estimates**

1. Standardization of the cost estimate preparation process
2. Alignment of objectives between the customer and team
3. Selection of estimate methodology commensurate with the desired level of accuracy
4. Collection of project data and confirmation of historical cost information
5. Organizing the estimate into the desired format
6. Documentation and communication of estimate basis, accuracy, etc.
7. Review and checking of estimate
8. Feedback from project implementation

miscommunications. It also enables establishment of the estimate work plan and staffing requirements. The estimate kick-off meeting provides an excellent forum for establishing alignment.

A successful process provides a clear understanding of the work to be performed and the products that will be produced. The level of scope definition for early estimates is low compared to later estimates. There must be a mutual understanding between the business unit and the estimator regarding the level of scope definition. The estimator must communicate to the business unit the expected range of accuracy based on the level of scope definition.

Prior to starting the estimate, a work plan for preparing the estimate should be developed. The work plan can be developed after alignment and scope definition. The estimate work plan identifies the work that is needed to prepare the estimate including who is going to do it, when it is to be done, and the budget for preparing the estimate. The plan also includes the tools and techniques that are appropriate for the level of scope definition and the expected accuracy of the estimate. The leader of the estimating team is responsible for developing the estimate work plan which is discussed in the following pages.

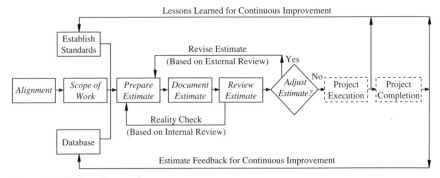

Figure 4-4 Estimating work process.

While preparing an estimate, there must be two-way communications between the estimating team and the party that requested the estimate. The estimating team must keep the requesting party informed of the work being performed, and the requesting party must respond to questions that may arise from the estimating team. The estimating process can assist the requesting party in identifying areas of uncertainty and additional information that may be needed or assumptions that must be made in lieu of definitive information about the project.

After the estimate is completed, a document should be prepared that defines the basis of the estimate. Estimate documentation is essential for presentation, review, and future use of the estimate. The documentation for an estimate improves communications among project participants, establishes a mechanism for estimate reviews, and forms a basis for early project cost control. The estimating team should develop a standard cost estimate presentation format that is easily understood by internal business and engineering management.

Contingency is the amount of money that must be added to the base estimate to account for risk and uncertainty. Contingency is a real and necessary component of an estimate. Assessing risk and assigning contingency to the base estimate is one of the most important tasks in preparing early estimates. Typically, risk analysis is a prerequisite to assigning contingency. Based on the acceptable risks and the expected confidence level, a contingency is established for a given estimate. The lead estimator for a project must assess the uniqueness of each project and select the technique of risk analysis that is deemed most appropriate.

No estimating process is complete without the continuous feedback loops shown in Figure 4-4. To improve early estimates, the estimating process must be a continuous cycle. Actual cost information from completed projects must be captured in a feedback system that can be integrated into the cost database for use in preparing future estimates. Lessons learned during project execution must also be documented and incorporated into estimating standards and procedures. The lessons learned during construction must be communicated back to the estimating team, to enable them to establish better standards for preparing future project estimates.

Importance of Team Alignment in Preparing Early Estimates

Early communication between the team and the customer is essential to the success of any estimate, particularly *early estimates*. This early communication is necessary to ensure a clear understanding of the customer's expectations and the team's ability to meet those expectations. Table 4-2 is a list of benefits of team alignment.

TABLE 4-2   Benefits of Team Alignment

---

1. Establishes a clear understanding between the customer and the team of the project's parameters
2. Assists in determining the level of effort required of the estimating team to deliver the estimate
3. Enables the estimating team to establish a work process and staffing plan to provide the deliverables required to meet the customer's expectations
4. Highlights issues that might not otherwise have been considered in the development of the estimate
5. Improves and documents the level of scope definition and the information that is known about the project
6. Assists the customer's understanding of what is included in the estimate and what is not included in the estimate
7. Establishes the responsibility of all project team members and the customer in the preparation of the estimate
8. Serves to establish a cohesiveness between the project team and the customer

---

To achieve alignment, a special effort must be made to resolve all issues that can impact the team's work and the customer's understanding of the contents of the estimate. This can be accomplished by two-way, open communications before starting the estimating process.

The level of scope definition of the project is one of the key issues that must be resolved early, because the accuracy of the estimate is dependent on scope definition. The customer must provide the level of accuracy and detail that is expected in the estimate. The estimating team must clearly communicate to the customer what will be provided in the estimate. The customer must also define the required deliverables of the team and the type of decisions that will be made based on the estimate. Critical questions that must be addressed early are shown in Table 4-3.

The estimate kick-off meeting is an effective method of achieving alignment by addressing the issues shown in Table 4-4. The estimate kick-off meeting between the customer and the team provides an exchange of information regarding the customer's expectations and the team's ability to meet those expectations. Regularly scheduled progress meetings after the kick-off meeting ensure continued alignment throughout the estimating process.

Project management should initiate open communication between the customer and the team to assist in identifying and documenting the issues to be resolved. Early involvement of the customer reduces the potential

TABLE 4-3   Critical Questions for Preparing Early Estimates

---

1. How thorough is the scope definition of the project?
2. What level of accuracy and detail is the customer expecting?
3. What deliverables are required from this effort?
4. What decisions will be made based on this estimate?

TABLE 4-4   Checklist of Issues for the Estimate Kick-Off Meeting

1. What are the customer's driving principles and expectations?
2. What is the level of scope definition of the project?
3. What level of accuracy and detail is the customer expecting?
4. What deliverables are required from this effort?
5. What decisions will be made based on the estimate?
6. Does the project have unique or unusual characteristics?
7. What is the estimate due date and the anticipated project start/completion date?
8. What level of confidentiality is required by the team?
9. Who are the customer's contacts with the team?
10. What other organizations need to interface with the team?
11. Are there other information sources that can aid the estimating team?
12. What is the budget for developing the estimate and who is paying for it?
13. Have similar projects/estimates been developed previously?
14. What customer-furnished items are to be excluded from the estimate?
15. What customer-furnished costs are to be included in the estimate?
16. Are there specific guidelines to be used in preparing the estimate?
17. Are there special permitting requirements that may affect the cost and schedule?
18. Are there any special funding requirements that might influence the final total installed cost?
19. Are there other issues that could affect the cost or schedule of the project?
20. What level of effort is required to meet the desired accuracy?

of giving conflicting instructions and directions to the team. Alignment requires a cooperative effort between the team and the customer. Common pitfalls in alignment of early estimates are shown in Table 4-5.

Scope Definition and Early Estimates

Good scope definition is extremely important in preparing estimates. However, early estimates are usually prepared based on very limited scope definition and scant information regarding specific needs of the proposed project.

TABLE 4-5   Common Pitfalls in Team Alignment for Preparing Early Estimates

1. Early estimates that are heavily influenced by a preconceived number developed by individuals outside the estimating team
2. Failure to include the business decision maker early in the estimating process
3. Failure to resolve issues early
4. Excessive constraints on the estimating team, such as inadequate time for preparing the estimate or lack of cost data
5. Failure to hold an estimate kick-off meeting
6. Failure to identify information required of the estimating team to prepare the estimate
7. Inadequate understanding of the expected accuracy based on the level of information, estimate methodology, or other factors that impact the estimate
8. Failure to identify costs and scope to be captured in the estimate
9. Failure to identify the scope that is not to be included in the estimate
10. Failure to communicate the estimating methodology to be used in preparing the estimate

It is common knowledge that the accuracy of any estimate depends on the amount of information that is known about the project when the estimate is prepared. Any cost estimate usually is assigned a range of accuracy (+/– percentage). These ranges narrow as the quantity and quality of information increase through the life of a project. This infers that estimate accuracy is a function of available information (scope definition), a generally accepted fact in engineering and construction.

During the past thirty years, numerous papers have been published, emphasizing the importance of scope definition. The lack of scope definition has been identified as the root cause of cost overruns, late completion dates, excessive rework, unnecessary disputes, poor team alignment, and other problems associated with engineering and construction projects.

It should be recognized that determining the level of scope definition is a progressive activity. It starts at the inception of the project, when the project is only an idea of the project sponsor for a product to be produced. As engineering progresses, the level of scope definition increases. Consequently, early estimates are often subject to high variability.

Although good scope definition is important in preparing estimates, the skills and experience of the project team and the estimating procedure also play an important role. Figure 4-5 illustrates the importance

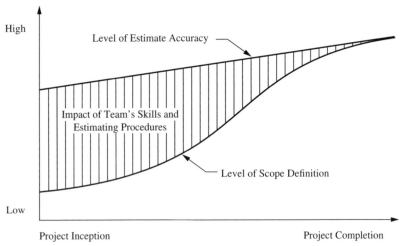

Figure 4-5   Relative impact of team's skills and estimating procedures on estimate accuracy.

of having the team involved early in a project's life, when the level of scope definition is low. The business unit must rely on the experience and skills of the team to produce accurate early estimates, because early in the project the level of scope definition is low and it is often poorly defined. The estimator must address limited scope definition and clearly communicate to the customer the level of scope definition that was used in preparing the estimate.

Preparing Early Estimates

Issues that should be discussed, defined, and documented when preparing estimates are shown in Table 4-6. As the estimate is being prepared it is important to perform periodic "reality checks" to make sure the costs developed are within reason. Based on estimator experience and familiarity with the project, this may include

- Simple "intuitive" checks for reasonableness
- Comparisons with similar projects
- Comparisons with industry data ($/square foot, cost/megawatt, indirect/direct costs, etc.)
- Check ratios such as lighting costs/fixture, fire protection, costs/sprinkler, etc.

Once the estimate is complete, a detailed review should be made of the entire estimate package, including the backup materials, assumptions, unit prices, and productivity rates. The estimate should also be checked against the project schedule requirements to ensure they are compatible, such as overtime rates assumed during outages, and price escalation.

TABLE 4-6   Issues for Preparing Early Estimates

1. Work plan for preparing the estimate
2. Costs/scope to be included, or excluded, in this estimate
3. Estimating methodology, tools, and techniques
4. Expected accuracy of the estimate
5. Impact of time allowed for preparing the estimate
6. Information needed by the estimating team to prepare the estimate
7. Roles and responsibilities for preparing the estimate
8. Format for presenting the estimate to the customer
9. Schedule for preparing the estimate that includes
   a. Kick-off meeting, estimate reviews and approvals
   b. Milestones for delivery of information and deliverables

Organizing to Prepare Estimates

The lead project estimator is responsible for initiating and leading the effort to develop a plan for preparing the estimate. In almost all situations, higher-quality estimates can be produced by professional cost estimators with an engineering or technical background. Like any technical specialty, estimating requires specific skills, training, and experience. Involvement of the estimating team early in the project is essential in the business development process.

Cost estimates for projects in the engineering and construction industry are prepared by individuals with many different job titles, responsibilities, and functions. Depending on the size and needs of each company, those preparing cost estimates may be working alone or as part of a group. They may be centralized in one location or in multiple locations. In some situations they may be integrated with different organizations or they may work in one homogeneous group.

There are advantages and disadvantages to centralizing or decentralizing the estimating staff. *Where* an estimate is prepared is not as important as *who* is preparing the estimate and the *process* used in preparing the estimate. It is important to implement and maintain effective control over the estimating process. Procedures must be in place for

- Disseminating knowledge and sharing expertise among the estimating staff
- Assigning and sharing the work load among estimators to improve efficiency
- Reviewing, checking, and approving work for quality control

Timely exchange of information is critical to ensure current price data, databases, and feedback. Preparing estimates requires expertise from multiple disciplines. An effective organization includes members of key disciplines, estimators, and management personnel who are knowledgeable in estimating. An effective team must be organized to prepare, review, check, and approve the work. This same team must also capture lessons learned to improve the estimating process and improve efficiency.

Establishing an Estimate Work Plan

Effective management of the estimating effort requires planning, scheduling, and control. The leader of the estimating team is responsible for developing an estimate work plan for the project. The estimate work plan is a document to guide the team in preparing accurate estimates and improving the estimating process. It identifies the work that needs to be accomplished to prepare the estimate: who is going to do it, when it is to be done, and the budget for preparing the estimate.

The estimate work plan is unique for each project, based on specific project parameters and requirements. Figure 4-6 illustrates the

---

Estimate Work Plan

Project Name: _____

Project Number: _____

Customer's Name: _____

Type of Estimate Required
    Desired Level of Accuracy
    Level of Effort Required
    Deliverables of Estimate
Estimating Services to Be Provided
    Deliverables of Estimate by In-House Resources
    Deliverables of Estimate by Outside Resources
Budget for Preparing Estimate
    Anticipated work-hours for estimating staff
    Dollars budget for non-salary estimating work
Required Staffing for Preparing Estimate
    Principle Estimate (leader of estimating team)
    In-House and Outside Resources
    Availability of Personnel for Staffing
Schedule for Preparing Estimate
    Anticipated Start Date
    Requirements of Review Date
    Customer Due Date
Estimating Methodology
    Tools
    Technique
    Method
    Procedures
Estimate Control
    Level of Scope Definition
    Checklists
    Review Process
Presentation
    Format for Presenting Estimate
    Audience of Presentation

Figure 4-6   Typical information to be addressed in the estimate work plan.

type of information that should be included in an estimate work plan. The work plan should contain sufficient detail to allow all members of the estimating team to understand what is expected of them. After the work plan is finalized, it serves as a document to coordinate the estimating work and as a basis to control and maintain the estimating process.

In preparing early estimates, the skill level of the estimator and his or her experience with the type of facility to be estimated is extremely important. The quality of any estimate is governed by the following major considerations:

- Quality and amount of information available for preparing the estimate
- Time allocated to prepare the estimate
- Proficiency of the estimator and the estimating team
- Tools and techniques used in preparing the estimate

Typically, the technical definition and the completion date for an estimate is determined by others outside of the estimating team. Therefore, these two elements may be beyond the control of the estimator. However, the estimator does have control over the selection of the tools and methodology to be used in preparing the estimate. The approach to selecting the method of estimating should be commensurate with the owner's expected level of accuracy and constraints of time.

The estimating team should develop a standard cost estimate presentation format that includes the level of detail and summary of engineering design, engineered equipment, bulk materials, construction directs and indirects, owner's costs, escalation, taxes, and contingency. Computer methods, including spreadsheets or estimating programs, provide consistent formats for preparing and presenting estimates. Uniform formats provide the following benefits:

- Reduces errors in preparing estimates
- Enhances the ability to compare estimates of similar projects
- Promotes a better understanding of the contents of an estimate
- Provides an organized system for collecting future cost data

Presentation of the estimate is important. The estimating team must develop a format that is easily understood by business managers and engineering managers, as well as by external clients. Using a standard format for presentations promotes better communication among all participants in the project and a better understanding of what is included in the estimate. This understanding is necessary so good decisions can be made based on the estimate.

Methods and Techniques

Selection of the methods for preparing early estimates depends on the level of scope definition, time allowed to prepare the estimate, desired level of accuracy, and the intended use of the estimate. For projects in the process industry, the following methods are commonly used:

- Cost capacity curves
- Capacity ratios raised to an exponent
- Plant cost per unit of production
- Equipment factored estimates
- Computer-generated estimates

Cost-capacity curves

A cost-capacity curve is simply a graph that plots cost on the vertical axis and capacity on the horizontal axis. These curves are developed for a variety of individual process units, systems, and services. The minimum information needed to prepare an estimate by cost-capacity curves is the type of unit and capacity. For example, the type of unit may be a coker unit or hydrogen unit and the capacity may be barrels per day or cubic feet per hour. Examples of additional information that can enhance the quality of the estimate may include adjustments for design pressure, project location, and project schedule.

Cost-capacity curves are normally prepared by a conceptual estimating specialist who develops, maintains, and updates the cost-capacity curves on a regular basis. These curves are developed and updated utilizing return cost data from completed jobs. This information is normalized to a location, such as the U.S. Gulf Coast, and for a particular time frame expressed as a baseline, such as December of a particular year.

The estimated cost is determined by locating the capacity on the horizontal x-axis and then following a straight line up to the point of intersection with the curve. The estimated cost is then read from the vertical y-axis by a straight line from the y-axis to the point where the x-axis intersects with the curve. The total installed cost derived from the curve may be adjusted for escalation to the present day or to some point in the future and may be further adjusted to reflect other geographic locations.

Example 4-1   The figure shows cost-capacity curves for process units in a chemical plant. What is the estimated cost for a project that has a process unit C with a capacity of 3,000 barrels per day?

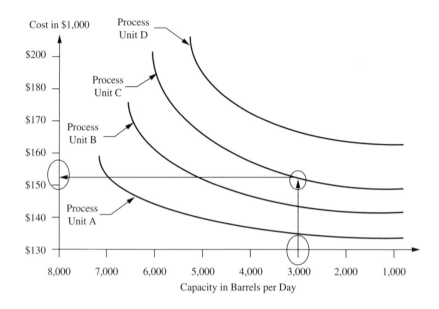

Locating the 3,000 barrels per day capacity along the abscissa, draw a vertical line upward until it intersects the process unit C cost curve. Then, draw a horizontal line to the left to read the estimated cost as $153,000.

Capacity ratios raised to an exponent

Capacity ratios raised to an exponent is another estimating technique for conceptual estimating. This approach takes into account the effect of economy of scale on the total installed cost. For example, if the cost of process unit B with capacity B is known, then the estimated cost of process unit A is calculated by multiplying the cost of process unit B times the ratio of the process unit capacities raised to an exponent (X) as shown in the following equation:

$$\text{Cost of process unit A} = (\text{Cost of process unit B}) \times \left(\frac{\text{Capacity of process unit A}}{\text{Capacity of process unit B}}\right)^{X}$$

Essentially, this method is a mathematical solution to the cost-capacity curves presented earlier in this section, which was a graphical technique. The exponent represented by X is mathematically derived from historical records from completed projects. It represents the non-linear relationship between cost with size, based on economies of scale.

Historical data can be captured from completed projects and a least-squares fit of the data or other methods of curve fitting can be used to determine an appropriate value of X for similar types of projects. Thus, the exponent distinguishes the curve of one process unit from another. Typically the range of the exponent X is between 0.55 and 0.88, depending on the type of process unit. When utilizing this equation to develop a cost estimate, if the exponent for the particular process unit is unknown, an exponent of 0.6 is used, which represents a standard or typical exponent for process plants.

Example 4-2   The cost of a 410 cubic feet per hour (ft^3/h) process unit is $6,750,000. From historical cost records, the capacity ratio exponent of a process unit is 0.72. Estimate the cost of a similar process unit with a capacity of 575 ft^3/h.

$$\text{Cost of process unit A} = (\text{Cost of process unit B}) \times \left(\frac{\text{Capacity of process unit A}}{\text{Capacity of process unit B}} \right)^{X}$$

$$= \$6,750,000 \times \left(\frac{575\ ft^3/h}{410\ ft^3/h} \right)^{0.72}$$

$$= \$6,750,000 \times 1.27523252$$

$$= \$8,611,132$$

Plant cost per unit of production

This conceptual estimating technique is used to estimate the total plant cost based on the average plant costs per unit of production on previously completed projects. This is a very simple and broad estimating approach where the only information available is the product description and the plant capacity. For example, cost records may show the average cost per unit for cogeneration facilities to be $1,000 per kilowatt ($1,000/kW) of production. Thus, for a future 300-megawatt (MW) cogeneration facility, the estimated cost would be calculated by multiplying the $1,000/kW times the 300 MW of power to derive a total estimated cost of $300,000,000.

This estimating technique assumes that the relationship between plant cost and production capacity is linear and, therefore, would apply best within a fairly narrow range. Ideally, average plant costs per unit of production capacity are best developed over various capacity ranges so that the estimator can select the relationship that is applicable for his or her estimate.

This method of preparing early estimates is similar to the square-foot estimating method used for projects in the building sector. The total estimated cost of a particular building project is determined by multiplying the average cost per square foot of previous projects by the total square feet in the proposed building.

Equipment Factored Estimates

For the process industry, equipment factored estimates are derived by applying various factoring techniques to estimated equipment costs. The factors used are developed and updated utilizing return cost data from completed projects. This information is normalized to a location, usually the U.S. Gulf Coast, and for a particular time frame, such as December of a particular year. The estimated total installed cost of a normalized unit is defined to include the following costs:

- Direct equipment costs
- Direct bulk material costs
- Subcontract costs
- Construction labor costs
- Construction indirect costs
- Home office services costs

One example of the factoring technique is the "equipment cost" to "total installed cost" (TIC) factor. This factoring technique is relatively simple for projects where equipment costs have been estimated. As the name implies, TIC factors are developed by dividing the equipment costs of a particular process unit into the total installed cost of that unit. The estimated cost of the project is determined by multiplying the equipment costs by the TIC factor, or multiplier. The factors for process plants generally range between 2.5 to 6.0, depending on the nature of the process unit. Conditions that affect the equipment to TIC factors are

- Equipment sizes
- Pressure
- Metallurgy
- Degree of prefabrication
- Site conditions
- Equipment costs
- Special conditions (large structures, pits, buildings, etc.)
- Explanation of engineering costs included

Another equipment factored estimating technique develops equipment costs manually or by utilizing commercially available computer software systems. Bulk material costs are factored from the estimated equipment costs, using historical cost data for the same or similar type units. Field labor work-hours are estimated for each individual equipment item, and the work-hours for installing bulk materials are estimated as a percentage of the bulk material costs for each category of materials. The resultant field labor work-hours are adjusted for productivity and labor costs by applying local labor rates to the estimated construction work-hours. Construction indirect costs are developed for the major categories by percentages of direct labor costs. Home office costs are estimated as a percentage of the total installed cost. The equipment factored estimating techniques described can be utilized when there is sufficient technical definitions available consisting of the following:

- Process flow diagram
- Equipment list
- Equipment specifications
- Project location
- General site conditions (assumed if not specified)
- Construction labor information
- Project schedule

Computer-generated estimates

There are numerous commercially available computer software systems for estimating capital costs for a number of different types of industries, including the process industry, building construction industry, and the heavy/highway infrastructure industries. These systems can be simple or very sophisticated. Most of the software packages can operate on a personal computer and are furnished with a cost database, which is updated on an annual basis. The more flexible systems allow the purchaser to customize the database.

Sophisticated software packages are available to assist the estimator in generating detailed material quantities as well as equipment and material costs, construction work-hours and costs, field indirects and engineering work-hours and costs. The detailed quantity and cost output allows early project control, which is essential in the preliminary phases of a project, before any detail engineering has started. The accuracy of an estimate can be improved because some systems allow vendor costs, takeoff quantities, project specifications, site conditions, etc., to be introduced into the program. To maximize the benefits of these software programs, the use of system defaults should be minimized and replaced with the following definitions:

- Specifications, standards, basic practices, and procurement philosophy
- Engineering policies
- Preliminary plot plans (if available) and information relating to pipe-rack, structures, buildings, automation and control philosophy, etc.

- Adequate scope definition
- Site and soil conditions
- Local labor conditions relating to cost, productivity, and indirect costs
- Subcontract philosophy

To become proficient in the use of computer software programs, frequent usage is required, and the user should compare the computer-generated results with other estimating techniques to determine the limitations and shortcomings of the programs. Once the shortcomings are known, corrective action to eliminate or minimize the shortcomings can be taken. To maximize the benefits of the use of software systems in developing early estimates, the following should be considered:

- Index or benchmark the unit costs and installation work-hours in the computer software's databases to match the company's cost databases.
- Establish system defaults that correspond to the company's engineering and design standards.
- Create a program that allows conversion of the output of the software programs to the company's account codes and format.

By adopting the above recommendations, confidence in the output of estimating software systems will improve. This will result in more consistency and reliability of computer-generated estimates.

There are other non-commercial computer software systems that are used in preparing early estimates, in particular spreadsheet programs. Some of these systems, although not commercially available, have been developed by owner/operator companies and contractors.

Estimate Checklists

Checklists are valuable tools to reduce the potential of overlooking a cost item. Checklists act as reminders to the estimator by

- Listing information required to prepare the early estimate
- Listing miscellaneous other costs that may be required in the estimate
- Listing the project scope that may be required but not identified in the definition provided for the estimate

A listing of information required to prepare an estimate in the process industry may include type of unit, feed capacity, and project location. For a computer-generated estimate, the required information includes soil and site data, building requirements, plot plan dimensions, and other specific engineering requirements. For projects in the building sector, a listing of information to prepare an estimate may include type of building, functional use of building, number of occupants, and project location.

TABLE 4-7   Checklist for an Early Estimate in the Process Industry

1. Process unit description (delayed coker, hydrogen plant, etc.)
2. Process licenser
3. Feed capacity
4. Production capacity
5. Product yield
6. Utility levels at process unit location
7. Feedstock specifications
8. Integration of multiple units
9. Process pressure and temperature operating levels
10. Provision for future expansion of capacity
11. Provision for processing multiple or different feedstocks
12. Single train vs. multi-train concept
13. Project location
14. Miscellaneous costs (spare parts, training, chemicals, etc.)
15. Other items, such as unusually high or low recycle rate

Typical examples of miscellaneous cost items for projects in the process industry may include spare parts, catalyst and chemicals, permits, and training. Typical scope items that may be required, but not identified in the definition provided for the estimate, may include certain utility and auxiliary systems. Examples are special steam systems, refrigeration, lube and seal oil systems, and flare systems.

Checklists are useful during initial client/customer meetings where they serve as agenda items for discussion. Checklists also assist the estimator in preparing an *estimate work plan* by identifying important points to emphasize in the write-up for the execution of the estimate. Table 4-7 is an illustrative example of an early estimate checklist for a project in the process industry.

Estimate Documentation

Effective communication is necessary during the estimating process. A support document should be developed and available for presentation, review, and future use of the estimate. A thorough documentation of the estimate forms a baseline for project control, so decisions during project execution can be made with a better awareness of the budget, thereby improving the overall outcome of the project.

Inaccurate cost estimates are often the result of omissions in the estimate, miscommunications of project information, or non-aligned assumptions. Documenting the estimate will minimize these inaccuracies by

- Improving communications among all project participants
- Establishing a mechanism for review of the estimate
- Forming a solid basis for project controls

As the estimate is being developed, the act of preparing documentation facilitates communications among the parties involved: estimators, scope developers, project managers, and customers. Estimate documentation improves the outcome of the estimate through

- Sharing information
- Identifying items that require clarification
- Helping the estimator obtain and organize information needed for the estimate
- Avoiding confusion over what is covered and not covered by the estimate
- Providing useful information for future estimates
- Highlighting weak areas of the estimate
- Increasing the credibility of the estimate

A portion of the documentation may be developed by sources other than the estimator. For example, written scopes are developed by those who are defining the project, quotes may be obtained by procurement personnel, and labor information may be obtained from field personnel. However, the estimator has overall responsibility for collecting and organizing this information. Reviewing and clarifying the information with the originator improves the estimate accuracy.

A standard default format or outline should be developed to organize and prepare documentation for the cost estimates. A different standard can be developed for different types of estimates. The process of developing, utilizing, and storing the documentation for future use should be built into the cost estimate work process. The items that should be documented are shown in Table 4-8.

TABLE 4-8   Recommended Documentation of Early Estimates

1. Standard format for presenting cost categories (codes)—summary and backup levels
2. Basis of estimate—clear understanding of what constitutes the estimate
3. Level of accuracy—expected for the estimate
4. Basis for contingency—risk analysis, if applicable
5. Boundaries of the estimate—limitations of the estimate
6. Scope of work—the level of scope definition used in preparing the estimate
7. Labor rates—breakdown and basis of labor rate
8. Assumed quantities—conceptualized, etc.
9. Applied escalation—dates and basis of escalation
10. Work schedule—shifts, overtime, etc., to match the milestones (not contradictory)
11. Other backup information—quotes, supporting data, assumptions
12. Checklists used—a list of completed checklists
13. Description of cost categories—codes used in preparing the estimate
14. Excluded costs—list of items excluded from the cost estimate

Estimate Reviews

Well-executed estimate reviews will increase the credibility and accuracy of the estimate. They also help the team and project management to know the level of scope definition and the basis of the estimate. The review of estimates is an important part of the estimating process because it helps the customer to understand the contents and level of accuracy of the estimate, allowing the customer to make better business decisions.

The number of reviews will vary depending on the size of the project, type of estimate, length of time allowed for preparing the estimate, and other factors. For any estimate, there should be at least two reviews: an internal review during development of the estimate and a final review at or near the completion of the estimate.

About halfway through the development of the estimate, a "reality check" should be scheduled. The purpose of the midpoint check is to avoid spending unnecessary time and money in pursuing an estimate that may be unrealistic or based on assumptions that are no longer valid.

The internal midpoint estimate review is brief. Typically the lead estimator, engineer, and project manager are involved. There may be times when it is advantageous to include the customer. This review is intended as a reality check of the data being developed to assess whether to proceed with the estimate. This is a "go–no-go" point where the results of the review will guide the estimator and the team to one of the following two steps:

1. Recycle back to the scope of work because the capital or scope have gotten outside of the boundaries established as a target for the project.
2. Give the team the "go ahead" to proceed with the remaining estimate process to complete the estimate.

The final estimate review is a more structured process. The depth of the review depends on the type or class of estimate that is being prepared. The meeting is intended to validate assumptions used in preparing the estimate, such as construction sequence, key supplier selection, and owner's cost. Engineering and the customer must accept ownership of the scope that is represented in the estimate.

The final estimate review may be a lengthy meeting. For a final estimate review, the attendees should include the lead estimator, process engineer, discipline engineer, operations/maintenance representative, engineering manager, and constructability leader. To be effective, the final estimate review meeting should be conducted with a written agenda. The meeting should be documented with written minutes that are distributed to all attendees. The estimator must

TABLE 4-9   Presentation Items for Review of an
Early Estimate in the Process Industry

1. Product mix, volume, and quality requirements
2. Facility location
3. Scope of work
4. Simplified flow sheet
5. Key assumptions used
6. Major undecided alternatives
7. Historical data used
8. Estimate exclusions
9. Estimator's experience and track record
10. Checklists used to prepare estimate

come to the review meeting prepared with the following information for comparisons:

- Historical data used in preparing the estimate

- Actual total installed costs (TIC) of similar projects

- Percent of TIC on key cost accounts

Comparisons of the estimate with the above information provide useful indicators for the estimate review. The estimator needs to assess each estimate to determine the appropriate checks that should be included in an estimate review meeting.

In some situations it may be desirable to use outside assistance for estimate reviews. For example, it may be helpful to obtain a review of the estimate by an experienced peer group to validate assumptions, key estimate accounts, construction sequence, potential omissions, etc. In other situations, it may be advantageous to engage a third party to perform an independent review. This will provide a check to compare the estimate with past similar estimates from the perspective of a different team.

Estimate reviews should focus on the big picture and follow Pareto's law, separating the significant few from the trivial many. Generally an estimate is prepared bottoms up, whereas the review is conducted top down. Table 4-9 is an illustrative example of presentation items for review of an early estimate for a project in the process industry.

Risk Assessment

Assessing risk and assigning contingency to the base estimate is one of the most important tasks in preparing early estimates. Risk assessment is not the sole responsibility of the estimators. Key members of the project management team must provide input on critical issues that should be addressed by the estimators in assessing risk.

Risk assessment requires a participatory approach with involvement of all project stakeholders including the business unit, engineering, construction, and the estimating team.

The business unit is responsible for overall project funding and for defining the purpose and intended use of the estimate. Engineering design is responsible for providing input on the design criteria and factors that are susceptible to changes that may impact the cost of the project. The estimator is responsible for converting the information from the business unit and engineering into an appropriate procedure for assessing risk and assigning contingency. The estimator must communicate the risk, contingency, and level of accuracy that can be expected of the final estimate.

Risk Analysis

Typically, risk analysis is a prerequisite to assigning contingency. Based on the acceptable risks and the expected confidence level, a contingency is established for a given estimate. Risk analysis and the resultant amount of contingency help the business unit to determine the level of economic risk involved in pursuing a project. The purpose of risk analysis is to improve the accuracy of the estimate and to instill management's confidence in the estimate.

Since the owner's organization has overall project funding responsibility, it must consider both the contractor's and owner's risks. The owner's contingency should cover the entire project risk, after adjusting for any risk already covered by contractors.

Numerous publications have been written to define risk analysis techniques. Generally, a formal risk analysis involves either a Monte Carlo simulation or a statistical range analysis. There are also numerous software packages for risk analysis. The lead estimator for a project must assess the uniqueness of each project and select the technique of risk analysis that is deemed most appropriate. For very early estimates, the level of scope definition and the amount of estimate detail may be inadequate for performing a meaningful cost simulation.

Contingency

Contingency is a real and necessary component of an estimate. Engineering and construction are risk endeavors with many uncertainties, particularly in the early stages of project development. Contingency is assigned based on uncertainty and may be assigned for many uncertainties, such as pricing, escalation, schedule, omissions, and errors. The practice of including contingency for possible scope expansion is highly dependent on the attitude and culture, particularly that of the business unit, toward changes.

In simple terms, contingency is the amount of money that should be added to the base estimate in order to predict the total installed cost of the project. Contingency may be interpreted as the amount of money that must be added to the base estimate to account for work that is difficult or impossible to identify at the time a base estimate is being prepared. In some owner or contractor organizations, contingency is intended to cover known unknowns. That is, the estimator knows there are additional costs, but the precise amount is unknown. However, sometimes an allowance is assigned for known unknowns and a contingency is assigned for unknown unknowns.

CII Source Document 41 defines contingency as "A sum of money to cover costs which are forecast but are difficult or impossible to identify when proposing." AACE International Document 18R-97 defines contingency as "An amount of money or time (or other resources) added to the base estimate to (a) achieve a specific confidence level or (b) allow for changes that experience shows will likely be required."

Traditional Methods of Assigning Contingency

The most effective and meaningful way to perform risk analysis and assign contingency is to involve the project management team. Estimators have insights and can assess imperfections in an estimate to derive an appropriate contingency. However, the interaction and group dynamics of the project management team provide an excellent vehicle to assess the overall project risk. The integration of the project management team's knowledge and the estimator's ability to assign contingency provides management with an appreciation and confidence in the final estimate. The end result is an estimate that represents the judgment of the project management team, not just the estimator's perspective.

Figure 4-7 illustrates the risk assessment process. The estimator must select the method deemed most appropriate for each project, based on information provided by the project management team and based on the intended use of the estimate by the business unit. The estimator must communicate the method selected, risk, accuracy, and contingency for the estimate.

Percentage of base estimate

For some situations, contingency may be assigned based on personal past experience. A percentage is applied to the base estimate to derive the total contingency. Although this is a simple method, the success depends on extensive experience of the estimator and historical cost

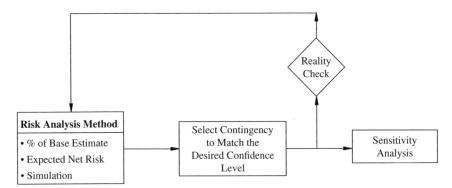

Figure 4-7 Risk assessment process.

information from similar projects. It is less accurate than other more structured methods.

Some organizations use standard percentages for contingencies based on the class of estimate. This method is governed by company policy rather than by a numerical analysis. Typically, the percentage used is based on the level of scope definition or on the stage of project development.

In some situations, contingency is determined as a percentage of major cost items rather than as a percentage of the total base estimate. This method typically relies on the personal experience and judgment of the estimator, but the percentage can also be from established standard percentages based on historical data. This method has the advantage of considering risk and uncertainty at a lower level than that used when contingency is based on a percentage of the total base estimate.

The personal experience and judgment of the estimators and engineers should not be overlooked in the process of assigning contingency. Even the most advanced computers are not a substitute for the knowledge and experience of the human mind. Estimators with many years of experience with a particular type of facility can often be quite accurate in assigning contingency based on how they "feel" about the level of uncertainty and risk associated with a project, the cost data used in preparing the estimate, and the thoroughness of the effort in preparing the base estimate.

Expected net risk

The estimator may determine contingency based on expected maximum risk and likelihood. After the evaluation of normal contingency of each estimate element, an individual element may also be evaluated

TABLE 4-10   Expected Net Risk Analysis

Estimate Item	Base estimate	Maximum cost	Maximum risk	Percentage probability	Expected net risk
1	$40,000	$50,000	$10,000	20%	$2,000
2	8,000	12,000	4,000	40%	1,600
3	100,000	150,000	50,000	30%	15,000
4	250,000	320,000	70,000	50%	35,000
5	72,000	95,000	23,000	70%	16,100
6	237,000	320,000	83,000	60%	49,800
7	12,000	28,000	16,000	10%	1,600
8	94,000	135,000	41,000	30%	12,300
9	730,000	870,000	140,000	40%	56,000
10	43,000	72,000	29,000	80%	23,200
11	572,000	640,000	68,000	50%	34,000
12	85,000	97,000	12,000	20%	2,400
	$2,243,000	$2,789,000	$546,000		$249,000

for any specific unknowns or potential problems that might occur. The first step involves determining the maximum possible risk for each element, recognizing that it is unlikely that all the risk will occur for all elements.

The next step involves assessing the percentage probability that this risk will occur. The expected net risk then becomes a product of the maximum risk times the probability. The sum of all the expected net risks provides the total maximum risk contingency required. Table 4-10 illustrates an expected net risk analysis.

The risk analysis data in Table 4-10 shows a base estimate of $2,243,000 with a maximum anticipated cost of $2,789,000, a difference of $546,000. However, it is unlikely that all the risks will occur for each item in the estimate. Therefore, the expected net risk for each bid item is calculated by multiplying the maximum risk by the probability percentage of each item. The total expected net risk for the project is $249,000, which is calculated as the sum of the net risk for each item in the estimate. The $249,000 represents a contingency markup of 11.1% on the base estimate of $2,243,000, calculated as [($249,000/$2,243,000) = 11.1%]. For this project the final estimate is calculated as the base estimate plus the total expected net risk, $2,243,000 + $249,000 = $2,492,000.

Simulation

A formal risk analysis for determining contingency is usually based on simulation. A simulation of probabilistic assessment of critical risk elements can be performed to match the desired confidence level. Monte Carlo simulation software packages are useful tools for performing simulation.

However, a knowledge of statistical modeling and probability theory are required to use these tools properly.

Range estimating is a powerful tool that embraces Monte Carlo simulation to establish contingency. Critical elements are identified that have a significant impact on the base estimate. For many estimates there are less than 20 critical elements. The range of each critical element is defined and probability analysis is used to form the basis of simulation. Using this method, non-critical elements can be combined into one or a few meaningful elements.

Range estimating is probably the most widely used and accepted method of formal risk analysis. In range estimating the first step requires identification of the critical items in the estimate. The critical items are those cost items that can affect the total cost estimate by a set percentage, for example +/–4%. Thus, a relatively small item with an extremely high degree of uncertainty may be critical whereas a major equipment item for which a firm vendor quote has been obtained would not be considered critical. Typically, no more than 20 critical items are used in the analysis. If more than 20 critical items are identified, the set percentage can be increased to reduce the number of critical items.

Once the critical items are identified, a range and a target are applied to each item. For example, the range may include a minimum value so there is only a 1% chance that the cost of the item would fall below that minimum. Similarly, an upper value may be established so there will be only a 1% chance of going over that value. The target value represents the anticipated cost for that item. The target value does not need to be the average of the minimum and maximum values. Usually the target value is slightly higher than the average.

After the critical items have been identified and ranged, a Monte Carlo simulation is performed. The Monte Carlo analysis simulates the construction of the project numerous times, as many as 1,000 to 10,000, based on the ranges given to the critical items and the estimated values of the non-critical items. The results of the simulation are rank ordered and then presented in a cumulative probability graph, commonly called an S-curve. The cumulative probability graph typically shows the probability of underrun on the horizontal x-axis and either the total project cost or contingency amount on the vertical y-axis. The decision maker can then decide the amount of contingency to add based on the amount of risk.

Caution also must be exercised because it is possible to seriously underestimate the cost of a project when using range estimating. There is a risk of understating the true risk of a project due to statistical interdependencies among the critical items in the analysis. Whenever two or more cost items are positively correlated, meaning they increase together or decrease

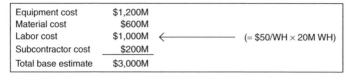

Figure 4-8   Base estimate summary. (M = 1 million, WH = work-hours.)

together, the Monte Carlo simulation may cause one to be high and the other low, thus canceling each other out. Thus, the true risk would be understated. Also, underestimating the ranges on the critical items can have a profound impact on the results, also leading to an understatement of the true risks inherent in the design and construction of the project.

When used properly, formal risk analysis using Monte Carlo simulation range estimating can be an extremely valuable tool because it requires a detailed analysis of the components of the estimate, a process that can identify mistakes and poor assumptions. However, precautions must be used when using simulation methods for early estimates. For many early estimates there is not enough detailed information or an adequate number of cost items for a valid simulation.

Assessing estimate sensitivity

The contingency added to an estimate includes the combined impact of all risk elements. The accuracy of an estimate can be improved by assessing high-cost impact factors, increasing the level of scope definition, or a combination of both. A sensitivity analysis can be performed to illustrate how a specific risk element can impact the total estimate.

The sensitivity analysis evaluates the impact of only one risk element at a time. It is frequently used in conjunction with an economic analysis. During the process of determining contingency, risk elements that have the maximum impact on the total installed cost are prime candidates for sensitivity analysis. Figures 4-8 and 4-9 show a sample sensitivity analysis for a $3,000M base estimate.

Risk element	% Change from estimate	New base estimate ($M)
Labor rate ($50/WH)	0	$3,000
Labor rate	+10	3,100
Labor rate	−5	2,950
Total work-hours (20M WH)	0	$3,000
Total work-hours	+15	3,150
Total work-hours	−7	2,930
Equipment ($1,200M)	0	$3,000
Equipment	+5	3,060
Equipment	−5	2,940

Figure 4-9   Sensitivity analysis. (M = 1 million, WH = work-hours.)

For any estimate, it is necessary to add a contingency to the base estimate. The method used for assigning contingency will vary depending on analysis of risk and other factors that can impact the cost of a project. This section has presented traditional methods used for assigning contingency.

Estimate Feedback for Continuous Improvement

It is unfortunate, but many people have the perception that the estimator's involvement with a project is over when the estimate is finished. In reality, it is an advantage to the management of a project, and in the best interest of the customer, for the estimator to remain connected to the project during execution.

The estimator can be an important asset to project management during the execution phase of a project. Involvement of the estimator during project execution allows the estimator to stay in touch with the project and provide an early warning for any potential cost overruns. Including the estimator in the distribution list of monthly project reports can provide input to the project management team members to enable them to make good decisions related to costs.

During project execution the estimator can also be a valuable resource for recasting the cost estimate into work/bid packages and for analyzing actual bids with the recast estimate. The estimator can also assist in management of changes during project execution, by assessing the impact of changes on cost.

No estimating process is complete without the continuous feedback loops shown in Figure 4-4. Feedback from project execution provides lessons learned to the estimator that allows the estimating team to modify estimating standards and practices. Feedback from project completion also allows the estimating team to update the database for improving the accuracy of future estimates. Terminating the estimator's involvement when the estimate is finished prevents continuous improvement of the estimating process.

To provide meaningful feedback, the estimator must explore how the cost will be tracked during project execution. An estimate should be prepared with cost breakdowns in a format that allows easy future cost tracking. A standard code of accounts enables an organization to simplify the estimating process, update the database, and facilitate cost control. This benefits both the estimating team and the project management team.

A final project cost report is an extremely valuable document for improving estimates because it provides a real feedback to compare with the original cost estimate. Pitfalls for future estimates can be eliminated or minimized. Both the original estimate and the final project cost reports should be maintained at a central location. A cost reference with reports sorted by project location, type, size, etc., can be used to update the cost database for future estimates.

The best source of data for estimators to develop and enhance the estimating tools and techniques is their own organization. There is an abundance of project data that is available from completed projects and definitive estimates. The key to success is the establishment of a mechanism to capture and retrieve this information in a format that can be useful in developing statistical relationships, such as percentage of breakdowns of TIC by cost category, TIC to equipment cost ratios, and construction indirect costs to direct labor cost ratios. When a project is completed, the actual TIC can be added to the database. Estimate feedback is an integral part of the estimating process. It is not an add-on feature. A process for providing feedback loops is necessary for improving the accuracy of early estimates.

References

1. Alroomi, A., Jeong, H. S., and Oberlender, G. D., "Cost Estimating Competencies Using Criticality Matrix and Factor Analysis," *Journal of Construction Engineering and Management*, American Society of Civil Engineers, Reston, VA, 2013.
2. *Cost Estimate Classification System As Applied in Engineering, Procurement, and Construction for the Process Industry,* Recommended Practice No. 18R-97, AACE International, Morgantown, WV, 1997.
3. Diekmann, J. E., Sewester, E. E., and Taher, K., *Risk Management in Capital Projects,* Source Document 41, Construction Industry Institute, Austin, TX, October 1988.
4. Gibson, G. E., Jr. and Dumont, P. R., *Project Definition Rating Index,* Research Report 113-11, Construction Industry Institute, Austin, TX, April 1998.
5. O'Connor, J. T. and Vickroy, C. G., *Control of Construction Project Scope,* Source Document 6, Construction Industry Institute, Austin, TX, March 1986.
6. Peurifoy, R. L. and Oberlender, G. D., *Estimating Construction Costs,* 6th ed., McGraw-Hill Book Company, New York, NY, 2014.
7. *Richardson General Construction Estimating Standards*, Richardson Engineering Services, Inc., Mesa, AZ, published annually.

5

Project Budgeting

Project Budgets

The discussion of budgeting in this chapter is an extension of Chapter 4 on preparing early estimates. It is also closely related to Chapter 7 on design proposals, which presents the process of determining the cost of engineering design services for projects. The cost of construction work is a major portion of the total cost of any project. Much of the construction work on many projects is performed by subcontractors who are awarded a contract by the general contractor. This chapter presents an overview of construction costs. A more detailed coverage of estimating construction costs is presented in *Estimating Construction Costs,* 6th ed., published by McGraw-Hill, Inc.

The budget for a project is the maximum amount of money the owner is willing to spend for design and construction to economically justify the project. Estimating is a prerequisite to project budgeting. Chapter 4 presented the process of preparing early estimates of projects. As presented in that chapter, after the base estimate is completed, a risk assessment must be performed. The purpose of the risk assessment is to determine an appropriate amount of contingency funds that should be added to the base estimate in order to reasonably predict the final cost of the project. Thus, the budget can be considered to be the base estimate plus contingency.

Development of Project Estimates for Budgeting

The preparation of estimates, assessing risk, and assigning contingency for budgeting is one of the most difficult tasks in project management because it must be done before the work is started. It is a process that involves a series of successive approximations beginning with the owner's feasibility study and continuing through design development and construction.

The preparation of cost estimates for budgeting is important to each party because the decision to proceed, at each phase in the project, is based on the estimated cost that was determined in the preceding phase. The owner's organization must determine a realistic maximum and minimum cost of the entire project, which includes the cost of design and construction. The designer's organization must determine the cost of performing design tasks and producing the contract documents. It must also determine the probable cost of construction as a part of the design process. The construction contractor's organization must determine the cost of all material, labor, and equipment to build the project on the job-site.

Each contractor on a project must develop a base estimate, consider risk, and assign contingency for the work they will be performing on a project. Since the owner's organization has overall project funding responsibility, the owner's management must consider both the contractor's and owner's risks in order to determine the overall budget for the project.

Project estimating and budgeting begins with the owner during the study of needs, priorities, and scope. As discussed in Chapter 3, the project budget is derived from scope definition; therefore, a special effort should be made early in the development of a project to define the scope as detailed and accurately as possible. The control of project scope growth and cost overruns can be greatly enhanced if the owner obtains the early advice and expertise of experienced design and construction professionals, who have the knowledge of construction costs. All parties must realize that the estimated cost, at any time, is based upon the amount of information that is known about the project when the estimate was prepared. Too often this concept is not fully recognized. A project manager can play an important role as mediator in the early stages of the development of a project by testing, scrutinizing, and identifying the variances that should be applied to an estimate.

The owner's organization must prepare estimates to determine the overall project budget, which includes the approved cost for design and construction. If the scope is not well defined or the owner's organization does not have the expertise to perform such an estimate, the owner can enlist a designer to perform these services on a cost-reimbursable basis. Because this budget is prepared prior to any detailed design work, it should include a reasonable amount of contingency funds to allow some flexibility in decision making during design development.

The designer's organization must prepare a budget based on the estimated costs to provide design services. In addition, as a part of the design process the designer must prepare the estimated construction costs of the various design alternatives that are being evaluated to meet the owner's needs for the project. This is necessary before completion of the contract documents. It is the designer's responsibility to keep design costs and estimated construction costs within the owner's overall approved project budget. This requires extensive cooperation and involvement with the owner because the scope must sometimes be readjusted to meet the owner's approved budget, or the budget must be readjusted to meet the owner's needs. This decision must be made by the owner's organization.

The construction contractor's organization must prepare a bid that is submitted to the owner, based on the estimated costs to build the project in accordance with the contract documents. For competitive-bid projects, the contractor is not obligated to a cost that is within the owner's approved budget because this information is usually not known to the contractor. For negotiated cost-reimbursable projects the contractor's organization works closely with the owner to determine construction alternatives with costs that are within the owner's overall approved budget.

Levels of Accuracy

A range of accuracy, usually a plus or minus percentage, should be assigned to any estimate by the estimator based on his or her best assessment of the project's true cost. There is no industry standard that has been agreed on regarding the amount of plus or minus percentage that should be applied to an estimate. To discuss this issue it is helpful to divide projects into two general categories: building projects and industrial projects.

Building projects generally have two types of estimates: approximate estimates (sometimes called preliminary, conceptual, or budget estimates) and detailed estimates (sometimes called final, definitive, or contractor's estimates). For large owner organizations the approximate estimate is prepared by the owner during the feasibility study of the project's needs, priorities, and scope definition. For small owner organizations it is usually prepared in cooperation with the design organization that is contracted by the owner to design the project and prepare the contract documents. The level of accuracy of the approximate estimate can vary significantly, depending upon the amount of information that is known about the project. With no design work it may range from +50% to −30%. After preliminary design work, it may range from +30% to −20%. On completion of detailed design work it may range from +15% to −10%.

For building projects, the detailed estimate is prepared by the construction contractor from a complete set of contract documents prior to submittal of the bid or formal proposal to the owner. The detailed estimate is important to both the owner and the contractor because it represents the bid price, the amount of money the owner must pay for completion of the project, and the amount of money the contractor will receive for building the project. For a building project that has a complete set of well-defined contract documents and no unusual features, the competitive bidding of numerous contractors will often result in less than a 1% variation in the lowest two bids.

For petrochemical and processing projects, estimating is difficult because of the wide range of variations in the number and sizes of piping, instrumentation, equipment, and other components that are required to process the product that the plant is built to produce. Because of the complexity of the project, estimating is done in stages as the design progresses and more information becomes known about the project.

Although there is no industry agreement, the petrochemical and processing industry generally develops project budget estimates in stages. For example, the feasibility estimate is the first estimate and is usually done within an owner's organization as a part of the feasibility plan. Estimates at this stage are commonly referred to as *order of magnitude* cost estimates. The estimate is prepared as a ratio of costs of previously completed similar projects, contractor quotes, or owner cost records, such as cost per horsepower, cost per barrel of throughput, or cost per pound of finished product. The level of accuracy is usually ±50%.

After the major equipment is identified and process flow sheets are developed, an *equipment factored* estimate can be prepared. This estimate is based on applying factors to in-house priced major equipment in order to compensate for piping, instrumentation, electrical, and other construction costs that are required to complete the cost estimate. The level of accuracy at this stage is usually ±35%.

After completion of piping and instrumentation drawings, a preliminary *control estimate* can be developed. The documents and data for this estimate usually include equipment sizing and layout, process flow sheets, piping and instrumentation drawings, building sizes, and a milestone schedule. The level of accuracy is usually ±15%.

The final estimate is performed near the end of engineering design when most of the costs have been identified and is called the definitive estimate, commonly referred to as an Approved for Expenditure estimate or AFE Definitive estimate. It is based on process flow sheets, mechanical flow sheets, equipment layout, isometrics, and building plans. The level of accuracy of an AFE Definitive estimate is usually ±10%.

Owner's Estimate for Budgeting

Every project must be shown as economically feasible before it is approved by the owner's management. Economic feasibility is determined by an economic analysis for projects in the private sector or by a benefit/cost ratio for projects in the government sector. An economic analysis can be performed once an owner's estimate has been prepared.

Estimating costs during the inception of a project by the owner, prior to any design, is difficult because only limited detailed information is known about the project. However, this cost estimate is important because it is used to set the maximum project budget that will be approved for design and construction. At this stage of project development the only known information is the number of units or size of the project, such as number of square feet of building area, number of cars in a parking garage, number of miles of 345-kilovolt (kV) transmission line, or number of barrels of crude oil processed per day. At some point in time an estimate has to be frozen and converted to a project budget.

Preparation of the owner's estimate requires knowledge and experience of the work required to complete the project. Cost information from professionals who are knowledgeable about design and construction is essential. Cost information for preparation of the owner's budget is usually derived from one of two sources: cost records from previous projects of similar type and size, or pricing manuals that are published annually by several organizations.

For buildings, public works, and heavy construction projects, the *Means Cost Guide* is commonly used. The Richardson's manual for construction estimating is a common reference for petrochemical and processing projects. These pricing manuals provide costs per unit for various types of projects, such as cost per square foot of building area for offices, warehouses, and maintenance buildings. The costs are derived from previous projects that have been completed at numerous geographic locations. Figure 5-1 illustrates examples of information that is available for several types of buildings. It shows the low, average, and high cost per square foot, based on the level of quality. The budget for a proposed project can be calculated by multiplying the cost per square foot by the total square feet in the project. The cost of land, permits, and design fees should be added to the calculated cost of construction. A reasonable percentage multiplier should also be applied for contingency since the design is not prepared for the project during the owner's budgeting process. Adjustments for time and location should also be made as discussed in the following paragraphs.

The costs in Figure 5-1 do not include site-work or utilities outside of the building.

Weighted Unit Cost Estimating

The other source of cost information is company records from previous projects. Although the total cost of previously completed projects will vary between projects, unit costs can be calculated to forecast the cost

Component	Office buildings			Motels		
	Low $/SF	Median $/SF	High $/SF	Low $/SF	Median $/SF	High $/SF
Foundation	3.95	4.00	4.80	0.90	1.40	1.60
Floors on grade	3.10	3.15	3.90	3.95	5.00	5.40
Superstructure	14.90	16.90	20.25	10.95	13.30	21.70
Roofing	0.20	0.25	0.30	2.40	3.40	3.45
Exterior walls	4.90	9.75	13.00	2.80	4.45	5.55
Partitions	4.35	5.30	7.05	2.60	3.65	5.25
Wall finishes	2.35	3.75	5.00	0.75	2.60	2.75
Floor finishes	2.05	3.90	5.15	2.40	3.55	4.55
Ceiling finishes	1.55	2.80	3.75	2.05	4.60	4.90
Conveying systems	5.55	6.70	8.25	1.15	1.80	2.35
Specialties	0.65	0.80	2.65	1.10	1.35	4.00
Fixed equipment	1.05	2.80	3.75	1.15	1.65	1.95
Heat/vent/air cond.	8.85	9.50	12.20	3.10	5.55	6.25
Plumbing	3.50	3.80	4.85	4.45	5.40	6.15
Electrical	4.60	4.75	6.25	4.20	7.45	8.20
Total $/SF	$61.55	$78.10	$101.15	$43.95	$65.15	$84.05

Component	Secondary schools			Hospitals		
	Low $/SF	Median $/SF	High $/SF	Low $/SF	Median $/SF	High $/SF
Foundation	1.35	1.85	2.70	4.35	4.80	6.65
Floors on grade	3.65	4.40	6.00	0.30	0.40	0.60
Superstructure	10.95	12.30	17.25	17.05	18.55	25.50
Roofing	1.70	2.05	2.45	3.25	3.70	5.20
Exterior walls	3.75	5.55	8.00	16.00	18.55	25.10
Partitions	5.90	6.55	8.50	7.20	11.00	24.70
Wall finishes	3.05	3.40	5.15	6.75	7.95	11.10
Floor finishes	3.10	3.95	5.25	2.60	2.75	4.00
Ceiling finishes	3.20	3.65	4.65	2.15	2.20	3.55
Conveying systems	0.00	0.00	0.00	12.95	13.00	19.55
Specialties	1.70	1.90	2.60	3.10	3.25	4.60
Fixed equipment	2.85	3.35	6.00	5.20	5.25	7.65
Heat/vent/air cond.	9.05	10.45	14.45	21.65	25.50	36.05
Plumbing	5.05	6.00	9.20	9.10	10.65	16.45
Electrical	10.25	12.00	16.50	13.45	17.50	24.40
Total $/SF	$69.55	$77.40	$108.70	$125.10	$145.05	$215.10

Figure 5-1   Illustrative examples of cost-per-square-foot information available from pricing manuals.

of future projects. The term of *weighting* is commonly used to refer to the procedure of analyzing historical cost data to determine a unit cost for forecasting future project costs. A unit cost should be developed that emphasizes the average value, yet accounts for extreme maximum and minimum values. Equation (5-1) can be used for weighting cost data from previous projects:

$$UC = \frac{A + 4B + C}{6} \qquad (5\text{-}1)$$

where UC = forecast unit cost
A = minimum unit cost of previous projects
B = average unit cost of previous projects
C = maximum unit cost of previous projects

Example 5-1 illustrates the weighting procedure. The procedure can be applied to other types of projects and their parameters. Examples are apartment units, motel rooms, miles of electric transmission line, barrels of crude oil processed per day, and square yards of pavement.

Example 5-1 Cost information from eight previously completed parking garage projects is given in the following table.

Project	Total cost	No. cars	Unit cost	
1	$1,387,500	150	$ 9,250	
2	896,000	80	11,200	
3	1,797,000	120	14,975 ←	highest value
4	1,107,000	90	14,975	
5	590,400	60	12,300	
6	1,903,000	220	8,650 ←	lowest value
7	889,000	70	12,700	
8	1,615,500	180	8,975	
			Total = $79,034	

Average cost per car = $ 9,879 ← average value

From Eq. (5-1) the forecast unit cost can be calculated as

$$UC = \frac{\$8,650 + 4(\$9,879) + \$14,975}{6} = \$10,524$$

For a project with 135 cars the estimated cost can be calculated as

135 cars @ $10,524 = $1,420,673

Adjustments for Time, Size, and Location

It is necessary to adjust the cost information from previously completed projects for differences in size, time, and location. The previous example illustrates adjustment relative to size. Time adjustments represent variation in costs due to inflation, deflation, interest rates, etc.

Location adjustments represent variation in costs between locations due to geographical differences in costs of materials, equipment, and labor.

An index can be used to adjust previous cost information for use in preparation of the owner's estimate. Various organizations publish indices that show economic trends. The *Engineering News Record* (ENR) annually publishes indices of construction costs for both time and location. Example 5-2 illustrates the combination of adjustments of cost estimates for size, time, and location. *Estimating Construction Costs,* 6th ed., published by McGraw-Hill, Inc., presents a comprehensive discussion of estimating project costs.

Example 5-2   Use the time and location indices below to calculate the forecast cost for a building with 52,000 square feet of floor area. The building is to be constructed 3 years from now in City B. A similar type building with 41,000 square feet of floor area that cost $4,510,000 was completed 2 years ago in City D.

Time	Index	Location	Index
4 yr ago	3780	City A	1251
3 yr ago	3821	City B	1372
2 yr ago	3915	City C	1452
1 yr ago	4150	City D	1589
Current year	4287	City E	1426

There is no method to predict future cost escalation with absolute certainty. For this example a compound interest is calculated based on the trend in cost indices over the past 4 years. Using the equation compound interest equation $F/P = (1 + i)^n$, an interest rate can be calculated as follows:

$$\frac{4287}{3780} = (1+i)^4 \rightarrow i = 3.2\%$$

The combined adjustments of costs for time, location and size can be calculated as follows:

Estimated cost		Previous cost		Time adjustment		Location adjustment		Size adjustment
"	=	$4,510,000	×	$(1 + 0.032)^5$	×	(1372/1589)	×	(52,000/41,000)
"	=	$4,510,000	×	1.17057	×	0.86344	×	1.26829
"	=	$5,781,289						

Parametric Estimating

Parametric estimating is commonly used to prepare conceptual cost estimates of building construction projects in the early stages of project development, after preliminary design is completed. Parametric estimating relates the total cost of a project to a few physical measurements, or "parameters." For example the gross square floor area of a building is a typical overall parameter for a building project. Sometimes parametric estimating is referred to as *square foot* estimating, or simply *unit cost* estimating.

Unit costs for parametric estimating may be derived from internal company records of past similar projects, or from national pricing manuals that are available from several sources, such as *Engineering News-Record* (ENR) and *RS Means*. The cost data used in parametric estimating should be current with respect to time and the cost data should be from projects that are similar in type and size as the project that is being estimated. It may be necessary to adjust the unit costs based on time and size as presented in subsequent sections of this chapter.

Figure 5-2 illustrates data for parametric estimating of a building, based on an office building with 10,000 square feet (SF) of floor space, 2 stories, and 8-ft walls. Some of the parameter *unit costs* are expressed in the gross enclosed square feet of floor area (such as electrical work), while some unit costs are expressed in square foot of the building components (such as square feet of roof), and some unit costs are expressed in the unit of measure of an individual components, such as linear feet (LF) of partition wall.

In parametric estimating, the unit costs of all components are converted to equivalent costs per square foot of the building. For example Figure 5-2 shows the unit cost of partition walls as $11.25 per linear foot of wall based on 1 linear foot of partition wall for each 6 square feet of floor area in the building. The unit cost of this component of partition walls is converted to square feet of building as: [10,000 SF/6 LF/SF] × [$11.25/LF/10,000 SF] = $1.88/SF. Thus, the $11.25/LF of partition wall is equivalent to $1.88/SF of floor area of the building.

Another example is interior doors, which are measured as "each" with a unit cost of $260/door. Figure 5-2 shows 1 door for 75 square feet of building. The unit cost of interior doors at $260 each is converted to cost per square foot of building as: [10,000 SF/75 SF/door] × [$260/door/10,000 SF] = $3.46/SF. An alternate calculation is [$260/door/75 doors/SF] = $3.46/SF. Thus, the unit cost of $260/door is equivalent to $3.46/SF of floor area of the building.

The interior wall finish is based on painting 2 sides of the linear feet of 8-ft high interior partition walls. The unit cost for this building component is calculated as: [1 LF of wall/6 SF of building] × [8-ft × 2 sides × $1.25/SF] = $3.33/SF. Thus, the $1.25/SF of wall painting is equivalent to $3.33/SF of floor area of the building.

The parametric estimate is prepared by multiplying the square feet of the proposed building by the cost per square foot of each component. The sum of all components is the estimated cost. This method of estimating is commonly used by the owners for budgeting during the economic

Office building, 10,000 SF, 2-story, 8-ft walls

System/Component	Specification	Unit	Unit cost	Cost per SF	% of total
Foundation					
Piers and footings	Poured concrete footings	SF ground	4.36	2.18	
Slab on grade	Concrete, grading, and fill	SF slab	6.95	3.48	7.3
Excavation	Excavation and backfill	SF ground	2.75	1.38	
Exterior Closure					
Exterior walls	Concrete block with stucco	SF wall	12.27	8.59	
Elevated floors	Concrete, precast double tees	SF floor	9.96	4.98	25.7
Roof	Concrete, precast double tees	SF roof	14.55	7.28	
Exterior doors	Wood/glass (1,250 SF ground/door)	each	2,145.00	0.86	
Exterior windows	Aluminum/glass (300 SF flr/window)	each	724.00	2.41	
Interior work					
Partition walls	Studs/gyp board (6 SF flr/LF partition)	LF wall	11.25	1.88	
Interior doors	Hollow core (75 SF floor/door)	each	260.00	3.46	
Wall finishes	Paint, 2 coats	SF wall	1.25	3.33	16.4
Floor finishes	Carpet	SF floor	37.98	4.11	
Ceiling finishes	Suspended acoustic ceiling	SF ceiling	2.58	2.58	
Mechanical					
Plumbing	Fixtures/feeds (160 SF flr/fixture)	each	1,450.00	9.06	
Heating/cooling	HVAC system	SF floor	14.15	14.15	26.8
Fire protection	Sprinkler system	SF floor	1.98	1.98	
Electrical					
Service/distribution	400 amp, panel board, leads	SF floor	5.47	5.47	
Lighting & wiring	Fixtures, feeds, switches, wiring	SF floor	15.82	15.82	23.6
Special electrical	Alarm system	SF floor	0.83	0.83	
	Subtotal of system/components:			$93.83	100.0
	Add 15% for contingency:			$18.77	
	Total building cost = $107.90/SF				

Figure 5-2 Unit costs for parametric estimate of a building.

feasibility phase of a project and by architects and engineers during the design phase to evaluate the cost of design alternatives.

With good historical cost records on comparable structures, the parametric estimating method can give reasonable accuracy for conceptual cost estimates. With this method an experienced estimator with access to well-documented records can quickly prepare an estimate and budget that will help in making decisions during the early phases of a project.

Economic Feasibility Study

Regardless of its size or type, a project must be economically feasible. There are at least two ways to determine economic feasibility, depending on whether the owner is in the private sector or government sector. For a private project the economic feasibility can be determined by an economic analysis of the monetary return on the investment to build the project. For a public government project the economic feasibility is usually determined by a benefit/cost ratio.

There are three methods that are commonly used by the private sector to evaluate the monetary return on a potential investment: capital recovery, pay back period, and rate of return. Each method uses the fundamental equations of time value of money. There are numerous books that have been published that describe the development and use of these equations for economic analysis. Below is a brief introduction to the equations.

Money can earn value when invested over time. For example, $100 invested today at 7% interest will be worth $107 one year from now. Thus, the time value of money means that equal dollar amounts at different points in time do not have equal value. For example, the present worth of $100 today is equivalent to $107 one year from now provided the interest is 7%. Engineering economic analysis is the process of evaluating alternatives. The basic variables used in a time value analysis for economic feasibility are

P = present-worth amount (the value of money today)
F = compound amount (the future value of money after n periods of time at i interest)
i = interest rate per interest period (usually one year)
n = number of periods of time (usually in years)

Single payments

The following calculations are provided to illustrate the relationship between P and F for a given i and n.

End of year	Interest earned during the year	Compound amount at end of year (F = future value of money)
0		$= P$
1	Pi	$P + Pi = P(1 + i)$
2	$P(1 + i)i$	$P(1 + i) + P(1 + i)i = P(1 + i)^2$
3	$P(1 + i)^2 i$	$P(1 + i)^2 + P(1 + i)^2 i = P(1 + i)^3$
n	$P(1 + i)^3 i$	$P(1 + i)^3 + P(1 + i)^3 i = P(1 + i)^n$

Thus, $F = P(1 + i)^n$. Note that F is related to P by a factor that depends only on i and n. This factor, $(1 + i)^n$, is called the single-payment compound-amount factor, which makes F equivalent to P. Rearranging terms,

$P = F/(1+i)^n$. The factor $1/(1+i)^n$ is known as the present-worth compound-amount factor, which makes P equivalent to F. These two equations provide equivalent single payments, now and in the future. There are four parameters: P, F, i, and n. Given any three parameters, the fourth can easily be calculated.

Uniform payment series

Instead of a single amount of money of P today and F in the future as illustrated previously, a series of equal payments may occur at the end of succeeding annual interest periods. The sum of the compound amounts of the series of equal payments may be calculated by using the single-payment compound-amount factor. If A represents a series of n equal payments, then the sum of these payments after n years at i interest can be calculated as in the following equation:

$$F = A(1) + A(1+i) + \ldots + A(1+i)^{n-2} + A(1+i)^{n-1}$$

Multiplying both sides of the equation by $(1+i)$ gives the following equation:

$$F(1+i) = A(1+i) + A(1+i)^2 + \ldots + A(1+i)^{n-1} + A(1+i)^n$$

Subtracting the first equation from the second gives the compound amount of the uniform series of payments as follows:

$$F(1+i) - F = -A + A(1+i)^n$$

$$F = A\left[\frac{(1+i)^n - 1}{i}\right]$$

Note that F is related to A by a factor that depends only on i and n. This factor, $[(1+i)^n - 1]/i$, is called the uniform-series compound-amount factor, which makes the series of A payments equivalent to F. Rearranging terms, $A = F\{[i/(1+i)^n] - 1\}$. The factor $[i/(1+i)^n] - 1$ is known as the uniform-series sinking-fund factor. It represents the amount of money A that must be invested over the uniform series of payments to equate to the future amount F.

Substituting $P(1+i)^n$ for F in the uniform-series sinking-fund equation gives the following equation:

$$A = \frac{p(1+i)^n i}{(1+i)^n - 1}$$

$$= p\left[\frac{i(1+i)^n}{(1+i)^n - 1}\right]$$

The equation $A = P\{[i(1+i)^n]/[(1+i)^n - 1]\}$ gives a uniform series of payments A that is equivalent to the present worth of P. Again, note that A is related to P by a factor that depends only on i and n. This factor, $[i(1+i)^n]/[(1+i)^n - 1]$, is called the uniform-series capital recovery. Rearranging terms, $P = A\{[(1+i)^n - 1]/[i(1+i)^n]\}$. This gives the present worth P of the uniform series of A payments. The factor, $[(1+i)^n - 1]/[i(1+i)^n]$, is called the uniform-series present-worth factor.

For convenience, each of the six basic equations for economic analysis are shown in Eqs. (5-2) to (5-7).

Fundamental equations of time value of money

P = Present Worth
F = Future Sum
A = Equal Payment Series
i = Annual Interest Rate
n = Study Period (years)

Single Payment Series:

Compound Amount $\quad F = P[(1+i)^n] = P(^{F/Pi-n})$ $\qquad$ (5-2)

Present Worth $\qquad P = F\left[\dfrac{1}{(1+i)^n}\right] = F(^{P/Fi-n})$ $\qquad$ (5-3)

Equal Payment Series:

Compound Amount $\quad F = A\left[\dfrac{(1+i)^n - 1}{i}\right] = A(^{F/Ai-n})$ $\qquad$ (5-4)

Sinking Fund $\qquad A = F\left[\dfrac{i}{(1+i)^n - 1}\right] = F(^{A/Fi-n})$ $\qquad$ (5-5)

Present Worth $\qquad P = A\left[\dfrac{(1+i)^n - 1}{i(1+i)^n}\right] = A(^{P/Ai-n})$ $\qquad$ (5-6)

Capital Recovery $\quad A = P\left[\dfrac{i(1+i)^n}{(1+i)^n - 1}\right] = P(^{A/Pi-n})$ $\qquad$ (5-7)

The factor designation at the right of the preceding interest equations represents an abbreviated form of the equation illustrated in the following:

$(^{F/Pi-n})$ means find the Future Sum, F, from a known Present Worth, P, and Interest Rate, i, during a Study Period of n.

In general, an economic analysis involves the process of solving for one of the variables. For example, capital recovery solves for A, pay back period solves for n, and rate of return solves for i. A more complex analysis using one or more of the basic six equations is required when multiple sums of money are distributed over the study period, or when tax advantages are included.

The capital recovery method evaluates the amount of annual money, A, that must be obtained throughout the study life, n, of a project in order to obtain a recovery on the original capital investment, P. A simple illustration is given in Example 5-3. There may be other considerations that should be evaluated, such as tax advantages, non-uniform generation of revenue, and the disposal of the facility after its useful life. The pay back period method evaluates the number of years, n, that a project must be operated in order to obtain an interest rate, i, for a given investment, P, with an annual generated income, A. This method is illustrated in Example 5-4. A rate-of-return analysis evaluates the interest rate, i, that equates the initial investment, P, to the yearly net cash flow as illustrated in Example 5-5. A trial-and-error solution is required in this example because the annual payments are not uniform. All these examples are simple illustrations of the methods that can be used to determine the economic feasibility of a project.

Example 5-3   This example illustrates the capital recovery method for determining economic feasibility. Suppose the feasibility estimate for a project is $7.0M with an expected operating life of 12 years. Annual maintenance and operating expenses are forecast as $560K per year. Using a 10% interest rate, what net annual income must be received to recover the capital investment of the project?

$$A = P\left[\frac{i(1+i)^n}{(1+i)^n - 1}\right] + \$0.56\text{M}$$

$$= \$7.0\text{M}\left[\frac{0.10(1+0.10)^{12}}{(1+0.10)^{12} - 1}\right] + \$0.56\text{M}$$

$$= \$1.588\text{M per year}$$

If the project can be built for $7.0M, operated for $560K per year, and earn $1.5876M per year, the project is economically feasible, neglecting any tax advantages. If the project can be built for less than $7.0M, or can be built to operate more efficiently than $560K per year, it is even more economically attractive.

Example 5-4   This example illustrates the pay back period method for determining the economic feasibility of a project. The initial investment for a project is $18.0M. A net annual profit of $3.5M is anticipated. Using a 15% desired rate of return on the investment, what is the pay back period for the project?

$$P = A\left[\frac{(1+i)^n - 1}{i(1+i)^n}\right]$$

$$\$18.0M = \$3.5M\left[\frac{(1+0.15)^n - 1}{0.15(1+0.15)^n}\right]$$

$$n = 10.5 \text{ years}$$

Example 5-5 This example illustrates the rate-of-return method for determining the economic feasibility of a project. An initial project investment of $1.05M is being considered for a 5-year study period. It is anticipated the project will be sold after the 5-year period for $560K. Determine the rate of return with the anticipated net profit shown below.

End of year	Net profit ($1,000)
0	0
1	−350
2	−120
3	+420
4	+735
5	+680

Using Eq. (5-2) to transfer the costs from each year to an equivalent present worth:
Try $i = 15\%$.

$$P = [-\overset{P/F15\text{-}1}{\$350(0.8695)} - \overset{P/F15\text{-}2}{\$120(0.7562)} + \overset{P/F15\text{-}3}{\$420(0.6575)} + \overset{P/F15\text{-}4}{\$735(0.5718)}$$
$$+ \overset{P/F15\text{-}5}{\$680(0.4972)} + \overset{P/F15\text{-}5}{\$560(0.4972)}] \times 1{,}000$$

$1.05M > $0.92M; therefore try a lower rate of return.
Try $i = 10\%$.

$$P = [-\overset{P/F10\text{-}1}{\$350(0.9090)} - \overset{P/F10\text{-}2}{\$120(0.8264)} + \overset{P/F10\text{-}3}{\$420(0.7513)} + \overset{P/F10\text{-}4}{\$735(0.6830)}$$
$$+ \overset{P/F10\text{-}5}{\$680(0.6209)} + \overset{P/F10\text{-}5}{\$560(0.6209)}] \times 1{,}000$$

$1.05M < $1.17M; therefore try a higher rate of return.
Try $i = 12\%$.

$$P = [-\overset{P/F12\text{-}1}{\$350(0.8929)} - \overset{P/F12\text{-}2}{\$120(0.7972)} + \overset{P/F12\text{-}3}{\$420(0.7118)} + \overset{P/F12\text{-}4}{\$735(0.6355)}$$
$$+ \overset{P/F12\text{-}5}{\$680(0.5674)} + \overset{P/F12\text{-}5}{\$560(0.5674)}] \times 1{,}000$$

$1.05M ≈ $1.06M; therefore the rate of return is slightly over 12%.

It is important for the project manager and his or her team to realize that an owner's economic study, similar to one of those illustrated, is used by the owner to approve the project budget, which is the capital investment, P. When a project exceeds its budget, then the economic justification that was used by the owner to proceed with the project is impaired.

A popular method for deciding on the economic justification of a public project is to compute the benefit/cost ratio, which is simply the ratio of the benefits to the public divided by the cost to the government. If the ratio is 1, the equivalent benefits and equivalent costs are equal. This represents the minimum justification for an expenditure by a public agency. Generally, the first step is to determine the benefits that can be derived from a project. This is in contrast to the consideration of profitability as a first step in evaluating the merits of a private enterprise. The second step in evaluating a public project involves an analysis of cost to the governmental agency. When a public project is being considered, the question is: Will this project result in the greatest possible enhancement of the general welfare in terms of economic, social, environmental, or other factors that serve the needs of the general public? The measurement of benefits is sometimes difficult because they cannot always be expressed in dollars.

Many government agencies have a list of projects that are waiting for approval, but for which funds are not yet available. The decision as to which project to approve may be based upon the amount of money that is allocated in a fiscal year, rather than economic feasibility.

Design Budgets

The design organization has a difficult task of estimating the cost of providing design services and/or producing contract documents for the project before the design and construction phases begin. For many projects the magnitude of work that is required by the designer cannot be fully anticipated, because design is a creative process that involves the evaluation of numerous alternatives. The evaluation of design alternatives is a necessary part of the design process required to select the best design that satisfies the owner's need for the project.

Compensation for design services is usually priced by one of the following methods: lump-sum, salary cost times a multiplier, cost plus a fixed payment, or percent of construction. The method that is used depends on the accuracy of the scope definition that is provided to the design organization.

For projects that have a well-defined scope with no unusual features, and are similar to projects that a designer has handled in the past, a lump-sum design contract is commonly used. Preparation of the design budget can be developed by defining tasks, and grouping of tasks, in a work breakdown structure. The development of a project work breakdown structure and design work packages is discussed in Chapter 6. A design work package, shown in Figure 6-7, can then be prepared for each task. Based upon the past experience of the designer with similar projects, the estimated labor-hours of design calculations, number of drawings, labor-hours per drawing, travel, and other expenses can be estimated for each task. The total cost for design can be calculated by adding the cost of all design work packages. The final design budget is usually broken down by discipline with the labor-hours based on the number of drawings to be produced. Figure 5-3 illustrates the summary of a design engineering budget. Figure 5-4 shows an example of time distribution for design calculations as well as the development of drawings and specifications for a project. Chapter 7 presents the process of developing design proposals and design budgets. Reference Figures 7-5 and 7-6 for gathering the information necessary to prepare the budget forms shown in Figures 5-3 and 5-4.

The salary cost times a multiplier method is used for projects when it is difficult to accurately define the scope of work at the time the designer is retained for the project. For these types of projects, preliminary services, such as process studies, development of alternate layout plans, or other services are required to establish information that is needed for the final design. The designer provides a fee schedule to the owner that lists the classification and salary costs of all personnel, and a rate schedule for all other costs that are directly chargeable to the project. Work is then performed based on actual time expended in the design effort. A multiplier, usually within a range from 2.0 to 3.0, is applied to direct salary costs that compensate the design organization for overhead, plus a reasonable margin for contingencies and profit. A larger multiplier may be used for unusual projects that require special expertise, or for projects of short duration or small size. Travel, subsistence, supplies, and other direct non-salary expenses are generally reimbursed at actual costs, plus a 10% to 15% service charge.

The cost plus a fixed payment method is used for projects that have a general description or statement of the scope of contemplated work, such as the number, size, and character of buildings or other facilities and the extent of utilities. The design organization is reimbursed for the actual cost of all salaries, services, and supplies plus a fixed fee that is agreed on between the designer and owner. The fixed fee usually varies from 10% for large projects to 25% for small projects that are short in duration.

Design work may also be compensated based on a percentage of the construction costs of a project, although this method is not as common today as it was in the past. Generally the percentage is on a sliding scale that decreases as the construction cost increases. The percentage also varies depending on the level of design services that are provided, such as design only, design and preparation of drawings, or full design

		HOURS							
									DATE:___/___/___
								PROJECT BUDGET FORM	
	Project Name								
Dept. Number	Department	ADMIN	MTGS	SHED	SPECS	CALCS	DWGS	TOTAL	DOLLARS
0100	Project Management							12,000	$840,000
		3,000	2,400	900				6,300	
		1,500	1,200	3,000				5,700	
0200	Architecture		350	100	350	250	800	1,850	$111,000
0300	Mechanical		350		480	360	1,760	2,950	$177,000
0400	Electrical		350		2,100	3,200	7,450	13,100	$786,000
0500	Structural							26,360	$1,581,600
	Project Engineer	2,500	750					3,250	
	CADDS Coord.		160				2,800	2,960	
	Department		550		1,300	6,700	11,600	20,150	
0600	Environmental	400						400	$28,000
0700	Civil							7,140	$428,400
	Turb./Gen. Spec				340			340	
	Department		190	100	1,310	2,000	32,00	6,800	
0800	CADDS		100	100			1,000	1,200	$60,000
0900	Clerical	7,000	400		600			8,000	$200,000
1000	Document Control	1,000			200		800	2,000	$50,000
1100	Reproduction				200		800	1,000	$26,000
1200	Project Control	1,000	200	500				1,700	$102,000
1300	Management	400						400	$36,000
1400	Subcontractor A	100	100		86	114	200	600	$60,000
1500	Subcontractor B	50	100		50		200	400	$40,000
1600	Record Drawings					1,000	2,000	3,000	$150,000
1700	Support Buildings		100		400	500	1,500	2,500	$150,000
WORK–HOUR SUBTOTAL		16,950	7,300	4,700	7,416	14,124	34,110	84,600	$4,826,000
Task #	Description								
1800	Contingency								$500,000
1900	General Expenses								$24,000
2000	Travel								$100,000
2100	Office Budget								$50,000
EXPENSES SUBTOTAL									$674,000
TOTAL									$5,500,000

Figure 5-3   Illustrative design engineering budget.

DRAWINGS

	50% Drawings	20% Drawing Review	20% Drawing Submittals	10% Final Record	Total
Architecture	400	160	160	80	800
Mechanical	880	352	352	176	1,760
Electrical	3,725	1,490	1,490	745	7,450
Structural	5,800	2,320	2,320	1,160	11,600
Civil	1,600	640	640	320	3,200
	12,405	4,962	4,962	2,481	24,810

CALCULATIONS

	70% Drawings	20% Drawing Review	10% Drawing Submittals	Total
Architecture	175	50	25	250
Mechanical	252	72	36	360
Electrical	2,240	640	320	3,200
Structural	4,690	1,340	670	6,700
Civil	1,400	400	200	2,000
	8,757	2,502	1,251	12,510

DRAWINGS

	25% Drawings	25% Drawing Review	25% Drawing Submittals	25% Final Record	Total
Reproduction	200	200	200	200	800
Doc. Control	200	200	200	200	800
	400	400	400	400	1,600

SPECIFICATIONS

	80% Specs	20% Review	Total
Architecture	280	70	350
Mechanical	384	96	480
Electrical	1,680	420	2,100
Structural	1,040	260	1,300
Civil	1,048	262	1,310
Turbine/Gen.	272	68	340
Consultants	108	28	136
Clerical	480	120	600
	5,292	1,324	6,616

SPECIFICATIONS

	50% Specs	50% Review	Total
Reproduction	100	100	200
Doc. Control	100	100	200
	200	200	400

Figure 5-4 Example of time distribution for design calculations, drawings, and specifications.

services which include design, preparation of drawings, and observation during construction. The percentage generally will range from 5% to 12% of the anticipated construction cost.

The percentage data given in the above paragraphs are not fixed, nor should the ranges be considered as absolute maximums or minimums. Instead, they are presented as a guide to establish the approximate costs that may be incurred for the design of a project.

Contractor's Bid

Most of the cost of a project is expended during the construction phase when the contractor must supervise large work forces who operate equipment, procure materials, and physically build the project. The cost of construction is determined by the contractor's bid that has been accepted by the owner before starting the construction process. Depending on the completeness of the design and the amount of risk that is shared between the owner and contractor, there are many methods that have been developed to compensate the construction contractor.

The pricing format for providing construction services can be divided into two general categories: fixed price and cost reimbursable. Fixed price contracts usually are classified as lump-sum, unit-price, or a combination of lump-sum and unit-price. Cost-reimbursable contracts can be classified as cost plus a fixed fee or cost plus a percentage. An incentive is often built into cost-reimbursable contracts to control the total cost of a project. Examples of incentives are "target price" and "guaranteed maximum price."

For projects with a complete set of plans and specifications that have been prepared prior to construction, the contractor can prepare a detailed estimate for the purpose of submitting a lump-sum bid on the project. Only one total-cost figure is quoted to the owner, and this figure represents the amount the owner will pay to the contractor for the completed project, unless there are revisions in the plans or specifications. The contractor's bid is prepared from a detailed estimate of the cost of providing materials, labor, equipment, subcontract work, overhead, and profit.

The preparation of lump-sum detailed cost estimates generally follows a systematic procedure that is developed by the contractor for his or her unique construction operations. Building construction contractors usually organize their estimates in a format that closely follows the Construction Specifications Institute (CSI) numbering system, which divides work into major divisions. Each major division is subdivided into smaller items of work. The numbering system is developed for architects writing specifications that are unique to each project. However, contractors typically use CSI numbering system as a checklist and guide for quantity takeoff, price extensions, and summary of costs for the final lump-sum bid.

For years, contractors used the CSI 16 division numbering system, which now has been expanded into a 50 division number system. It doesn't matter which numbering system is used as long as the information that is needed to assist in the bidding process is included and can be found easily.

An illustrative example of the cost summary for a building construction project is shown in Figure 5-5, which follows the CSI 16 division numbering system. This figure shows the summary costs of material, labor, and subcontract for Item 2, Site-work, which has a total cost of $181,389. Figure 5-6 shows a breakdown of the $181,389 site-work cost, which includes clearing, excavation, compaction, handwork, termite control, drilled piers, foundation drains, landscape, and paving. Preparation of

Item	Division	Material	Labor	Subcontract	Total
1	General requirement	$ 16,435	$ 36,355	$ 4,882	$ 57,672
2	Site-work	15,070	20,123	146,196	181,389
3	Concrete	97,176	51,524	0	148,700
4	Masonry	0	0	212,724	212,724
5	Metals	212,724	59,321	0	272,045
6	Woods and plastics	38,753	10,496	4,908	54,157
7	Thermal and moisture	0	0	138,072	138,072
8	Doors and windows	36,821	32,115	0	68,936
9	Finishes	172,587	187,922	0	360,509
10	Specialties	15,748	11,104	9,525	36,377
11	Equipment	0	0	45,729	45,729
12	Furnishings	0	0	0	0
13	Special construction	0	0	0	0
14	Conveying systems	0	0	0	0
15	Mechanical	0	0	641,673	641,673
16	Electrical	0	0	354,661	354,661
	Total direct costs	$605,314	$408,960	$1,558,370	$2,572,644
	Material tax (5%)	30,266			2,602,910
	Labor tax (18%)		73,613		2,676,523
	Contingency (2%)			53,530	2,730,053
	Bonds/Insurance			34,091	2,764,144
	Profit (10%)			276,414	3,040,558
				Bid price =	$3,040,558

Figure 5-5 Example of building construction project bid summary using the CSI organization of work.

Cost code	Description	Quantity	Material	Labor	Subcontract	Total
2110	Clearing	L.S.	$ 0	$ 0	$ 3,694	$ 3,694
2220	Excavation	8,800 yd³	0	11,880	9,416	21,296
2250	Compaction	950 yd³	0	2,223	722	2,945
2294	Handwork	500 yd²	0	1,750	0	1,750
2281	Termite control	L.S.	0	0	3,475	3,475
2372	Drilled piers	1,632 lin ft	14,580	2,800	14,534	31,914
2411	Foundation drains	14 ea.	490	1,470	0	1,960
2480	Landscape	L.S.	0	0	8,722	8,722
2515	Paving	4,850 yd²	0	0	105,633	105,633
			$15,070	$20,123	$146,196	$181,389

Figure 5-6   Division 2 estimate for site-work.

the estimate would involve further breakdown of the items in Figure 5-6. For example, the cost of excavation work would be broken down into labor and equipment costs for the crew performing excavation work.

Heavy engineering construction projects generally are organized in a Work Breakdown Structure (WBS) that is unique to the project to be constructed. A WBS organizes costs into major divisions of work. Each division is divided into subdivisions and each subdivision is organized into more detailed components.

An example of a WBS for an electrical power construction project is illustrated in Figures 5-7 to 5-9. Figure 5-7 is the top division level in the WBS, which shows the summary of costs for each major facility in the project. For example no. 2100 is the major facility of Transmission Line A. The subdivision of costs for Transmission Line A are shown in Figure 5-8. Cost Item 2370 in Figure 5-8 shows the summary of costs for tower foundations. A more detailed breakdown of the tower foundation costs are shown in Figure 5-9.

The unit-price bid is similar to the lump-sum bid, except that the contractor submits a cost per unit of work in place, such as a cost per cubic yard of concrete. The contract documents define the units the owner will pay the contractor. The final cost is determined by multiplying the bid cost per unit by the actual quantity of work that is installed by the contractor. Thus, the price that the owner will pay to the contractor is not determined until the project has been completed, when the actual quantities are known.

Cost-reimbursable contracts for construction may be used for several reasons; to start construction at the earliest possible date, to allow the owner to make changes in the scope of work without substantial modifications in the contract, or because the project is unique with features that

Group-level report					
No.	Group	Material	Labor and equipment	Subcontract	Total
1100	Switch station	$1,257,295	$ 323,521	$3,548,343	$ 5,129,159
2100	Transmission line A	3,381,625	1,259,837	0	4,641,462
2300	Transmission line B	1,744,395	0	614,740	2,359,135
3100	Substation at spring creek	572,874	116,403	1,860,355	2,549,632
4200	Distribution line A	403,297	54,273	215,040	672,610
4400	Distribution line B	227,599	8,675	102,387	338,661
4500	Distribution line C	398,463	21,498	113,547	533,508
		$7,985,548	$1,784,207	$6,454,412	$16,224,167

†For large projects the costs are sometimes rounded to the nearest $100 or $1,000. Figures 5-7 to 5-9 show full dollars to illustrate the transfer of costs among the component, division, and group levels of an estimate.

Figure 5-7 Example of electric power construction bid summary using the WBS organization of work.†

prevent a reasonable approximation of the actual cost of construction. The estimate for this type of project is usually prepared by the contractor as an approximate estimate. The owner and contractor agree on a cost rate for all labor, equipment, and other services that may be charged to the project by the contractor. The contractor is reimbursed for all costs that are accrued during the construction phase of the project plus a percentage of the costs or a fixed fee.

To maintain some degree of control over the total cost of a project, an incentive is often placed on cost-reimbursable contracts. For example, the contract may be awarded on a cost plus basis with a guaranteed

DIVISION-LEVEL REPORT FOR TRANSMISSION LINE A					
Cost item	Description	Material	Labor	Equipment	Total
2100	TRANSMISSION LINE A				
2210	Fabrication of steel towers	$ 692,775	$ 0	$ 0	$ 692,775
2370	Tower foundations	83,262	62,126	71,210	216,598
2570	Erection of steel towers	0	144,141	382,998	527,139
2620	Insulators and conductors	2,605,588	183,163	274,744	3,063,495
2650	Shield wire installation	0	78,164	63,291	141,455
Total for 2100		$3,381,625	$467,594	$792,243	$4,641,462

Figure 5-8 Example of electric power construction estimate using the WBS organization of work.

COMPONENT-LEVEL REPORT FOR TOWER FOUNDATIONS						
Cost Item	Description	Quantity	Material	Labor	Equipment	Total
2370	TOWER FOUNDATIONS					
2372	Drilling foundations	4,196 lin ft	$ 0	$25,428	$44,897	$ 70,325
2374	Reinforcing steel	37.5 tons	28,951	22,050	15,376	66,377
2376	Foundation concrete	870 yd^3	53,306	13,831	10,143	77,280
2378	Stub angles	3,142 lb	1,005	817	794	2,616
Total for 2370			$83,262	$62,126	$71,210	$216,598

Figure 5-9 Example of electric power construction estimate using the WBS organization of work.

maximum price, commonly called a GMP contract. For a GMP contract, the owner and contractor agree on a guaranteed maximum price prior to the start of construction. They also agree on the distribution of costs that each will incur if the final cost is above or below the guaranteed maximum price. Then, during construction, the contractor is reimbursed for actual costs plus a fixed fee or percentage of actual costs. If the actual final cost is above or below the guaranteed maximum price, then the predetermined distribution of the difference of costs is distributed between the owner and contractor. To illustrate, if the cost exceeds the guaranteed maximum price, the contractor may pay 70% of the cost and the owner pays 30% of the cost. If the cost is less than the guaranteed maximum price, the contractor receives 60% of the reduced costs and the owner receives 40%.

References

1. Ahuja, H. N. and Campbell, W. J., *Estimating: From Concept to Completion,* Prentice-Hall, Englewood Cliffs, NJ, 1987.
2. *Building Estimator's Reference Book,* The Frank R. Walker Company, Chicago, IL.
3. *Consulting Engineering: A Guide for the Engagement of Engineering Services,* Manual No. 45, American Society of Civil Engineers, Reston, VA, 1988.
4. *Current Practice in Cost Estimating and Cost Control,* Conference Proceedings, American Society of Civil Engineers, Reston, VA, 1983.
5. *Engineering News Record,* a publication of the McGraw-Hill Companies, published weekly, New York, NY.
6. *Means Cost Guide,* Robert Snow Means Company, Duxbury, MA, published annually.
7. Neil, J. M., *Construction Cost Estimating for Project Control,* Prentice-Hall, Englewood Cliffs, NJ, 1982.
8. Palmer, W. J., Coombs, W. E., and Smith, M. A., *Construction Accounting and Financial Management,* 5th ed., McGraw-Hill, Inc., New York, NY, 1995.
9. Parker, A. D., Barrie, D. S., and Snyder, R. N., *Planning and Estimating Heavy Construction,* McGraw-Hill, Inc., New York, NY, 1984.
10. Peurifoy, R. L. and Oberlender, G. D., *Estimating Construction Costs,* 6th ed., McGraw-Hill, Inc., New York, NY, 2014.
11. "The Richardson Rapid System," *Process Plant Construction Estimating Standards,* Richardson's Engineering Services, Inc., Mesa, AZ, published annually.
12. White, J. A, Agee, M. H., and Case, K. E., *Principles of Engineering Economic Analysis,* Wiley, New York, NY, 1989.

6

Development of Work Plan

Project Manager's Initial Review

The discussion of developing the project work plan in this chapter is based on handling the project in its early stage of development, prior to design. It is presented from this perspective because the ability to influence the overall quality, cost, and schedule of a project can best be achieved during design. Most books and articles discuss project management during the construction phase, after design is completed. At this time in the life of a project the scope of work is fully defined, the budget is fixed, and the completion date is firm. It is then too late to make any significant adjustments in the project to improve quality, cost, or schedule to benefit the owner.

When a project manager is assigned to a project, his or her first duty is to gather all the background material that has been prepared by the sponsoring organization. This includes the owner's study and the contract that has been signed by the project manager's organization. These documents must be thoroughly reviewed to be certain there is a well-defined scope, an approved budget, and a schedule that shows major milestones for the project, in particular the required completion date.

The purpose of this initial review process is to become familiar with the owner's objectives, the overall project needs, and to identify any additional information that may be needed to begin the process of developing a work plan to manage the project. To organize the review process it is best to divide the questions into the three categories that define a project: scope, budget, and schedule. To guide this initial review, the project manager should continually ask questions like those shown in Table 6-1.

TABLE 6-1   Guidelines for Project Manager's Initial Project Review

Scope
1. What is missing?
2. Does it seem reasonable?
3. What is the best way to do this?
4. What additional information is needed?
5. What technical expertise is needed?
6. How is the best way to handle construction?
7. What is the owner's expected level of quality?
8. What codes and regulations are applicable?

Budget
1. Does the budget seem reasonable?
2. How was the budget determined?
3. Who prepared the budget?
4. When was the budget prepared?
5. Should any portion of the budget be rechecked?
6. Has the budget been adjusted for time & location?

Schedule
1. Does the schedule seem reasonable?
2. How was the schedule determined?
3. When was the schedule prepared?
4. Who prepared the schedule?
5. How firm is the completion date?
6. Are there penalties or bonuses?

Owner's Orientation

After the project manager has performed the initial project review and become familiar with the project, the owner's authorized representative should be identified and a meeting scheduled to set up the necessary coordination arrangements with the owner. The owner's representative serves two roles in a project: as a participant in providing information and clarifying project requirements, and as a reviewer and approver of all team decisions. The owner must be considered an integral part of the project team, beginning at the start of the project and continuing through all phases until completion.

During this initial meeting, the owner's authorized representative should set priorities for the project. There are four elements of concern for a project: quality, scope, time, and cost. It is understood that quality is an element that must be satisfied. The owner should set the level of quality that is expected in the project. There must be a mutual understanding of quality between the project manager and the owner's representative. Scope is the fixed quantity of work to be performed. It may be expanded or reduced by the owner as the project proceeds, generally depending on the costs. The priority of time or cost is set by the owner. Frequently, time is initially set as a priority over cost. However, cost may take precedence over time if the market for the product changes or other conditions arise. If a priority is not set, the project manager must attempt to optimize time and cost.

The level of involvement required by the owner's representative must be determined at the beginning of the project. If he or she wants to sign everything, then the project manager must include time in the project schedule and cost into the budget for the owner's involvement. Two-way communication is an absolute requirement. The project manager should also inform the owner's representative of how he or she plans to create a project team that will be coordinated to represent each part of the project.

This initial meeting also gives both the owner's representative and the project manager the opportunity to meet each other. At this meeting it may be desirable to visit with others in the owner's organization that may be concerned with the project. Issues to discuss might include clarification of goals and requirements, desired level of quality, any uniqueness about the project, financing, regulatory agencies, and approval process.

In some instances this meeting may be the project manager's first introduction to the owner's representative. Because many owners expect an all-knowing project manager, some precautions need to be taken. Since the project team has not yet been formed, all discussions should focus on the work to be performed rather than on work that has been completed. Ideally, the project manager should have assisted in the proposal preparation that was approved by the owner to proceed with the project. This gives the project manager a better understanding of the history behind the project and previous contact with the owner's representative.

Organizational Structures

Each project manager is affected by the environment in which he or she works. The organization of a company can have a large impact on the ability to manage a project. Figures 6-1 through 6-5 show various organizational structures of companies. A project manager may work for a company that is organized as shown in these figures, or he or she may manage a project for a client whose company organization is similar to one of these organizational structures.

If a company is product oriented, it will be organized around manufacturing and marketing of the product, with the priority of decisions focused on products. A company that is service oriented will be organized around providing customer service. The design and construction of a project is a means to an end for the company to provide a product or service and does not represent the primary function of that company. This secondary emphasis on a project can hamper the work of a project manager.

The organizational structure shown in Figure 6-1 is an example of a business with an emphasis on manufacturing and marketing of products. The engineering portion of the company exists to support the manufacturing operation. Manufacturing exists to produce the product for the marketing group to sell. Questions related to the engineering/ construction of a project for this company would typically be directed to

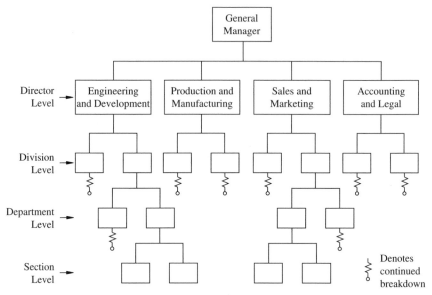

Figure 6-1 Traditional management organization (production line/business oriented).

the engineering department. However, the answers to these questions often come from the manufacturing department, which in turn may have to obtain input from the marketing group. This requires a channel of communications between various parties that can cause misinterpretation of information and a delay in obtaining answers. A project manager performing work for a company that is organized as shown in Figure 6-1 should include a contingency in the project schedule for delays of owner responses, and should be alert to the potential for scope growth.

An example of a functional organization is illustrated by the electrical power company shown in Figure 6-2. The company emphasis is on generation, transmission, and distribution of electrical power services. Utilities and governmental agencies are usually organized in functional departments. This type of organization is efficient for the design and construction of projects involving a single function, such as the design and construction of a transmission line or a substation. However, if a project involves design and construction of a unit of a power station, plus two transmission lines and a substation, it can be difficult to identify the project within the organization. There is a tendency for the project to pass from one department to another if a single project manager is not assigned overall responsibility. This can lead to lost information and schedule delays. Even if a single project manager is assigned, coordinating across departments lines can be difficult.

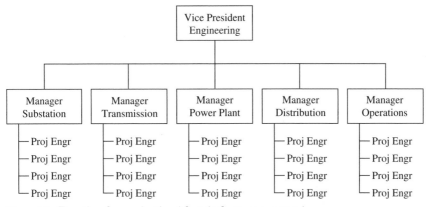

Figure 6-2 Functional organization (electrical power company).

Figure 6-3 shows a typical work environment of a consulting engineering company that provides design services for projects. The company emphasis is discipline oriented, involving a group of specialists who share knowledge and technical expertise. Overemphasizing separate disciplines can encourage competition and conflicts at the expense of the whole organization, resulting in focus on internal department operations rather than external relations and project work. When emphasis is focused on internal departments, decision making and communication channels tend to be vertical, rather than horizontal, with little attention paid to costs, schedules, and coordination.

Many consulting engineering companies are organized as shown in Figure 6-3. For small projects with short durations this type of organization is efficient. However, project management can be hindered because

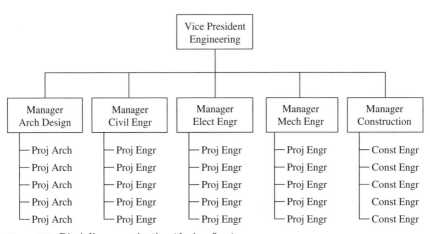

Figure 6-3 Discipline organization (design firm).

some of the engineers have a dual role, as both a designer and a project manager. As the number of disciplines increases, coordination of complex projects becomes more difficult. For example, a complex project may involve architectural, civil, structural, mechanical, and electrical engineering work. The work may begin with the architectural layout, followed by the various engineering designs. As the work moves from discipline to discipline, the project identity can be lost and it becomes difficult to know where the project is or what its status is. By the time the project reaches the last discipline there may not be enough budget left to complete the work. Discipline organizations develop a strong resistance to change.

Another type of organizational structure for a consulting engineering company is shown in Figure 6-4. The company is organized into functional departments: buildings, heavy/civil, process, and transportation. The disciplines are dispersed among the functional departments and serve on design teams for projects that are assigned to the department. Lead designers are appointed as team leaders to manage the design effort. Each designer remains in his or her functional department to provide technical expertise for the project. However, if there is a decline in the number of projects in one or more departments, one or more designers may be transferred to another functional department. This can be disruptive to the management of projects.

To increase emphasis on project cost, schedule, and general coordination, a matrix organization as shown in Figure 6-5 is often used. The objective is to retain the design disciplines in their home departments so technical expertise is not lost, and to create a projects group that is responsible for overall project coordination. To accomplish this the designer has two channels of communications, one to the technical supervisor and another to the project manager. Issues related to technical expertise are addressed vertically while issues related to the project are addressed horizontally.

The matrix organization provides a work environment with emphasis on the project. Each project is defined by a horizontal line on the matrix.

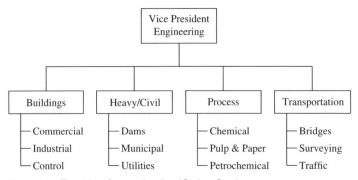

Figure 6-4 Functional organization (design firm).

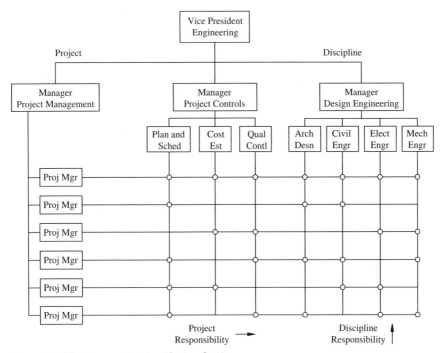

Figure 6-5 Matrix organization (design firm).

The project manager is responsible for overall project coordination, interfacing of disciplines, client relations, and monitoring of overall project costs and schedules. The various design disciplines are responsible for providing technical expertise, quality performance, and the cost and schedule for their particular part of the project. No one person works for the other on the project team; instead everyone works for the project. The project manager is the leader of the team and serves as a focal point for integrating responsibility.

A matrix defines lines of communications but does not indicate the authority for conflict resolution. A matrix may be defined as a "strong matrix," where project managers have the authority to decide what is good for the overall project. At the opposite end of the spectrum is the "weak matrix," where discipline managers have the authority in decision making. A discipline supervisor may be more concerned with his or her technical area than the overall project. Designers are usually concerned with producing the best design possible, sometimes at the expense of project cost or schedule and without regard to the effect on other departments.

The success of project management in the matrix organization depends on the philosophy of the company and the attitude of the employees.

Too much emphasis on disciplines can lead to time and cost problems. Likewise, too much emphasis on projects can lead to inefficiencies and quality problems due to losing control of and contact with the technical departments. Therefore, there must be a balance between managing the project and providing technical expertise. Mutual respect among disciplines is essential. The project manager relies on the expertise of each team member and recognizes that everyone is a key player on the team of a successful project. A "can do" attitude must exist, with a drive to complete a quality project in an efficient manner that meets the needs of the owner. What is good for the project is good for the entire company. Effective communications among team members is a must.

As a project moves from the design phase to the construction phase, a work structure must be developed around the work that must be accomplished in the field. A project organization must be developed that is matched to the project to be constructed. Management of the project is best performed in the field, where the actual work is being performed.

Work Breakdown Structure

For any size project, large or small, it is necessary to develop a well-defined Work Breakdown Structure (WBS) that divides the project into identifiable parts that can be managed. The concept of the WBS is simple; in order to manage a whole project, one must manage and control each of its parts. The WBS is the cornerstone of the project work plan. It defines the work to be performed, identifies the needed expertise, assists in selection of the project team, and establishes a base for project scheduling and control. Chapters 8 and 9 show how the WBS is used in project scheduling and tracking.

A WBS is a graphical display of the project that shows the division of work in a multi-level system. Figure 6-6 is a simple illustrative example of a WBS for a project that has three major facilities: site-work, utilities, and buildings. Each major facility is subdivided into smaller components. For example, the major facility of buildings is subdivided into three buildings: office, maintenance, and warehouse. The project is further broken down so the components at each level are subsets of the next higher level. The number of levels in a WBS will vary depending upon the size and complexity of the project. The smallest unit in the WBS is a work package. A work package must be defined in sufficient detail so the work can be measured, budgeted, scheduled, and controlled. Development of work packages is discussed later in this chapter.

The development of the WBS is a continuing process that starts when the project is first assigned to the project manager and continues until all work packages have been defined. The project manager starts the process of developing the WBS by identifying major areas of the project.

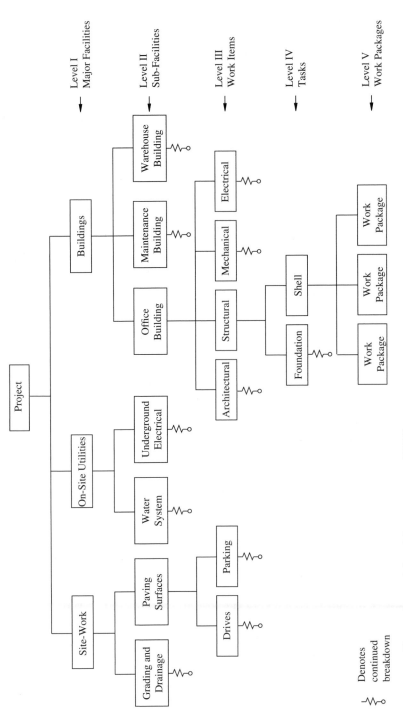

Figure 6-6 Illustrative work breakdown structure (WBS).

Level I
Major Facilities

Level II
Sub-Facilities

Level III
Work Items

Level IV
Tasks

Level V
Work Packages

Project

Site-Work

On-Site Utilities

Buildings

Grading and Drainage

Paving Surfaces

Water System

Underground Electrical

Office Building

Maintenance Building

Warehouse Building

Drives

Parking

Architectural

Structural

Mechanical

Electrical

Foundation

Shell

Work Package

Work Package

Work Package

⌐Ｗ⌐ Denotes continued breakdown

As members of the project team define the work to be performed in more detail, the WBS is adjusted accordingly. Thus, the WBS is used from the start to the finish of the project for planning and controlling. It is an effective means of defining the whole project, by parts, and providing effective communication channels for exchange of information that is necessary for management of the project.

The WBS is the foundation of a project management system. Code numbers can be used to relate the WBS to the Organizational Breakdown Structure (OBS) for management of people. Code numbers can also be used to relate the WBS to the Cost Breakdown Structure (CBS) for management of costs. Similarly, code numbers can relate the WBS to the Critical Path Method (CPM) schedule to manage time. Thus, the WBS provides a systematic approach for identifying work, compiling the budget, and developing an integrated schedule. Since the WBS is developed jointly by the project team, the people that will actually perform the work, it is an effective tool for relating work activities to ensure that all work is included and that work is not duplicated. Most importantly, it provides a basis for measurement of performance.

Formats for Work Breakdown Structures

A WBS can be developed in a graphic format or outline format. Figure 6-7 is an illustrative example of a graphic format WBS for a project and Figure 6-8 is a WBS in an outline format for the same project. The WBS in these figures represents the Engineering/Procurement/Construction (EPC) for a maintenance facility. The WBS for this EPC project has three major divisions: engineering, procurement, and construction. The engineering division is subdivided into contract design and in-house design. Procurement is subdivided into purchase of major equipment and securing contractor bids. Construction is subdivided into site-work and two buildings. The project management of this type of project is further discussed in Chapters 8 and 9 and the Appendix.

The graphic format of a WBS is easy to read and gives a good overall picture of the total project. However, most projects have many items in the WBS, which makes it difficult, or impossible, to display all items in the WBS on one page. For presentations to upper management or clients it is usually desirable to display the WBS on a large size drawing. Another option is to develop the graphic format of the WBS on three smaller size drawings. For example, for an EPC project all of engineering is displayed on one page, procurement on the second page, and construction on the third page.

Most project teams develop the WBS in the outline format. It is easier to transmit the WBS in a paper size sheet than handling a large drawing. The outline format can be easily modified with little effort and becomes a working document to guide the work of the project team.

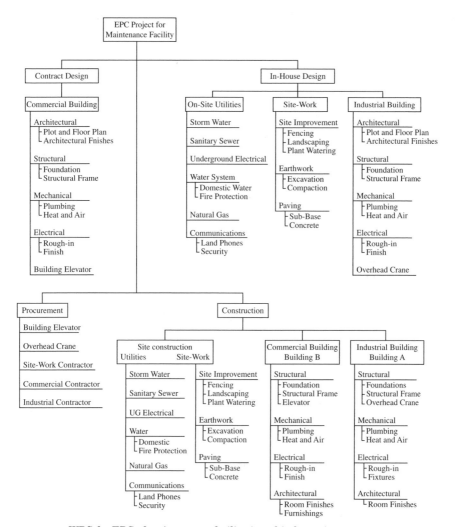

Figure 6-7 WBS for EPC of maintenance facility (graphic format).

The numbering system shown in Figure 6-8 provides a good system to show the divisions and subdivisions in the WBS. For some projects, it may be desirable to develop the graphic format for upper management and the outline format for the project team.

Forming the Project Team

A key concept in project management is to organize the project around the work to be accomplished. After review of all backup material from the owner's study and all other information that is known about the project, the project manager should develop a preliminary WBS that identifies

**WBS for Engineering-Procurement-Construction (EPC)
of Maintenance Facility**

1.0 Engineering design

1.1 Contract design
   1.1.1 A/E for Building B
      1.1.1.1 Plot and floor plan
      1.1.1.2 Architectural finishes
   1.1.2 Structural
      1.1.2.1 Foundations
      1.1.2.2 Structural frame
   1.1.3 Mechanical
      1.1.3.1 Plumbing
      1.1.3.2 Heat and air
   1.1.4 Electrical
      1.1.4.1 Rough-in
      1.1.4.2 Finish
   1.1.5 Building Elevator

1.2 In-house design
   1.2.1 On-site utilities
      1.2.1.1 Storm water
      1.2.1.2 Sanitary sewer
      1.2.1.3 Underground electrical
      1.2.1.4 Water system
         1.2.1.4.1 Domestic water
         1.2.1.4.2 Fire protection
      1.2.1.5 Natural gas
      1.2.1.6 Communications
         1.2.1.6.1 Land phones
         1.2.1.6.2 Security
   1.2.2 Site-work
      1.2.2.1 Site improvement
         1.2.2.1.1 Fencing
         1.2.2.1.2 Landscaping
         1.2.2.1.3 Plant watering
      1.2.2.2 Earthwork
         1.2.2.2.1 Excavation
         1.2.2.2.2 Compaction
      1.2.2.3 Paving
         1.2.2.3.1 Sub-base
         1.2.2.3.2 Concrete
   1.2.3 A/E Building A
      1.2.3.1 Architectural
         1.2.3.1.1 Plot and floor plan
         1.2.3.1.2 Architectural finishes
      1.2.3.2 Structural
         1.2.3.2.1 Foundations
         1.2.3.2.2 Structural frame
      1.2.3.3 Mechanical
         1.2.3.3.1 Plumbing
         1.2.3.3.2 Heat and Air
      1.2.3.4 Electrical
         1.2.3.4.1 Rough-in
         1.2.3.4.2 Finish
      1.2.3.5 Overhead crane

Figure 6-8 WBS for EPC of Maintenance Facility (outline format)

2.0 Procurement

2.1 Building elevator
2.2 Overhead crane
2.3 Site-work contractor
2.4 Commercial contractor bid
2.5 Industrial contractor bid

3.0 Construction

3.1 Site construction
    3.1.1 Utilities
        3.1.1.1 Storm water
        3.1.1.2 Sanitary sewer
        3.1.1.3 Underground electrical
        3.1.1.4 Water
            3.1.1.4.1 Domestic
            3.1.1.4.2 Fire Protection
        3.1.1.5 Natural gas
        3.1.1.6 Communications
            3.1.1.6.1 Land Phones
            3.1.1.6.2 Security
    3.1.2 Site-work
        3.1.2.1 Site improvement
            3.1.2.1.1 Fencing
            3.1.2.1.2 Landscaping
            3.1.2.1.3 Plant watering
        3.1.2.2 Earthwork
            3.1.2.2.1 Excavation
            3.1.2.2.2 Compaction
        3.1.2.3 Paving
            3.1.2.3.1 Sub-base
            3.1.2.3.2 Concrete

3.2 Commercial Building B
    3.2.1 Structural
        3.2.1.1 Foundations
        3.2.1.2 Structural frame
        3.2.1.3 Elevator
    3.2.2 Mechanical
        3.2.2.1 Plumbing
        3.2.2.2 Heat and air
    3.2.3 Electrical
        3.2.3.1 Rough-in
        3.2.3.2 Finish electrical
    3.2.4 Architectural
        3.2.4.1 Room Finishes
        3.2.4.2 Furnishings

3.3 Industrial Building A
    3.3.1 Structural
        3.3.1.1 Foundations
        3.3.1.2 Structural frame
        3.3.1.3 Overhead crane
    3.3.2 Mechanical
        3.3.2.1 Plumbing
        3.3.2.2 Heat and air
    3.3.3 Electrical
        3.3.3.1 Rough-in
        3.3.3.2 Finish electrical
    3.3.4 Architectural
        3.3.4.1 Room finishes

Figure 6-8 *(Continued)*

the major tasks that must be performed. A detailed list of tasks should be prepared and grouped into phases that show the sequencing of tasks and the interdependences of work. This provides identity for the project to assist in selection of resources and the technical expertise that will be required of the project team. A time schedule should be attached to each task. All this preparatory work is required because the project manager cannot effectively form the project team until the work to be done is known. In essence the project manager must develop a preproject work plan, which should be reviewed by his or her supervisor. This plan will be expanded into a final project work plan after the project team is formed.

After the preparatory work is complete, the project manager is responsible for organizing the project team to achieve project objectives. The project manager and appropriate discipline managers are jointly responsible for selecting team members. This can sometimes be difficult because every project manager wants the best people on his or her team. Each project has a specific list of needs, but the overall utilization of all people in the company must be considered. It is not practical to shift key personnel from project to project; therefore compromise in the assignment of people is required. The assignment of appropriate staff for a project must take into consideration the special technical expertise needed and personnel available on a company-wide basis.

The project team consists of members from the various discipline departments (architectural, civil, structural, mechanical, electrical, etc.), project controls (estimating, planning and scheduling, quality control, etc.), and the owner's representative. The number of team members will vary with the size and complexity of the project. The project manager serves as leader of the team. All team members represent their respective discipline's area of expertise and are responsible for early detection of potential problems that can have an adverse effect on the project's objectives, cost, or schedule. If a problem occurs, each team member is to notify his immediate supervisor and the project manager.

It is important that each team member clearly understands the project objectives and realizes his or her importance in contributing to the overall success of the project. A cooperative working relationship is necessary between all team members. Although the project manager is the normal contact person for all discipline departments involved in the project, he or she may delegate contact responsibility to lead members of the team. Since the initiative and responsibility to meet project objectives, costs, and schedules rest with the project manager, he or she should be kept fully advised and informed.

The project manager must organize, coordinate, and monitor the progress of team members to ensure the work is completed in an orderly manner. He or she should also maintain frequent contact with the owner's representative.

Kick-Off Meeting

After formation of the project team, the project manager calls the first team meeting, commonly called the kick-off meeting. It is one of the most important meetings in project management and is held prior to starting any work. The purpose of the kick-off meeting is to get the team members together to identify who is working the project and to provide them with the same base of knowledge about the project so they will feel like they are a part of the team. It is important for the project manager to fully understand the project objectives, needs, budget, and schedule and to transmit this information to team members early in the project. In particular the scope of work must be closely reviewed.

The kick-off meeting allows the team to set priorities, identify problem areas, clarify member responsibilities, and to provide general orientation so the team can act as a unit with a common set of goals. At the meeting the project manager should present project requirements and the initial work plan, discuss working procedures, and establish communication links and working relationships. Every effort should be made to eliminate any ambiguities or misunderstandings related to scope, budget, and schedule. These three elements of a project cannot be changed without approval of both the project manager and the owner's representative.

Prior to the meeting the project manager should prepare general project information data, including the project name (that will be used for all documents and correspondence), project location, job account number, and other information needed by the project team. Standards, CADD requirements, policies, procedures, and any other requirements should also be presented. It is important to provide this information to key people on the project so they know the project is approved for work and feel that they are a part of the team. The project manager should visit with key team members prior to the meeting to identify and resolve any peculiar problems and clarify any uncertainties.

In general the meeting is short in duration, but it is the first step in understanding what needs to be done, who is going to do it, when it is to be done, and what the costs will be. This is not a design meeting but an orientation meeting. The project manager must keep the meeting moving and not get overly involved in details. Minutes of the meeting must be recorded and distributed to team members. In particular, there should be documentation of the information that is distributed, the agreements among the team members, and the identification of team concerns or questions that require future action by the project manager or team members.

There are three important purposes of the kick-off meeting: to orient team members regarding project objectives and needs, to distribute the project manager's overall project plan, and to assign to each team member the responsibility of preparing work packages for the work required in his or her area of expertise. Work packages should be

TABLE 6-2   Kick-Off Meeting Checklist

1. Review the agenda and purpose of the meeting
2. Distribute the project title, account number, and general information needed by the project team
3. Introduce team members and identify their areas of expertise and responsibility
4. Review project goals, needs, requirements, & scope (including guidelines, limitations, problems)
5. Review the project feasibility estimate of the owner & the approved budget for the project team
6. Review the project preliminary schedule & milestones
7. Review the initial project work plan:
    How to handle design
    How to handle procurement
    How to handle construction
8. Discuss assignments to team members:
    Ask each member of the team ---> (who?)
    To review the scope of work required in their area ------------------> (what?)
    To develop a preliminary schedule for their work ------------------> (when?)
    To develop a preliminary estimate for their work ------------------> (how much?)
9. Ask each team member to prepare design work packages for their responsible work and report this information to the project manager within two weeks
10. Establish the next team meeting, write minutes of kick-off meeting, and distribute to each team member and management

prepared and returned to the project manager within two weeks of the kick-off meeting. To facilitate orderly conduct of the meeting and to ensure that important items are covered, the project manager should use a checklist for the kick-off meeting as illustrated in Table 6-2.

Work Packages

The project manager is responsible for organizing a work plan for the project; however he or she cannot finalize the project plan without extensive input from each team member. The kick-off meeting should serve as an effective orientation for team members to learn the project requirements and restrictions of budget and schedule. At that meeting the project manager assigns each team member to review the scope of work required of his or her respective expertise, to identify any problems, and to develop a budget and schedule required to meet the scope. This can be accomplished by preparing a design work package that describes the work to be provided.

Each team member is responsible for the development of one or more work packages for the work he or she is to perform. A work package provides a detailed description of the work required to meet project needs and to match the project manager's initial work plan. The work packages should be assembled by each team member and supplied to the project manager within two weeks of the kick-off meeting.

A work package is divided into three categories: scope, budget, and schedule. Figure 6-9 illustrates the contents of a work package.

Work Package

Title: ______________________

WBS Code: ______________________

1. Scope

Required Scope of Work: ______________________

Services to Be Provided: ______________________

Services not included in this Work Package, but included in another work package: ______________________

Services not included in this Work Package, but will be performed by: ______________________

2. Budget

Personnel Assigned to Job	Work-Hours	$-Cost	CBS Code Acct.	Computer Services Type	Computer Services Hours	Computer Services $-Cost

Total Work-Hours = _____ Personnel Costs = $ _____

Computer Hours = _____ Computer Costs = $ _____

Travel Expenses _____ + Reproduction Expenses _____ + Other Expenses _____ = $ _____

Total Budget = $-Labor + $-Computer + $-Other = $ _____

3. Schedule

OBS Code	Work Task	Responsible Person	Start Date	End Date

Work Package: Start Date: _____ End Date: _____

Additional Comments: ______________________

Prepared by: ______________________ Date: _____

Approved by: ______________________ Date: _____

Figure 6-9 Team member's design work package.

The scope describes the required work and services to be provided. It should be described in sufficient detail so other team members, who are providing related work, can interface their work accordingly. This is important because a common problem in project management is coordinating related work. There is a risk of the same work being done by two persons, or work not being done at all, because two people are each thinking the other person is providing the work. Team members must communicate among themselves during the process of preparing the work packages for a project.

A work package is the lowest level in the WBS and establishes the baseline for project scheduling, tracking, and cost control. The work package is extremely important for project management because it relates the work to be performed to time, cost, and people. As shown in the budget section of Figure 6-9, a code account number relates the work to the CBS. Likewise, the schedule section has a code number that relates the work to the OBS. The CBS is used for management of project costs and is further discussed in Chapter 9. The OBS code number identifies and links the work to the people. Many articles have been published that discuss the relationship of the work packages to the WBS, OBS, and CBS. An example of a design work package is shown in Figure 7-7 of Chapter 7.

The preparation of the budget portion of a work package requires a careful evaluation of all resources needed to produce the work. All work tasks and items must be budgeted, including personnel, computer services, reproduction expenses, travel, expendable supplies, and incidental costs.

Team members must consider their overall work load when they prepare the schedule portion of a work package. Since team members are generally assigned to one or more projects, their other assigned duties and future commitments to other projects must be considered when preparing a work package for a new project. The failure of team members to carefully integrate the schedule of all projects for which they are assigned is a common source of late completion of projects. Too often team members overcommit their time without making allowances for potential interruptions and unforeseen delays in their work. All tasks should be identified and scheduled.

Follow-Up Work

After the exchange of information at the kick-off meeting and a review of the required work by each team member, there may be a need to readjust the work breakdown structure of the initial project plan. A team member may have the capability to perform the work, but may determine that the magnitude of the work is in excess of what he or she can schedule because of prior commitment to other projects. Thus, a part or all of his or her portion of the project may require assignment to outside contract work. Another option would be a restaffing of the project based on the overall available resources of the project manager's organization. These situations should be resolved within two weeks of the kick-off meeting.

An accumulation of the budgets from all the team's work packages provides an estimated cost for the total project. If the estimated cost exceeds the approved budget, the project manager is made aware of this situation early in the project, within two weeks of the kick-off meeting. The team as a whole must then work together to determine alternative methods of handling the project to keep the estimated cost within the approved budget. If it cannot be resolved within the team, the project manager must work with his or her supervisor to determine a workable solution. If a solution cannot be found, the owner must be advised so an agreeable solution can be determined for a scope of work that matches the approved budget. It is important to resolve issues of this nature at the beginning of the project, when choices of alternatives can be made, rather than later when it is too late.

After receipt of all work packages, the project manager must integrate the schedules of all team members to develop a schedule for the entire project. If the project schedule exceeds the required completion date, the team as a whole must work together to determine alternative methods of scheduling the work. If the discrepancy between the planned schedule and required schedule cannot be resolved within the team, the project manager must then resolve the issue with his or her supervisor. If the required schedule cannot be achieved, then the owner must be advised so that acceptable agreements can be reached.

Issues related to project scope, budget, and schedule must be resolved early. Effective communication and cooperation among team members is necessary. The results of the team assignments and definitions of work packages allow the project manager to finalize the work breakdown structure that forms the foundation of the project work plan. After receipt of all information, the project manager can finalize the overall plan to manage the project.

Project Work Plan

The project manager must develop a written work plan for each project that identifies the work that needs to be done, who is going to do it, when it is to be done, and what the costs will be. The level of detail should be sufficient to allow all project participants to understand what is expected of them in each phase and time period of the project; otherwise there is no basis for control. It is important that a participatory approach be used and that team members understand project requirements, jointly resolve conflicts, and eliminate overlaps or gaps in related work. There must be agreements on priorities, schedule, and budget.

TABLE 6-3   Components of a Project Work Plan

Directory ←—————————————	*Who?*
Project title and number	
Project objectives and scope	
Project organizational chart	
Tasks ←—————————————	*Does what?*
Detailed listing of tasks	
Grouping of tasks	
Work packages	
Schedule ←—————————————	*When?*
Sequencing and interdependencies of tasks	
Anticipated duration of each task	
Calendar start and finish dates of tasks	
Budget ←—————————————	*How much?*
Labor-hours and cost of staff for each task	
Other expenses anticipated for each task	
Billing approach and anticipated revenue by month	
Risk Assessment ←—————————————	*What can go wrong?*
List of risks and their probability of occurrence	
Impact of each risk on quality, cost, and schedule	
Strategy to prevent and mitigate each risk	
Measurement ←—————————————	*What is accomplished?*
Accomplishment of tasks	
Completion of work packages	
Number of drawings produced	

Upon receipt of all the team members' work packages the project manager can assemble the final project work plan. Table 6-3 provides the basic components of a work plan: the directory, tasks, schedule, budget, risk management, and measurements of work. The project directory contains all pertinent information, such as project title, number, objectives, and scope. The project organization chart shows all participants, including the owner's representative. The detailed list of tasks, and grouping of tasks, is derived from the work breakdown structure. The sequencing and scheduling of tasks can be obtained by integrating the schedules of work packages provided by team members. Likewise, the budget can be obtained from a summary of the costs from all work packages.

Risk assessment is a vital part of the work plan and is too often overlooked by project managers and team members. Risk assessment and analysis address the question "What can go wrong?" Before commencing work on any project there needs to be a thorough evaluation of the risks associated with the project. Each risk needs to be identified and an evaluation should be made on the probability of occurrences of each risk. There also needs to be a plan to prevent risks and a strategy to mitigate a risk when it occurs.

Once the work plan is finalized it serves as a document to coordinate all work and as a guide to manage the overall effort of the project. It becomes the base for control of all work. The Appendix illustrates the components of a work plan for a project: work breakdown structure, project organizational chart, sample work package, and project schedule. Note the transfer of information from one component to another to form the integrated work plan.

The first step in organizing a project is development of a WBS. The WBS defines the work to be accomplished, but does not define who is responsible for performing the work. A successful project depends on people to make things happen. However, merely selecting good people is not enough. A key function in project management is to organize the project around the work to be accomplished and then select the right people to perform the work within the approved budget and schedule.

After the WBS is complete, the next step is to link the OBS from the company to the required work that is defined in the WBS. Figure 6-10 illustrates the linking of the WBS and OBS to identify the various disciplines that are responsible for each part of the WBS. The project manager, with the assistance of discipline managers, can then begin the process of selecting individuals from the various discipline departments who will form the project team.

The linking of the WBS and OBS establishes the project framework for management of the project. After the project framework is defined, a project schedule can be developed to guide the timing of activities and interface-related work. The time and cost required to accomplish each activity can be obtained from the work packages. The CPM technique is the most common network scheduling system that is used in the engineering and construction industry. Techniques for project scheduling are discussed in Chapter 8.

After completion of the project framework, a coding system, often referred to as a Cost Breakdown System, can be developed to identify each component of the WBS. The coding system provides a common code of accounts used by all participants in the project because it is directly related to the WBS, that is, the work to be performed. Coding systems are discussed in Chapters 8 and 9.

The integration of the WBS, OBS, and CPM forms the project plan, which is the base for project tracking and control. A code of accounts can be developed that relates the required work (defined in the WBS) to the people (shown on the OBS) who will do the work in accordance with the schedule (shown on the CPM) to complete the project. Thus, the WBS,

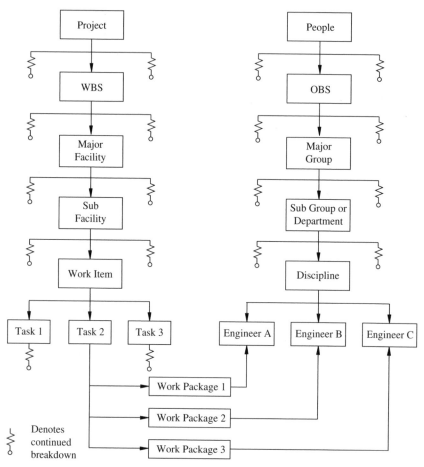

Figure 6-10 Linking the WBS and OBS.

OBS, and CPM must be linked together to form an all-encompassing project plan.

To be effective, a system of project management must integrate all aspects of the project; the work to be done, who is going to do it, when it is to be done, and what the cost will be. Actual work can then be compared to planned work, in order to evaluate the progress of a project and to develop trends to forecast at completion costs and schedules.

The development phases of the project work plan are illustrated in Figure 6-11. Topics related to project definition were discussed in Chapters 3 and 5. Project framework is presented in this chapter. Project scheduling and tracking is presented in Chapters 8 and 9, respectively.

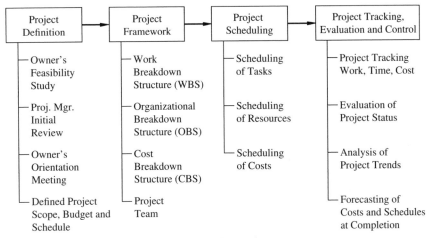

Figure 6-11   Phases of development of work plan.

References

1. Cable, D. and Adams, J. R., *Organizing for Project Management,* Project Management Institute, Newtown Square, PA, 1982.
2. Eldin, N. N., "Management of Engineering/Design Phase," *Journal of Construction Engineering and Management,* ASCE, Reston, VA, Vol. 117, No. 1, March 1991.
3. Emling, R. E., "A Code-of-Accounts from an Owner's Point of View," *Transactions of the American Association of Cost Engineers,* Morgantown, WV, 1970.
4. *Handbook for Preparation of Work Breakdown Structures,* NHB 5610, National Aeronautics Space Agency, Washington, DC, 1975.
5. Johnston, D. C., "Planning Engineering and Construction Projects," *Engineering Management Division Symposium,* ASCE, Reston, VA, 1986.
6. Neil, J. N., *Manual for Special Project Management,* Construction Industry Institute, Austin, TX, July 1991.
7. *Quality in the Constructed Project: A Guide for Owners, Designers, and Contractors,* Volume 1, Manual No. 73, ASCE, Reston, VA, 1990.

7

Design Proposals

Evolution of Projects

This chapter is written from the perspective of the project manager who is responsible for managing the design effort. Therefore, the use of project manager in this chapter refers to the design project manager. The material presented and discussed in this chapter on engineering design is an illustrative example of applying the principles and techniques of project management presented in Chapter 6. For example, *scope* refers to the scope of work required for the design effort, *budget* refers to the cost for providing design services, and *schedule* refers to the schedule for performing the design work.

A project is in a continual state of change as it develops from conception to completion. Because there are perpetual changes in a project, the design project manager should be involved at the beginning of the project and remain with the project until final completion. Continuity of the design project manager is crucial to a successful project. In all instances, the project manager is the prime contact with the sponsor of the project. Changes in a project are a major issue for the design team. Changes occur as a project progresses through the following phases:

- Sponsor's development phase
- Project organization phase
- Engineering phase
- Procurement phase
- Construction phase
- System testing and start-up phase
- Project completion and contract close-out phase

The sponsor's development phase usually ends in a request for proposal (RFP) to further develop the project. At this time in the development of a project, there must be a clear understanding of the proposal request and the sponsor's goals. Various names are given to the sponsor, including owner, business unit, operating group, client, customer, and end user. Essentially, the sponsor is the organization that requests the work and will use the work when it is completed. It is important for the design team to have a clear understanding of the desired outcome of the project and the reasons the project is being undertaken by the sponsor.

Project Execution Plan

The first step in developing a design proposal is development of the project execution plan to manage the design process. The plan must include the scope of work covered in the RFP and interfaces with others who will be involved in the project, including both in-house and contract personnel. Too often, RFPs have vaguely defined scopes of work that will later cause unforeseen additional work, which adversely impacts the budget and schedule.

The plan must also include a milestone schedule that shows major phases and areas of work, including critical due dates. An overall preliminary budget must be developed to guide the project to ensure there are no unexpected surprises as the project develops.

Project Definition

Although project definition is primarily the responsibility of the sponsor's organization, often the design organization is involved in assisting project definition. Early questions of the design team include: What do we know about the project? What are we trying to do? What work do we need to do? The answers to these questions depend on a good definition of the project. It is not possible to define the scope of design work without a good definition of the project. Project definition is a prerequisite to engineering design.

Poor project scope definition is the source of project changes, rework, schedule delays, and cost overruns. The preproject planning research team of the Construction Industry Institute (CII) developed the Project Definition Rating Index (PDRI) as a tool to measure the level of definition of a project. It allows a project team to quantify, rate, and assess the level of scope development on projects prior to authorization for detailed design or construction. The central premise of the research team effort was that teams must be working on the right project in a collaborative manner (alignment) and performing the right work (scope definition) during preproject planning.

Figure 7-1 is the PDRI for industrial projects, which applies primarily to process industry work. There is also a PDRI for building projects and a PDRI for infrastructure projects.

The PDRI for industrial projects consists of 70 elements, grouped into 15 categories, which are further grouped into three broad sections (see

I Basis of Project Decision
A. Manufacturing Objectives Criteria
    A1. Reliability Philosophy
    A2. Maintenance Philosophy
    A3   Operating Philosophy
B. Business Objectives
    B1. Products
    B2. Market Strategy
    B3. Project Strategy
    B4. Affordability/Feasibility
    B5. Capacities
    B6. Future Expansion Considerations
    B7. Expected Project Life Cycle
    B8. Social Issues
C. Basic Data Research & Development
    C1. Technology
    C2. Processes
D. Project Scope
    D1. Project Objectives Statement
    D2. Project Design Criteria
    D3. Site Characteristics Available vs. Required
    D4. Dismantling & Demolition Requirements
    D5. Lead/Discipline Scope of Work
    D6. Project Schedule
E. Value Engineering
    E1. Process Simplification
    E2. Design & Material Alternatives Considered/Rejected
    E3. Design for Constructability Analysis

II Front End Definition
F. Site Information
    F1. Site Location
    F2. Surveys & Soil Tests
    F3. Environmental Assessment
    F4. Permit Requirements
    F5. Utility Sources with Supply Conds.
    F6. Fire Protection & Safety Considerations
G. Process/Mechanical
    G1. Process Flow Sheets
    G2. Heat & Material Balances
    G3. Piping & Instrmt Diagrams (P&IDs)
    G4. Process Safety Mgmt. (PSM)
    G5. Utility Flow Diagrams
    G6. Specifications
    G7. Piping System Requirements

G. Process/Mechanical *(continued)*
    G8. Plot Plan
    G9. Mechanical Equipment List
    G10. Line List
    G11. Tie-in List
    G12. Piping Specialty Items List
H. Equipment Scope
    H1. Equipment Status
    H2. Equipment Location Drawing
    H3. Equipment Utility Requirements
I. Civil, Structural, & Architectural
    I1. Civil/Structural Requirements
    I2. Architectural Requirements
J. Infrastructure
    J1. Water Treatment Requirements
    J2. Loading/Unloading/Storage Facilities Requirements
    J3. Transportation Requirements
K. Instrument & Electrical
    K1. Control Philosophy
    K2. Logic Diagrams
    K3. Electrical Area Classifications
    K4. Substation Requirements/Power Sources Identified
    K5. Electric Single Line Diagrams
    K6. Instrument & Electrical Specifications

III Execution Approach
L. Procurement Strategy
    L1. Procurement Strategy Equipment & Materials
    L2. Procurement Procedures & Plans
    L3. Procurement Responsibility Matrix
M. Deliverables
    M1. CADD/Model Requirements
    M2. Deliverable Defined
    M3. Distribution Matrix
N. Project Control
    N1. Project Control Requirements
    N2. Project Accounting Requirements
    N3. Risk Analysis
P. Project Execution Plan
    P1. Owner/Approval Requirements
    P2. Engr./Const. Plan & Approach
    P3. Shut Down/Turn-Around Requirements
    P4. Pre-Commissioning Turnover Sequence Requirements
    P5. Startup Requirements
    P6. Training Requirements

Figure 7-1 Project definition rating index (PDRI) for industrial projects—sections, categories, and elements. (*Source:* Construction Industry Institute.)

Figure 7-1). To determine the PDRI, each of the elements are rated on a scale from 1 through 5. An element rating of 1 represents complete definition whereas a rating of 5 represents incomplete or poor definition of the element. The sum of the element weights is the composite weighted score of a project, which can range up to 1,000 points, with lower points as a better score and higher points as a worse score. Based on an analysis of 40 projects, the research team found that projects scoring less than 200, out of 1,000 total points, were significantly more successful than those that scored greater than 200.

Problems in Developing Project Definition

Too often, insufficient time is devoted to defining the requirements of projects. Project definition is usually performed by people outside engineering and construction. These individuals generally have job responsibilities involving finances or managing a business unit in the sponsor's organization. Their job responsibilities and expertise are often not related to defining requirements of a project in terms that can be converted to engineering design and construction.

Sometimes the only fixed known information about the project is the amount of money the owner has to spend, with only a vague idea of what the owner would like from the expenditure of money. The owner may have a wish list of what he or she would like, but the only firm information is the total amount of available funds. For this type of situation, the designer must work closely with the owner to identify the desired operational criteria of the project: what the owner wants to do with the project when it is completed. The designer must assist the owner in separating what he or she needs from what he or she wants. The designer must convert the owner's needs into engineering scope of work and the construction costs to produce the final project. A cost for each element in the project must be determined to ensure the project will not exceed the owner's available funds.

High turnover of people is another problem in defining objectives. Many owner organizations frequently promote and reassign personnel. Turnover of people can lead to changing priorities. By the time a project reaches the approval stage, the people who established the initial project definition may no longer be involved. Members of the current operating group that will use the project when it is completed should be involved in confirming that the established project definition will meet their goals and objectives. The goals and objectives must be adequately quantified and documented. This requires coordination between the sponsor's organization and the project team.

A well-defined definition of the project is a prerequisite to planning the work, because the team members must know what the project is before they can plan it. Too often there is a rush into the implementation phase even before there is a good understanding and agreement of project definition. Getting an early team agreement on definition prevents the project scope from creeping out of control.

Design Proposals

Upon receipt of a request for proposal (RFP), the design project manager should carefully review it to become familiar with the global issues related to environmental and community relations, hazardous waste, bidding strategy, required permits and regulations, expectations, and goals of the customer. Although these issues will be developed in more detail at a later date, the project manager must be aware of all aspects of the project.

The purpose of the proposal is to identify the scope and develop the budget and schedule for preparing the design. The project proposal may be as formal as a request for qualification for a new sponsor or as informal as a brief outline of scope for expansion of existing work.

During this early stage in project development, the design engineer must convert the sponsor's project definition into an engineering scope of work. However, the design engineer may feel the sponsor's definition is inadequate or may feel there is missing information. The sponsor should be contacted for clarifications. However, sometimes the sponsor is unable to fully clarify or respond to rectify the discrepancies. For these situations, the design engineer must define the scope of engineering work to the best of his or her ability and then develop a budget and schedule based on the designer's assumed scope of work. It is then mandatory to document and communicate to the sponsor the assumptions made and the impact of the work on the total project. This essentially locks in the scope of work at this time in the project. Then, at a later date when additional information is known, the assumed scope, budget, and schedule for that portion of the work can be adjusted appropriately.

Figure 7-2 is an illustrative example of a project proposal form. Project data should include a brief description of the work. If needed, the space under "Comments" may be used to relay information important to the proposal. All disciplines that will be involved in the project should be identified, including architectural, civil, electrical, mechanical, and structural, or other special expertise. If the budget or estimated fee for design is not known, then the magnitude of the project or estimated construction cost should be listed. After the top portion of the form is completed, it is submitted to management for review and approval. Analysis of the completed project proposal form will allow management to make decisions based on fundamental project information.

PROJECT PROPOSAL FORM

☐ Continuation of Existing Work   ☐ New Work

PROJECT DATA
Client Name: _____
Description of Work: _____
Location of Work: _____

Prepared By: _____   Date: _____

DISCIPLINES INVOLVED:

☐ Arch $ _____   ☐ Mechanical $ _____

☐ Civil $ _____   ☐ Structural $ _____

☐ Electrical $ _____   ☐ Other: _____

ESTIMATE

Fee: $ _____   Work-Hours: _____

Start Date: _____   Completion Date:

Proposal Required: ☐ No   ☐ Yes, Due Date: _____

Comments
_____
_____
_____
_____
_____
_____
_____
_____

APPROVAL:

☐ No Further Action Required   ☐ Further Discussion Required   ☐ Proceed

Date: _____

DISTRIBUTION
President   _____
Principal-in-Charge _____
Project Manager   _____
Marketer   _____
Document Control   _____

Figure 7-2   Project proposal form for design.

The design project manager is responsible for managing the overall coordination of the proposal effort. Specific duties include

- Defining the scope of work for the project
- Establishing a work plan, including budget and schedule, for the proposal effort

- Monitoring the work plan to ensure effective communication among team members
- Communicating with discipline managers to identify key personnel
- Assist in preparation of the proposal documents
- Attend the sponsor's interview
- Participate in establishing a rate schedule
- Assimilating the list of project deliverables

To accomplish these duties, the design project manager and his or her team need to address the following questions and subtopics:

Why?	Why are we doing this project—project objectives?
	▪ Expectations for the project—identify and document
	▪ Operating parameters—products to be produced or services to be provided
	▪ Business requirements—cost/schedule limitations and functional requirements
What?	What is needed to accomplish the end objective of the project?
	▪ Description of major systems required—plot plans, flow sheets, etc.
	▪ Identify applicable environmental and safety regulations—federal, state, local
	▪ Confirm standards to be used—company standards and sponsor's standards
How Much?	What is the cost to complete the project?
	▪ Identify costs—dollars, work-hours, or both
	▪ Description of the cost estimate method—historical records, vendor quotes
	▪ Organize estimate—develop estimate worksheets by disciplines
When?	How much time is required to complete the project?
	▪ Describe key milestone dates—design, procurement, construction, and close-out
	▪ Identify required advanced purchases—major equipment and special materials
	▪ Identify potential schedule risks—provide contingency options
Who?	Who will be doing the work?
	▪ Identify required expertise—technical and nontechnical
	▪ Determine availability of personnel—in-house personnel and outside personnel
	▪ Assign duties and roles—defined authority and responsibility

What can go wrong?	What are potential risks in the project?

- Assess risks—identify risks for cost, schedule, quality, and scope
- Analyze risks—probability of occurrence and impact of each risk
- Develop action plan—for prevention and/or mitigation of the risk

The discipline managers are responsible for providing technical support for the proposal. This support may include assigning personnel, establishing preliminary designs, reviewing sponsor information, and performing quality-control review of proposal documents. The discipline managers are also responsible for establishing the total work-hours needed for the project, to ensure that adequate technical expertise will be available when necessary to meet the project schedule.

Figure 7-3 illustrates a project proposal checklist. The project data can be the same as listed on the project proposal form. Prior to the sponsor's interview, the project manager should compile a list of proposed attendees, an agenda, and a list of materials that will be used for the presentation: boards, photographs, slides, or electronic media presentations, such as Power-point.

As a minimum, the general scope of work for the project is included in the space provided. Additional information, such as drawing lists, equipment lists, specification lists, or special sponsor requirements, can be attached to the form. The completed form, plus attachments, is submitted to management prior to preparation of the proposal.

Engineering Organization

Members of the design team should be involved in the project at the earliest possible date, preferably during the proposal preparation phase. The design engineers that will actually be doing the work can be extremely valuable for defining scope, identifying potential problems, and preparing realistic budgets and schedules for performing the work. Too often the designers are not a part of the proposal preparation or do not become involved until the proposal is presented to and approved by the owner. By then, the scope, budget, and schedule may be fixed, but it may not represent the actual work that is necessary to complete the project the way the owner expects. Getting input and early involvement of the design team is crucial to a successful project.

To effectively manage the design effort, an organizational chart should be developed for each project. The organizational chart is effective for defining the roles and responsibilities of the engineering manager and his or her team members during design. There must be a

---

PROJECT PROPOSAL CHECKLIST

Project Manager: _____   Marketer: _____

PROJECT DATA

Client Name: _____
Description of Work: _____
Location of Work: _____

Estimated Cost to Prepare Proposal: $_____

Estimated Work-Hours _____

PROPOSAL SUPPORT DOCUMENTATION

☐ Organization Chart   ☐ List of Special Graphics

☐ Rate Schedule   ☐ List of Travel Expenses

☐ List of Projects to Include   ☐ Proposal Sheets

☐ List of Employee Resumes   ☐ Other: _____

CLIENT INTERVIEW

Date of Interview: _____   Rehearsal Date: _____

☐ List of Attendees   ☐ Agenda

☐ Presentation Materials List   ☐ Other: _____

GENERAL SCOPE OF WORK

_____
_____
_____
_____
_____
_____
_____

☐ Drawing List   ☐ Specifications List

☐ Special Requirements List   ☐ Other: _____

APPROVAL:

_____ Date: _____   _____ Date: _____
President   Principal-in-Charge

Figure 7-3 Proposal checklist for design.

clear understanding of the reporting relationships of all members of the engineering team. If external consultants are involved, there must be a clear description of their reporting relationships as well as their roles and responsibilities. There must be a list of key members of the engineering team, including consultants where applicable, along with their telephone numbers, fax numbers, and e-mail addresses.

Figure 7-4 is an example of a list of technical expertise that may be needed for engineering design. For the list of particular technical expertise that is unique for each project, the project manager should develop an organizational chart that shows the interrelationships, roles, and responsibilities of each member on the project team.

Scope Baseline for Budget

Project proposal detail sheets as illustrated in Figure 7-5 help formulate a budget for estimated design services. These forms also assist in developing the final work breakdown structure (WBS) and organizational breakdown structure (OBS) for the project. Thus, the project proposal detail sheet forms the basis for setting up and managing the project.

The design project manager initiates one of these forms for each design discipline involved in the project. The scope definition and

Project Manager
- A. Project Engineer
- B. Schedule and Control Engineer
- C. Drafting Coordinator
- D. Disciplines:
    - 1. Architecture
        - a. Lead Architect
        - b. Architects
        - c. Draftsmen/CADD
    - 2. Civil
        - a. Lead Civil Engineer
        - b. Civil Engineers
        - c. Civil Technicians
        - d. Draftsmen/CADD
    - 3. Electrical
        - a. Lead Electrical Engineer
        - b. Electrical Engineers
        - c. Electrical Technicians
        - d. Draftsmen/CADD
    - 4. Mechanical
        - a. Lead Mechanical Engineer
        - b. Mechanical Engineers
        - c. Mechanical Technicians
        - d. Draftsmen/CADD
    - 5. Structural
        - a. Lead Structural Engineer
        - b. Structural Engineers
        - c. Structural Technicians
        - d. Draftsmen/CADD
- E. Clerical Personnel
- F. Document Control Personnel
- G. Reproduction Personnel
- H. Contract Administration

Figure 7-4 Illustrative example of expertise required for design.

estimated budgets should be as complete and accurate as possible. It is the responsibility of each design discipline to thoroughly examine the project definition from the sponsoring organization and convert project definition into engineering scope definition. A scope of work for the

---

PROJECT PROPOSAL DETAIL SHEET

Discipline: _____

PROJECT DATA

Project Manager: _____   Marketer: _____
Proposal Number: _____
Client Name: _____
Description of Work: _____
Location of Work: _____
Estimated Department Budget: _____ Work-Hours: _____

SCOPE DEFINITION

_____
_____
_____
_____
_____
_____
_____

CLIENT REFERENCES AND ATTACHMENTS

_____
_____
_____

PERSONNEL CATEGORIES REQUIRED WITH WORK-HOURS

	ADMIN	SPECS	ENGR	DRWGS
Department Manager				
Lead Engineer				
Senior Engineer				
Junior Engineer				
CADD/Drafting Coord.				
Total				

Do not include Clerical, Reproduction, or Other Disciplines. Return to PM by: _____

OUTSIDE COORDINATION/OTHER REQUIREMENTS

☐   Specification List

☐   Drawing List   Total Number of Drawings: _____

☐   Other: _____

Prepared by: _____   Date: _____

Figure 7-5 Detail sheet for design proposal.

discipline must be defined in sufficient detail to enable a reasonable estimate of the work-hours required to complete the design work.

After the project data shown on the top of Figure 7-5 is completed, each form and any supporting documentation should be reviewed with the appropriate discipline manager and the lead discipline representative. These two individuals should estimate the required design work-hours for each personnel category required to complete the scope of work requested. This estimate will require preparation of a detailed scope outline, a list of required specifications, and a preliminary drawing list. Preparation of mini-drawings as discussed later in this chapter is an effective method of preparing the drawing list. The number of drawings to be produced provides a basis for the number of required design work-hours.

The discipline manager should complete the project proposal detail sheet by summarizing the estimated design work-hours on the sheet. The personnel categories with design work-hours should be completed by the design manager. It is a summary of the hours required for each classification of work, including administration, specifications, engineering, and drawings. Depending on the uniqueness of the project it may be necessary to add to or delete some of the classifications shown in Figure 7-5. The completed form, plus all supporting documentation, is returned to the project manager. An example of this completed form is given in Figure 5-4.

After the design project manager has collected and reviewed the project proposal detail sheets from the design disciplines, the project budget form shown in Figure 7-6 can be completed. This form compiles all the information received from the discipline managers plus information supplied by the design project manager. The resource-hours for clerical, reproduction, project support, and project management are added to complete the design budget. An example of this completed form is given in Figure 5-3.

A dollar budget can be computed using one of several methods. For smaller projects, or projects without well-defined scopes, an average unit cost per resource-hour can be applied to all the identified hours. For larger projects, or projects with well-defined scopes, a more detailed dollar budget can be obtained by breaking down the hours by each personnel classification so the hourly rate can be applied more accurately and competitively, rather than using an average hourly rate. The total budget for resource-hours (labor costs and expenses) forms the basis for establishing the accounting system for the project during design.

During the budgeting process, checks should be made to reduce the possibility of major oversights in the budget and to ensure the design effort can be completed within the budget. Examples of checks may include number of expected drawings per building, anticipated number

DATE:____/____/____

PROJECT PROPOSAL BUDGET FORM

Project Name: _____

| Dept. Number | Department | RESOURCE HOURS | | | | | | | | DOLLARS |
| | | RATE | | | | | | | | |
		ADMIN	MTGS	SHED	SPECS	ENGR CALCS	DRWGS	RCD DRWG	TOTAL	
	Project Management								0	$0
	Architecture								0	$0
	Mechanical								0	$0
	Electrical								0	$0
	Structural								0	$0
	Environmental								0	$0
	Civil								0	$0
	CADDS								0	$0
	Clerical								0	$0
	Document Control								0	$0
	Reproduction								0	$0
	Project Control								0	$0
	Management								0	$0
RESOURCE-HOUR SUBTOTAL		0	0	0	0	0	0	0	0	0
FEE SUBTOTAL		$ -	$ -	$ -	$ -	$ -	$ -	$ -		0
Task #	Description									
	Contingency									
	General Expenses									
	Travel									
	Office Budget									
EXPENSES SUBTOTAL										$0
TOTAL										$0

Figure 7-6 Budget form for design proposal.

of design work-hours per drawing, expected time duration anticipated by design discipline, percent of design costs per total cost of project, and number of design hours per major piece of equipment. Simple checks of these ratios based upon data from past projects can be made to prevent significant errors in the design budget.

The design project manager must also communicate with the design managers of the various disciplines to ensure that the resources shown in the budget will be available when needed. A simple check of the average number of resources needed over the expected time duration of the design effort is an indication of the future work load expected of the design team. The discipline managers can compare this demand for resources with their normal availability of designers. Too often, a budget is established for a design effort and then later it is found the project cannot be completed in time or within budget because the resources are not available. Simple checks can reduce these types of problems in engineering design.

Design Work Package

After the design proposal is accepted by the sponsor, the project manager for the design firm can develop a complete work breakdown structure (WBS) for the design effort. As shown in Figure 6-6 the work package is the lowest level of the WBS. The work package links the work breakdown structure (WBS) to the organizational breakdown structure (OBS) of the design team.

Figure 7-7 is an example work package for design of foundations for an electrical transmission steel pole line project. The work package is divided into three categories: scope, budget, and schedule. The scope describe the required work and services to be provided. It should be described in sufficient detail so other team members, who will be providing related work, can interface their work accordingly. This is important because a common problem in project management is coordinating related design work. There is a risk of the same work being done by two persons, or work not being done at all, because two people are each thinking the other person is providing the work. Clarification of who does what is shown in the scope of Figure 7-7.

Information from the design proposal shown in Figures 7-5 and 7-6 are useful in preparing the budget for the work package. For example, the work hours for each person in the design of foundations are shown in the budget of Figure 7-7. Also, the list of work tasks, responsible person, and start and end date of each person involved in the foundation design are shown in the schedule of Figure 7-7. It should be noted that the overall start date and end date for the entire foundation design are shown at the bottom of the schedule.

WORK PACKAGE

Title: Foundation Design

WBS Code: 7-5-42-A10

1. Scope

Required Scope of Work: *Provide engineering analysis and design services for 31 foundations of the Spring Creek to Valley View steel pole transmission line project.*

Services to be Provided: *Review background material for project. Coordinate with structural engineer of steel pole manufacturer to obtain foundation ground line reactions without overload factors. Write soil investigation specification. Analyze soil report and borings, size diameter and depth of foundations, size vertical and horizontal reinforcing steel. Prepare drawings and write specification for construction.*

Services not included in this Work Package, but included in another work package: *Clearing of right-of-way in Site-work package. Staking of foundation locations for soil investigation firm in Surveying Work Package*

Services not included in this Work Package, but will be performed by: *Preparation of structural details by steel fabricator vendor. Construction inspection services to be provided by XYZ Consulting Services, Inc.*

2. Budget

Personnel Assigned to the Job	JobTitle	Work Hours	$-Cost	CBS Code Acct	Computer Services Type	Hours	$Cost
Fred Jones	Geotechnical Engineer	35	$1,925	8159	–	–	–
James Thomas	Foundation Engineer	180	$12,600	8172	Engr	45	$1,125
Ralph Smith	CADD Operator	68	$3,060	7080	CADD	85	$2,125
Support	Engineering Intern	20	$600	1054	Specs	25	$625
		303 hrs	$18,185			155 hrs	$3,850

Personnel Costs		Computer Expenses		Travel Expenses		Reproduction Expenses		Other Expenses		Total Budget
$18,185	+	$3,850	+	$3,200	+	$175	+	$280	=	$25,690

Schedule

OBS Code	Work Task	Responsible Person	Start Date	End Date
510	Review backup material	James Thomas	5/2/16	5/5/16
510	Obtain ground line reactions	James Thomas	5/4/16	5/10/16
530	Write soil investigation spec	Fred Jones	5/10/16	5/12/16
530	Evaluation of soil report	Fred Jones	6/3/16	6/9/16
510	Foundation Design	James Thomas	6/9/16	6/17/16
520	Development of Drawings	Ralph Smith	6/15/16	6/23/16
520	Write Construction Specifications	James Thomas	6/20/16	6/27/16
510	Design Review	James Thomas	6/28/16	6/30/16

Work Package: Start Date: 5/2/16   End Date: 6/30/16

Additional Comments: *Project engineer responsible for site-work needs to ensure right-of-way is cleared by May 10 to allow staking of foundations and to allow soil boring truck access to project. Project engineer responsible for surveying needs to ensure foundations are staked by May 15, so soil Investigation Company can begin soil borings. Need to notify steel supplier to ensure timely completion of structural steel details before completion of design drawings and writing of specifications.*

Prepared by: _____   Date: _____

Approved by: _____   Date: _____

Figure 7-7 Work package for design of foundations.

Design work packages provide the basis of design coordination, which is discussed in Chapter 10 of this book. Work packages establish a benchmark to monitor progress in work to ensure the design effort is completed on time and within the approved budget.

Mini-Drawings

For design proposals, mini-drawings are an effective and organized method of defining the design deliverables: the design drawings and specifications.

Mini-drawings, sometimes called cartoons or stick drawings, are sketches that are hand drawn on $8\frac{1}{2} \times 11$ paper to represent the full-size drawings that will be developed in the design effort. Since they are hand drawn, they can be prepared quickly.

A complete set of mini-drawings for the project includes title page, plan views, elevations, sections, schedules, and details in the order they will appear in the final drawings. Figure 7-8 is an illustrative example of one page from a set of mini-drawings. It shows five plan views for each floor in a building in the top portion of the page and provisions for text on the bottom portion of the sheet. In the full set of mini-drawings, there would be one page for each full-size drawing that is expected to be produced in the design effort. Development of the mini-drawings involves working through each drawing and blocking out to scale each anticipated sheet for size and location.

Design schedules, such as the room finish schedule, are also identified and assigned to locations on the drawings. Details can be determined, counted, and assigned locations within the typical detail sheet grid. Copies of the mini-drawings can be given, with instructions, to the draftsperson or CADD operator. As conditions change during design, the mini-drawings can easily be referred to, changed, and reorganized. This provides an effective means of managing the design effort.

The final set of mini-drawings can be submitted to the discipline managers for review and approval prior to starting design. The mini-drawings communicate the manner in which the design team intends to document the design and produce the design deliverables. Therefore, it is important that the mini-drawings be complete because they become a communications tool to management and the design team, for their reference throughout the design process.

Mini-drawings are living documents that may shift and change during the design process as changes occur. These documents must be kept up to date. Mini-drawings provide early establishment of drawing content and assist in defining drawing completeness. In addition, they provide the following benefits:

- Establish an efficient and orderly layout of the drawings
- Determine an early count of the anticipated number of drawings
- Identify the required areas of technical expertise needed for the design effort
- Assist in determining the work-hours required in design
- Provide a system for delegating drafting or CADD assignments
- Enable the department and discipline managers to schedule personnel

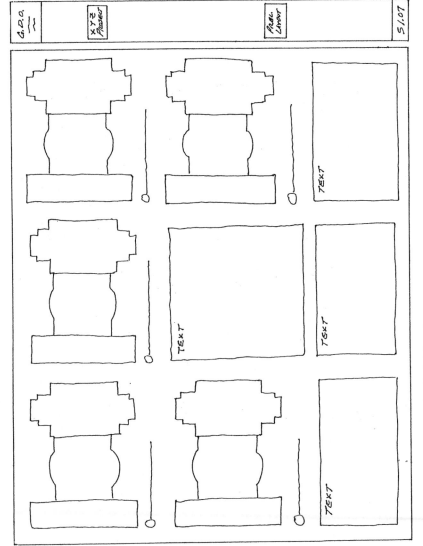

Figure 7-8 Illustrative example of mini-drawing.

After a set of mini-drawings are completed, the number of anticipated drawings are known, which provides a procedure for determining the cost for design. After the list of drawings is completed, a summary sheet can be prepared for each design discipline that shows the number of hours anticipated to produce the drawing. Based on the number of design-hours needed per drawing and the rate schedule for the designers, a design cost and budget can be determined. In addition to planning drawing content the mini-drawings can be used during actual design for project team communications and for keeping records of hours charged to the job.

Development of the Design Work Plan

Most design organizations work on billable hours. In simple terms, billable hours are hours that can be charged to, and are paid by, the sponsor's organization. Typical examples include design calculations, writing specifications, preparing drawings, conducting tests, or providing inspection. Many sponsoring organizations do not compensate the designer's organization for developing a design work plan. As a result, there may be little or no effort placed in developing an overall plan for design. However, this is a critical mistake because even a small amount of planning before starting work can prevent many future problems, including overspending on the design effort and late completion of design work. A good work plan for design helps also to reduce rework that can lead to errors in design. Every design effort should have a written work plan, even if the cost will not be reimbursed by the sponsoring organization.

The level of detail of the work plan depends on the time schedule and the allocated budget for preparing the design and contract documents. Preparation of the schedule should begin during proposal preparation or immediately after the contract award date. A CPM schedule as presented and discussed in Chapter 8 is recommended for larger or more complex projects because this format provides a better level of detail and a clearer definition of the order and interdependency of the work tasks. Using the CPM method of scheduling forces the user to think through and clearly define the interdependency and interrelationships of activities between the various design disciplines. This results in a more detailed design schedule.

For a small design project that must be completed in a short time a bar chart that shows each task in chronologic order is simple and easy to follow. The bar chart is less detailed and more useful for small, less complicated design projects. However, it may be advantageous to roll all the individual design bar charts into an overall master CPM diagram to schedule the entire design effort.

Chapter 8 presents a discussion and example of CPM scheduling for engineering design, and Chapter 9 presents integration of the CPM design schedule with procurement and construction. The Appendix also presents an illustrative example of linking the schedule of design, procurement, and construction of a project.

Regardless of the methods used, either CPM or bar charts, the schedule should incorporate all the required tasks, starting with a thorough review of all backup material that was used in preparing the proposal. In particular, the schedule should include review of backup material to identify conditions that can affect the design work, including the sponsoring organization's special requirements, applicable codes, and regulatory agencies. The schedule should also show key progress reviews, final checking and corrections, work to be performed by outside consultants, and any particular issue that may impact the successful completion of the design. The design schedule should be construction driven, because construction is the most costly component of most projects.

A common mistake in development of design schedules is the failure to include adequate contingency in the project schedule. Too often the design work plan includes everything that is known to be accomplished but fails to include a reasonable allowance to compensate for unforeseen delays that inevitably will arise during the design effort. Examples include delays in acquisition of permits, responses from regulatory agencies, reviews of design by clients, delays in responses from vendors, and requests for information (RFI) from outside organizations that supply information to the design team.

The design budget should be integrated with the design schedule. For design work it is more beneficial to define the budget in work-hours rather than in dollars of cost, because design is driven by hours. Integrating the total staff hours into the schedule provides a systematic method of managing the budget and schedule simultaneously.

After the work plan is established, a progress schedule can be developed using information in the work plan, including the mini-drawings that forecast the full-scale drawings. Each group of drawings should be allocated hours for completion determined by complexity, reusability from production design, and work-hour data based on past experience on similar projects. The progress schedule should be reviewed and updated on a regular reporting period, usually coinciding with the submittal of time cards. A review of actual progress, based on an analysis of the degree of completeness of each drawing or group of drawings or tasks, relative to the allocated hours versus estimated hours to complete, provides a projected outcome at the current level of effort. Regular reviews provide a consistent reporting of overall progress. This allows adjustments that may be necessary on a timely basis to ensure completion within the established time schedule and allocated budget.

After the work plan is established and mini-drawings are completed, the project manager must establish ground rules for the design team and outside consulting specialists. Standards for drawings and CADD work must be established, documented, and reviewed with the entire team. The American Institute of Architects (AIA) has layering standards that are commonly used for projects in the building sector. Sponsors or clients in the process industry sector have drawing standards that are required of engineering firms that perform design work on their projects. Many specify the particular CADD system that must be used for the design of their projects. The management system for handling the project should be reviewed with the team to ensure everyone knows what is expected of him or her. In particular, a system of checking design calculations and a procedure for checking drawings is necessary to ensure minimum errors and constructability.

The design budget is based on the contract for design services and related to the work plan for producing the contract documents. Careful monitoring of actual costs compared to the approved budget is necessary to ensure profitable completion of the construction documents.

Engineering Project Controls

For any design effort there must be a process to control changes in scope. The scope change process should ensure that when scope changes occur, the impact of the change on project cost and schedule is fully understood by all members of the project team, in particular the sponsor's organization. The adverse impact of late changes in scope must be communicated to the sponsor.

A system must be established for progress measurement and schedule control. The system should include the WBS for engineering, including the roles and responsibilities of the engineering manager and the engineering team with respect to progress measurement and schedule control. The roles and responsibilities of outside consultants should also be included in the system.

A system must also be established for cost control. The system should describe the roles and responsibilities of the engineering manager and team members, including outside consultants. The system should include the CBS for engineering. The cost contingency for engineering and how it will be managed is a critical factor in cost control of the design effort. A system for cost control must define the process that will be used for obtaining approval for changes in the engineering budget.

The cost control system must also include procedures for measuring productivity and reporting cost performance.

Progress Measurement of Engineering Design

The purpose of the project plan is to successfully control the project to ensure completion within budget and schedule constraints. Measuring the progress of a project supports management in establishing a realistic plan for execution of a project and provides the project manager and client with a consistent analysis of project performance. Progress measurement also provides an early warning system to identify deviations from the project plan and scope growth. To control engineering work, a drawing list, specification list, equipment list, and instrument list are used to determine the project status and continued planning.

After completion of the project proposal, the project manager should establish a system of measuring the progress of engineering design. The deliverables of design include performing calculations, producing drawings, and writing specifications for the project. These deliverables are defined in the scope of work, organized in the work breakdown structure, and used in determining the budget for design. Table 7-1 is an illustrative example of a progress measurement system for engineering design. The system is broken down into work packages by disciplines of engineering, including architectural, civil, structural, electrical, instrumentation, and mechanical engineering. The system includes the hours required to produce drawings as well as a breakdown of the percent of time for producing design calculations and drawings for each discipline. The information in this table forms the basis for measuring cost and schedule performance, which are discussed in Chapter 9.

Measuring progress of design is difficult because design is a creative process. Sometimes considerable time may be expended in the design effort without seeing any physical results. For example, numerous computer simulations may be necessary during a design analysis. Significant amounts of work and time may be expended in the simulation effort, but if the simulations are not completed, there will be no produced drawings to show any physical results of the design effort. Similarly, numerous drawings may be near complete, yet none of them completely finished. It is difficult to define a half-finished drawing. Thus, measuring design progress by counting the number of completed drawings may not fully measure the progress of the design effort.

Most sponsors request milestones in the design schedule for reviews. For example, there may be a 60% design complete milestone that is designated for review by the sponsor. There must be agreement between

TABLE 7-1   Example Progress Measurement System for Engineering Design Work Packages

---

A: Work Breakdown of Architectural Design Work Packages

Architectural design = 300 design-hours and 180 CADD hours
Drawings: 83 hours per drawing

Design = 40% of design effort

Design parameters identified	25%
Layout and methods established	20%
Ready to start related drawings	35%
Final drawings issued	20%
	Total = 100%

Drafting = 60% of design effort

Borders & basic layout established	15%
Review of completed information	10%
Related design work at 80% complete	25%
Quality-control review #1	3%
Drawing complete	39%
Quality-control review #2	3%
Record drawings	5%
	Total = 100%

Specifications: 8 total, 2 hours per specification

B: Work Breakdown of Civil Engineering Design Work Packages

Civil engineering design = 171 design-hours and 100 CADD hours
Drawings: 135 hours per drawing
   Site grading and drainage = 91 design-hours and 60 CADD hours
   Site-work details = 80 design-hours and 40 CADD hours

Drafting

Field survey	8%
Site base drawing	35%
Quality-control review #1	3%
Contract engineering	26%
Complete site drawing and details	20%
Quality-control review #2	3%
Record drawings	5%
	Total = 100%

Specifications: 7 total, 1 hour per specification

C: Progress Measurement of Structural Engineering Design Work Packages

Structural engineering = 480 design-hours and 240 CADD hours
Drawings: 80 hours per drawing

Engineering design = 40% of design effort

Design parameters identified	25%
Layout and methods established	20%
Ready to start related drawings	35%
Final drawings issued	20%

(Continued)

TABLE 7-1   Example Progress Measurement System for Engineering Design Work Packages (Continued)

C: Progress Measurement of Structural Engineering Design Work Packages (Continued)

Drafting = 60% of design effort

Borders and basic layout established	15%
Review of existing information	10%
Related design work at 80% complete	25%
Quality-control review #1	3%
Drawings complete	39%
Quality-control review #2	3%
Record drawings	5%

Specifications: 4 total, 1 hour per specification

D: Work Breakdown of Electrical Engineering Design Work Packages

Electrical engineering design = 72 design-hours and 42 CADD hours

Drawings: 18 hours per drawing

Engineering design = 40% of design effort

Design parameters identified	25%
Layout and methods established	20%
Ready to start related drawings	35%
Final drawings issued	20%
	Total = 100%

Drafting = 60% of design effort

Borders and basic layout established	15%
Review of complete information	10%
Related design work at 80% complete	25%
Quality-control review #1	3%
Drawing complete	39%
Quality-control review #2	3%
Record drawings	5%
	Total = 100%

Specifications: 4 total, 1 hour per specification

E: Progress Measurement of Instrumentation Design Work Packages

Instrumentation design = 124 design-hours and 84 CADD hours

Drawings: 121 hours per drawing

Engineering Design = 40% of design effort

Design parameters identified	25%
Layout and methods established	20%
Ready to start related drawings	35%
Final drawings issued	20%
	Total = 100%

Drafting = 60% of design effort

Borders and basic layout established	15%
Review of complete information	10%
Related design work at 80% complete	25%
Quality-control review #1	3%
Drawings complete	39%
Quality-control review #2	3%
Record drawings	5%
	Total = 100%

(Continued)

TABLE 7-1   Example Progress Measurement System for Engineering Design Work Packages (*Continued*)

F: Work Breakdown of Mechanical Engineering Design Work Packages

Mechanical engineering design = 496 design-hours and 235 CADD hours

Drawings

P&ID drawing of existing plant	56 design-hours and 20 CADD hours
Mechanical floor plan	60 design-hours and 25 CADD hours
Details—Piping, feed water	120 design-hours and 60 CADD hours
Boiler piping schematic sheet #1	60 design-hours and 30 CADD hours
Boiler piping schematic sheet #2	60 design-hours and 30 CADD hours
Boiler stack details	80 design-hours and 40 CADD hours
Miscellaneous details	20 design-hours and 10 CADD hours
Develop equipment list	40 design-hours and 20 CADD hours

Engineering design = 40% of design effort

Design parameters identified	25%
Layout and methods established	20%
Ready to start related drawings	35%
Final drawing issued	20%
	Total = 100%

Drafting = 60% of design effort

Borders and basic layout established	15%
Review of information complete	10%
Related design work at 80% complete	25%
Quality-control review #1	3%
Drawings complete	39%
Quality-control review #2	3%
Record drawings	5%
	Total = 100%

Specification

Boiler specification = 132 design-hours

Base specification marked up	25%
Related information reviewed	15%
Typed, ready for review	20%
Issue for client review	20%
Revise per client review	10%
Final issue	10%
	Total = 100%

Other specifications: 19 total, 1 hour per specification

---

the sponsor and the design team on how to define and measure the 60% design complete. For example 60% design complete may be defined as completion of all architectural work and structural work, completion of 50% of the electrical drawings, 25% of the mechanical drawings, and initiation of the specification writing. Chapter 9 presents methods of measuring design progress.

References

1. Anderson, S. D. and Tucker, R. L., *Assessment of Architecture / Engineering Project Management Practices and Performance,* Special Publication No. 7, Construction Industry Institute, Austin, TX, April 1990.
2. *Consulting Engineering: A Guide for the Engagement of Engineering Services,* Manual No. 45, American Society of Civil Engineers, Reston, VA, 1988.
3. Eldin, N. N., "Management of Engineering/Design Phase," *Journal of Construction Engineering and Management,* ASCE, Reston, VA, Vol. 117, No. 1, March 1991.
4. Vorba, L. R. and Oberlender, G. D., "A Methodology for Quality Design," *Project Management Symposium,* Project Management Institute, Newtown Square, PA, October 1991.
5. Bachner, J. P., *Practice Management for Design Professionals—A Practical Guide to Avoiding Liability and Enhancing Profitability,* John Wiley & Sons, New York, NY, 1991.

8

Project Scheduling

Project Planning and Project Scheduling

Project planning is the process of identifying all the activities necessary to successfully complete the project. Project scheduling is the process of determining the sequential order of the planned activities, assigning realistic durations to each activity, and determining the start and finish dates for each activity. Thus, project planning is a prerequisite to project scheduling because there is no way to determine the sequence or start and finish dates of activities until they are identified.

However, the terms of project planning and scheduling are often used synonymously because planning and scheduling are performed interactively. For example, a specific list of activities may be planned and scheduled for a project. Then, after the schedule is reviewed, it may be decided that additional activities should be added or some activities should be rearranged in order to obtain the best schedule of events for the project.

Planning is more difficult to accomplish than scheduling. The real test of the project planner/scheduler is his or her ability to identify all the work required to complete the project. The preceding chapters of this book focused on identifying work activities and grouping those activities into meaningful categories. For example, the process of developing a well-defined work breakdown structure (WBS) as presented in Chapter 6 results in a list of activities that must be performed to complete a project.

After the activities are identified, it is relatively easy to determine the schedule for a project. Many methods and tools have been developed for scheduling. The computer is universally used to perform the calculations for a project schedule. However, adequate attention must be given to both planning and scheduling. Sometimes a project schedule becomes non-workable due to too much emphasis on getting a computer-generated schedule. The planner/scheduler must give adequate time and think through the planning before turning to the computer to generate the schedule. In simple terms, it is better to be a good planner than to be proficient in computer applications. The material presented and discussed in the preceding chapters has set the stage for developing a good project schedule.

Desired Results of Planning

Project planning is the heart of good project management because it provides the central communication that coordinates the work of all parties. Planning also establishes the benchmark for the project control system to track the quantity, cost, and timing of work required to successfully complete the project. Although the most common desired result of planning is to finish the project on time, there are other benefits that can be derived from good project planning (see Table 8-1).

Planning is the first step to project scheduling. Planning is a process and not a discrete activity. As changes occur, additional planning is required to incorporate the changes into the schedule. There are many situations or events that can arise that can impact a project schedule. Examples are changes in personnel, problems with permits, change in a major piece of equipment, or design problems. Good planning detects changes and adjusts the schedule in the most efficient manner.

A common complaint of many design engineers is they cannot efficiently produce their work because of interruptions and delays. The cause of this problem is usually a lack of planning, and in some instances no planning at all. Planning should clearly identify the work that is required by each individual and the interface of work between individuals. It should also include a reasonable amount of time for the exchange of information between project participants, including the delay time for reviews and approvals.

TABLE 8-1   Desired Results of Project Planning and Scheduling

1. Finish the project on time
2. Continuous (uninterrupted) flow of work (no delays)
3. Reduced amount of rework (least amount of changes)
4. Minimize confusion and misunderstandings
5. Increased knowledge of status of project by everyone
6. Meaningful and timely reports to management
7. You run the project instead of the project running you
8. Knowledge of scheduled times of key parts of the project
9. Knowledge of distribution of costs of the project
10. Accountability of people, defined responsibility/authority
11. Clear understanding of who does what, when, and how much
12. Integration of all work to ensure a quality project for the owner

Another common complaint of many designers is the amount of rework they must do because of changes in the project. This also leads to confusion and misunderstandings that further hinder productive work. Planning should include a clear description of the required work before the work is started. However, it must be recognized that changes are a necessary part of project work, especially in the early development phases. If changes in the work are expected, or probable, then project planning should include provisions for a reasonable allowance of the anticipated changes. Too often people know that changes will occur, but fail to include them in the project planning.

Project planning and scheduling can serve as an effective means of preventing problems. It can prevent delays in work, a major cause of late project completion and cost overrun, which often leads to legal disputes. It can also prevent low worker morale and decline in productivity that is caused by lack of direction.

Benefits of Planning

One of the obvious purposes of project planning and scheduling is to complete a project on time. However, there are many benefits of planning projects. Planning is a centralized tool for communications that coordinates the work of all parties. The project plan provides a benchmark of work activities that can be used to measure progress. It keeps the work flowing in a continuous manner so everyone is better informed. The process of developing the plan requires project people to identify, in advance, the work that must be performed. After it is completed, the project plan provides accountability of people, with defined responsibility and authority. Organized work leads to improvement in the quality of the project.

Project planning also prevents problems. A well-developed project plan reduces the number of changes in a project and the amount of rework. A good project plan helps to prevent duplication of work, or overlaps in work. It helps in preventing interruptions and delays in work, which can improve productivity and quality of work. Project planning reduces confusion and misunderstandings between people on the project, which helps prevent low worker morale and declining productivity of workers. A good plan also helps to reduce and manage legal disputes that may arise in a project.

Principles of Planning and Scheduling

There must be an explicit operational plan to guide the entire project. The plan must include and link the three components of the project: scope, budget, and schedule. Too often, planning is focused only on schedule without regard to the important components of scope and budget.

TABLE 8-2   Key Principles for Planning and Scheduling

1. Begin planning before starting work, rather than after starting work
2. Involve people who will actually do the work in the planning and scheduling process
3. Include all aspects of the project: scope, budget, schedule, and quality
4. Build flexibility into the plan, include allowance for changes and time for reviews and approvals
5. Remember the schedule is the plan for doing the work, and it will never be precisely correct
6. Keep the plan simple, eliminate irrelevant details that prevent the plan from being readable
7. Communicate the plan to all parties; any plan is worthless unless it is known

To develop an integrated total project plan, the project must be broken down into well-defined units of work that can be measured and managed. This process starts with the WBS. Once this is completed, the project team members who have the expertise to perform the work can be selected. Team members have the ability to clearly define the magnitude of detail work that is required. They also have the ability to define the time and cost that will be required to produce the work. With this information a complete project plan can be developed.

The project plan and schedule must clearly define individual responsibilities, schedules, budgets, and anticipated problems. The project manager should prepare formal agreements with appropriate parties whenever there is a change in the project. There should be equal concern given to schedule and budget, and the two must be linked. Planning, scheduling, and controlling begin at the inception of the project and are continuous throughout the life of the project until completion. Table 8-2 lists key principles for planning and scheduling.

Responsibilities of Parties

The principal parties of owner, designer, and contractor all have a responsible role in project planning and scheduling. It is erroneous to assume this role is the responsibility of any one party. Each must develop a schedule for his or her required work and that schedule must be communicated and coordinated with the other two parties, because the work of each affects the work of the others.

The owner establishes the project completion date, which governs the scheduling of work for both the designer and contractor. The owner should also set priorities for the components that make up the project. For example, if the project consists of three buildings, the relative importance of the buildings should be identified. This assists the designer in the process of organizing his or her work and developing the design schedule to produce drawings that are most important to the owner. It also assists in the development of the specifications and contract documents that communicate priorities to the construction contractor.

The design organization must develop a design schedule that meets the owner's schedule. This schedule should include a prioritization of work in accordance with the owner's needs and should be developed with extensive input from all designers who will have principal roles in the design process. Too often, a design schedule is produced by the principal designer, or the project manager of the design organization, without the involvement of those who will actually do the work.

The construction contractor must develop a schedule for all construction activities in accordance with the contract documents. It should include procurement and delivery of materials to the job, coordination of labor and equipment on the job, and interface the work of all subcontractors. The objective of the construction schedule should be to effectively manage the work to produce the best-quality project for the owner. The purpose of construction scheduling should *not* be to settle disputes related to project work, but to manage the project in the most efficient manner.

For some projects, it may be desirable for one party to maintain the schedule and the other parties to participate in monitoring it. Ultimately each one of the parties will be responsible for his or her portion of the schedule. Maintaining one common schedule as a cooperative effort between parties can reduce problems associated with maintaining three separate schedules.

Planning for Multiple Projects

Many project managers are assigned the responsibility of simultaneously managing several small projects that have short durations. A small project is usually staffed with a few people who perform a limited number of tasks to complete the project. For projects of this type, there is a tendency for the project manager to forgo any formal planning and scheduling because each project is simple and well defined. However, the problem that the project manager has is not the management of any one project at a time; instead, the problem is managing all the projects simultaneously. The task of simultaneously managing multiple small projects can be very difficult and frustrating. Thus, the need for good planning and scheduling is just as important for the management of multiple small projects as it is for the management of a single large project.

To manage multiple small projects, the project manager must develop a plan and schedule that includes all projects for which he or she is assigned, even though the projects may be unrelated. This is necessary because the staffing of small projects requires assigning individuals to several projects at the same time so they will have a full-time work load.

Thus, their work on any one project affects the work on other projects. For this type of work environment the project manager must develop a plan and schedule that interfaces the work of each individual that is working on all the projects for which the project manager is responsible. In particular, the plan should clearly show how the work of each person progresses from one project to another.

A large project is commonly assigned to one project manager who has no other responsibilities than management of the single project at one time. It is staffed by persons who provide the diverse technical expertise that is required to accomplish the numerous tasks to complete the project. For projects of this type, the problem of the project manager is identifying and interfacing related tasks to ensure the work is accomplished in a continuous manner. He or she relies on the input of team members to develop the project plan and schedule. Much of the work of the project manager involves extensive communications with team members to ensure that work is progressing in a continuous and uninterrupted manner.

Regardless of the project size, large or small, planning and scheduling must be done. Perhaps the greatest mistake a project manager can make is to assume that planning and scheduling are not required for some reason, such as he or she is too busy, there will be too many changes, or the project is too small.

Techniques for Planning and Scheduling

The technique used for project scheduling will vary depending upon the project's size, complexity, duration, personnel, and owner's requirements. The project manager must choose a scheduling technique that is simple to use and is easily interpreted by all project participants. There are two general methods that are commonly used: the bar chart (sometimes called the *Gantt chart*) and the Critical Path Method (sometimes called CPM or network analysis system).

The bar chart, developed by Henry L. Gantt during World War I, is a graphical time-scale of the schedule. It is easy to interpret; but it is difficult to update, does not show interdependences of activities, and does not integrate costs or resources with the schedule. It is an effective technique for overall project scheduling, but has limited application for detailed construction work because the many interrelationships of activities, which are required for construction work, are not defined. Many project managers prefer the bar chart for scheduling engineering design work because of its simplicity, ease of use, and because it does not require extensive interrelationships of activities. However, it can require significant time for updating since the interrelationships of activities are not defined. A change in one activity on the bar chart will not automatically change subsequent activities. Also, the bar chart does not integrate costs with the schedule, nor does it provide resources, such as labor-hours, that are important for management of design.

Some designers argue that they cannot define the interrelationships between the activities that make up a design schedule. They use this argument to support the use of a bar chart. They will also argue that resources change constantly on a design project, resulting in a schedule that is too difficult to maintain. Either of these situations may occur at times on some projects. However, if these situations exist on every project, it is likely that the projects are not well planned, managed, or controlled.

The Critical Path Method (CPM) was developed in 1956 by the DuPont Company, with Remington Rand as consultants, as a deterministic approach to scheduling. The CPM method is commonly used in the engineering and construction industry. A similar method, Program Evaluation and Review Technique (PERT), was developed in 1957 by the U.S. Navy, with Booz, Allen, & Hamilton Management consultants, as a probabilistic approach to scheduling. It is more commonly used by the manufacturing industry; however, it can be used for risk assessment of highly uncertain projects. Both methods are often referred to as a network analysis system. The CPM provides interrelationships of activities and scheduling of costs and resources. It also is an effective technique for overall project scheduling and detailed scheduling of construction. However, it does have some limitations when applied to detailed engineering design work during the early stages of a project because it requires an extensive description of the interrelationships of activities.

Although the CPM technique requires more effort than a bar chart, it provides more detailed information that is required for effective project management. Using a network schedule to plan a project forces the project team to break a project down into identifiable tasks and to relate the tasks to each other in a logical sequence in much greater detail than a bar chart. This up-front planning and scheduling helps the project team to identify conflicts in resources before they occur. The project manager must use his or her own judgement and select the method of scheduling that best defines the work to be done and that communicates project requirements to all participants.

Network Analysis Systems

A network analysis system (NAS) provides a comprehensive method for project planning, scheduling, and controlling. NAS is a general title for the technique of defining and coordinating work by a graphical diagram that shows work activities and the interdependences of activities. Many books and articles have been written that describe the procedures and applications of this technique. It is not the purpose of this book to present the details of network methods because so much material has already been written. The following paragraphs and figures present the basic fundamentals of NAS to guide the project manager in the development of the project plan and schedule. The basic definitions shown in Figure 8-1 are presented to clarify the following paragraphs because there are variations in terminology used in network analyses.

Activity — The performance of a task required to complete the project, such as design of foundations, review of design, procure steel contracts, or form concrete columns. An activity requires time, cost, or both time and cost.

Network — A diagram to represent the relationship of activities to complete the project. The network may be drawn as either an "arrow diagram" or a "precedence diagram."

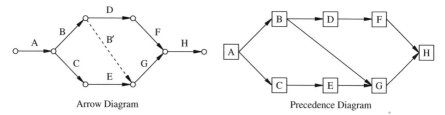

Arrow Diagram   Precedence Diagram

Duration (D) — The estimated time required to perform an activity. The time should include all resources that are assigned to the activity.

Early Start (ES) — The earliest time an activity can be started.

Early Finish (EF) — The earliest time an activity can be finished and is equal to the early start plus the duration.

$$EF = ES + D$$

Late Finish (LF) — The latest time an activity can be finished.

Late Start ((LS) — The latest time an activity can be started without delaying the completion date of the project.

$$LS = LF - D$$

Total Float (TF) — The amount of time an activity may be delayed without delaying the completion date of the project.

$$TF = LF - EF = LS - ES$$

Free Float (FF) — The amount of time an activity may be delayed without delaying the early start time of the immediately following activity. FFi 5 ESj 2 EFi, where the subscript i represents the preceding activity and the subscript j represents the following activity.

Critical Path — A series of interconnected activities through the network diagram, with each activity having zero, free and total float time. The critical path determines the minimum time to complete the project.

Dummy Activity — An activity (represented by a dotted line on the arrow network diagram) that indicates that any activity following the dummy cannot be started until the activity or activities preceding the dummy are completed. The dummy does not require any time.

Figure 8-1 Basic definitions for CPM.

For project management the CPM is the most commonly used NAS method. The concept is simple, the computations only require basic arithmetic; and a large number of computer programs are available to automate the work required of CPM scheduling. The most difficult task in the use of CPM is identifying and interfacing the numerous activities that are required to complete a project, that is, development of the CPM network diagram. If a well-defined WBS is developed first, the task of developing a CPM diagram is greatly simplified.

There are two basic methods of drawing CPM diagrams: the arrow diagram (sometimes called activity on arrow) and the precedence diagram (sometimes called activity on node). Although both methods achieve the same results, most project managers prefer the precedence method because it does not require the use of dummy activities. The precedence method can also provide the start-to-start, finish-to-finish, start-to-finish, and finish-to-start relationship of activities, which can significantly reduce the number of activities that are required in a network diagram. However, many individuals prefer to not use these relationships because of potential confusion in the network scheduling.

Figure 8-2 is a simple precedence diagram that is presented to illustrate the time computations for analysis of a project schedule by the CPM.

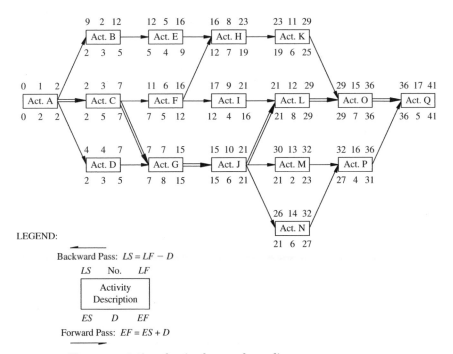

Figure 8-2 Time computations for simple precedence diagram.

Each activity is described by a single letter. The number at the top of the activity is the assigned activity number, and the number at the bottom of each activity represents the duration in working days. A legend is shown in the lower left hand corner to define the start and finish days. All calculations for starts and finishes are based on end-of-day.

After the CPM diagram has been prepared, the duration of each activity can be assigned and the forward-pass calculations performed to calculate the early start and early finish of each activity. The largest early finish of all preceding activities defines the early start of all following activities. For example, activity H cannot be started until activities E and F are both completed. Since the largest early finish of the two preceding activities is 12, the early start for activity H is 12. The forward-pass calculations are performed on all activities from the first activity A to the last activity Q. The early finish of the last activity defines the project completion, which is 41 days for this particular project. This project duration is a calculated value based upon the duration and interdependences of all activities in the project.

A backward pass can be performed to calculate the late start and late finish of each activity. The smallest late start of all following activities defines the late finish of all preceding activities. For example activities H and I cannot both be started until activity F is completed. Since the smallest early start of the two following activities is 16, the late finish for F is 16. The backward-pass calculations are performed on all activities from the last activity Q to the first activity A.

The difference between starts and finishes determines the amount of free and total float. For example, the total float for activity M is 9 days, the difference between its late start (30) and early start (21). The free float for activity M is 4 days, the difference between its early finish (23) and the early start (27) of the immediately following activity P.

The critical path, as noted by the double line on Figure 8-2, is defined by the series of interconnected activities that have zero total float. Since these activities have no float time available, any delay in their completion will delay the completion date of the project. Therefore, they are called critical activities.

Table 8-3 lists the basic steps to guide the process of developing a CPM diagram for project planning and scheduling. It is not always possible to complete a step without some readjustments. For example, the CPM diagram of step 2 may need readjusting after evaluation of the time and resources of steps 4 and 5. Some activities that were originally planned in a series may need to occur in parallel to meet a time requirement. Each project manager and his or her team must work together to develop a project plan and schedule that achieve the required project completion date with the resources that are available.

TABLE 8-3   Steps in Planning and Scheduling

1. Develop a work breakdown structure (WBS) that identifies work items (activities)
   a. Consider activities that require time
   b. Consider activities that require cost
   c. Consider activities that you need to arrange
   d. Consider activities that you want to monitor
2. Prepare a drawing (network diagram) that shows each activity in the order it must be performed to complete the project
   a. Consider which activities immediately precede each activity
   b. Consider which activities immediately follow each activity
   c. The interrelationship of activities is a combination of how the work must be done (constraints) and how you want the work to be done
3. Determine the time, cost, and resources required to complete each activity
   a. Review work packages of the WBS
   b. Obtain input from project team members
4. Compute the schedule to determine start, finish, and float times
   a. Perform a forward pass to determine early starts and finishes
   b. Perform a backward pass to determine late starts and finishes
   c. Determine the differences between start and finish times to determine float time and critical activities
5. Analyze costs and resources for the project
   a. Compute the cost per day for each activity and for the entire project
   b. Compute the labor-hours per day and/or other resources that are required to complete the project
6. Communicate the results of the plan and schedule
   a. Display time schedule for activities
   b. Display cost schedule for activities
   c. Display schedule for other resources

Development of CPM Diagram from the WBS

Table 8-3 provides the list of basic steps that can be used to guide the process of developing a network analysis system for project planning and scheduling.

The development of the WBS is an important first step that is often neglected. Attempting to develop the CPM logic diagram without a WBS usually leads to numerous revisions to the diagram. Each project manager and his or her team must work together to develop a project plan and schedule that achieves the required project completion date with the resources that are available.

Figure 8-3 is an example of a WBS for the design of a service facility project that consists of two buildings, site-work, and on-site utilities. As discussed in Chapter 6, a WBS can be developed in a graphical or outline format. Figure 8-3 shows the WBS in a graphic format and Figure 8-4 shows the same project in an outline format. A discussion of a typical owner's study for this type of project was presented in Chapter 3. To handle this project the contracting strategy is to use in-house personnel to design the on-site utilities, site-work, and the industrial maintenance building (denoted as Building A). This is commonly called performing

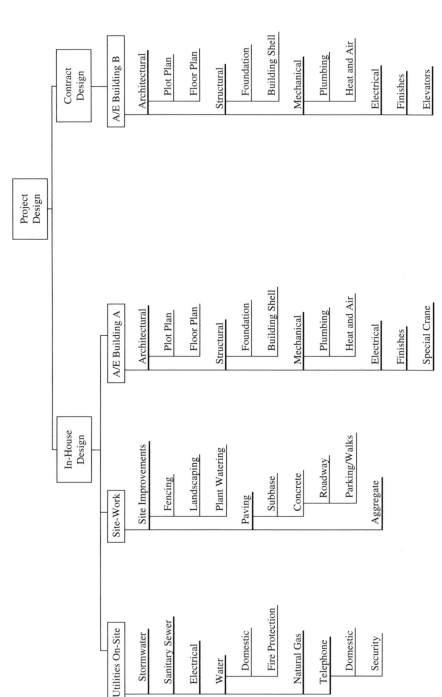

Figure 8-3 Work breakdown structure for design of service facility project (graphic format).

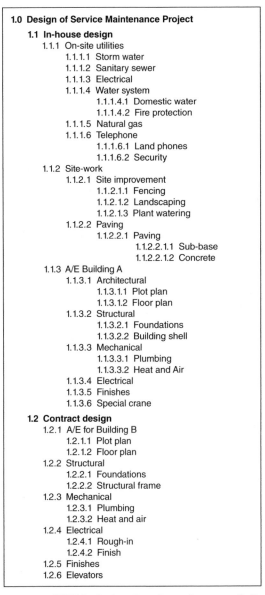

1.0 Design of Service Maintenance Project
  1.1 In-house design
    1.1.1 On-site utilities
        1.1.1.1 Storm water
        1.1.1.2 Sanitary sewer
        1.1.1.3 Electrical
        1.1.1.4 Water system
            1.1.1.4.1 Domestic water
            1.1.1.4.2 Fire protection
        1.1.1.5 Natural gas
        1.1.1.6 Telephone
            1.1.1.6.1 Land phones
            1.1.1.6.2 Security
    1.1.2 Site-work
        1.1.2.1 Site improvement
            1.1.2.1.1 Fencing
            1.1.2.1.2 Landscaping
            1.1.2.1.3 Plant watering
        1.1.2.2 Paving
            1.1.2.2.1 Paving
                1.1.2.2.1.1 Sub-base
                1.1.2.2.1.2 Concrete
    1.1.3 A/E Building A
        1.1.3.1 Architectural
            1.1.3.1.1 Plot plan
            1.1.3.1.2 Floor plan
        1.1.3.2 Structural
            1.1.3.2.1 Foundations
            1.1.3.2.2 Building shell
        1.1.3.3 Mechanical
            1.1.3.3.1 Plumbing
            1.1.3.3.2 Heat and Air
        1.1.3.4 Electrical
        1.1.3.5 Finishes
        1.1.3.6 Special crane
  1.2 Contract design
    1.2.1 A/E for Building B
        1.2.1.1 Plot plan
        1.2.1.2 Floor plan
    1.2.2 Structural
        1.2.2.1 Foundations
        1.2.2.2 Structural frame
    1.2.3 Mechanical
        1.2.3.1 Plumbing
        1.2.3.2 Heat and air
    1.2.4 Electrical
        1.2.4.1 Rough-in
        1.2.4.2 Finish
    1.2.5 Finishes
    1.2.6 Elevators

Figure 8-4 WBS for design of service maintenance facility (outline format).

work by the force-account method. A contract is assigned to an outside design organization for design of the commercial building (denoted as Building B), which is to be used as an employee's office building.

The WBS identifies the tasks and activities that must be performed, but does not provide the order in which they must occur. The CPM network diagram is prepared to show the sequencing and interdependences of the activities in the WBS. The diagram can be prepared by traditional drafting techniques or it can be prepared using the computer. The development on a computer can use either a computer-aided drafting and design (CADD) program, or a software package that is specifically written for CPM scheduling.

Regardless of the method that is used, the initial logic of the diagram must be arranged by the person who is developing the diagram. In simple terms, a person must tell a draftsman, or the computer, how to draw the diagram. An efficient way to accomplish this task is to record each activity on a 3×5 index card and to use a tack to post all activities on a bulletin board or office wall. The activities can then be easily rearranged and reviewed by key participants before development of the formal diagram.

Figure 8-5 is a CPM diagram that was developed from the WBS shown in Figure 8-3. Note that each activity on the CPM is derived from the work tasks that are shown on the WBS. Thus, the project manager plans the project around the work to be performed, which has been defined by the people who will perform the work. Activities that are related are grouped together and arranged in the order they are to be performed. For example, the architectural floor plans are developed before the structural, mechanical, and electrical designs. A careful planning of the interface of activities at the start of the project is necessary for successful management of a project.

The purpose of CPM is to plan the work to guide the progress of a project and provide a baseline for project control. Chapter 9 discusses linking the WBS to the CPM for project control by expanding Figures 8-3 and 8-5 to include procurement and construction activities.

Assigning Realistic Durations

The CPM network diagram defines the activities, and sequencing of activities, to be performed to accomplish the project; however, the anticipated time that is required to complete each activity must be determined in order to schedule the entire project. The durations that are assigned to activities are important because the critical path, timing of activities, distribution of costs, and utilization of resources are all a function of activity durations.

The assignment of the duration that is required to accomplish an activity will vary depending on many factors: quantity and quality of work, number of people and/or equipment that is assigned to the activity, level of worker skills, availability of equipment, work environment, effectiveness of supervision of the work, and other conditions. Although these variations exist, a special effort must be made to

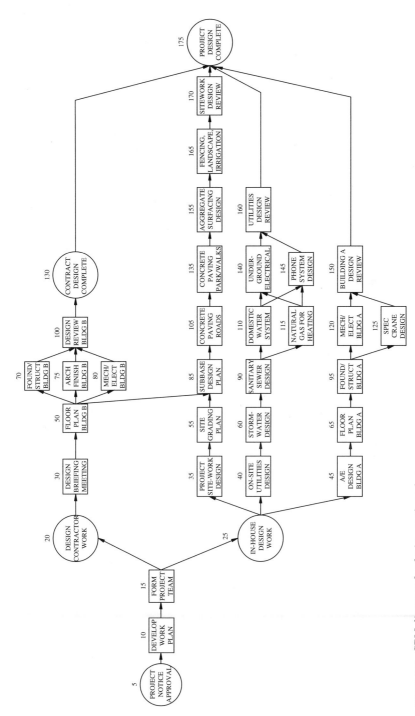

Figure 8-5 CPM diagram for design engineering.

determine a realistic duration for each activity because the duration that is assigned to activities in a CPM network diagram has a large impact on the schedule and overall management of a project.

Many activities in a project are routine in nature, which enables a reasonably accurate determination of the probable time of completion. For these types of activities the duration can be determined by dividing the total quantity of work by the production rate, which is a function of the number of individuals that are assigned to the activity. A common mistake that is made by many people is to calculate the time to accomplish an activity assuming a continuous flow of uninterrupted work. However, all work is subject to delays, interruptions, or other events that can impact time. Thus, a reasonable amount of time (allowance) must be added to the calculated time to determine a realistic duration for each activity.

Generally, the duration of an activity can be determined by one of three methods; by analyzing historical records from previously completed projects, by referencing commercially available manuals that provide costs and production rates for various types of work, or from the experience and judgement of the person who will be performing the work. It is often desirable to determine the probable duration by several methods so the results can be compared to detect any significant variations that may occur.

The schedule for the design work is the total time to produce the final drawings, including the overlap of design calculations and design drafting. As previously discussed most engineers prefer a bar chart for scheduling individual design tasks. However, for project control the individual bar charts must be developed into activities on the CPM diagram to develop the total project schedule. The start and finish of each activity of the CPM engineering design schedule is a composite of all tasks of the work package. The following illustrates the evaluation of overlapping tasks of the work package to determine the duration of an activity on the CPM diagram.

Tasks of Work Package	Duration
Project Engineering	5 days
Structural Design	16 days
CADD Operator	9 days
Total Design Days	30 days

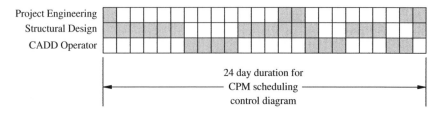

Computer Applications

The CPM network diagram, by itself, identifies the sequencing of activities but does not provide the scheduled start and finish dates, the distribution of costs, or the allocation of resources. This information can easily be determined by assigning the duration, cost, and resources that are required of each activity.

There are many CPM computer programs available to perform the numerous calculations necessary to determine the scheduled time, cost, and resources of activities. Although the number, type, and format of the computer-generated output reports vary widely, depending on the software, the basic input data required for each is the same information. The information required for the input data consists of activity number, description, duration, cost, and resources, such as labor-hours. The sequencing, or interrelationship of activities, is defined by the CPM network diagram. The input data are the same information that is compiled during preparation of the design work packages for the WBS, or during preparation of the estimate for a project by the construction contractor. Thus, the computer application of CPM is appropriate for both the design and construction phase of a project.

The information that must be assembled for a CPM computer analysis is illustrated in Figures 8-6 and 8-7. Figure 8-6 is a simple CPM precedence network diagram for a sewer and water lines construction project. Construction activities are selected for this illustrative example because they are easily recognizable by most readers. Each activity in Figure 8-6 is shown with its time and cost information. Resources are excluded for simplicity of this presentation and are discussed in the following section. A hand analysis of starts and finishes is shown in Figure 8-6 to illustrate the calculations and to relate them to the computer output reports discussed in the following paragraphs. The times that are shown on the diagram all represent end-of-day. A listing of the computer input data for this project is shown in Figure 8-7.

For this project, two surveying crews are available, which allows Activities 130 and 140 to occur simultaneously. Only one trenching machine is available; therefore trenching work for Water Line A, Activity 170, is planned before trenching of Water Line B, Activity 190. Other similar constraints are shown in the network to illustrate that planning must be done before project scheduling can be accomplished.

The input data required for a computer scheduling analysis are shown in Figure 8-7. The first part of the input data defines information related to each activity. The second part defines the order in which the activities are performed, that is, the sequencing or interfacing of activities. The project title is shown above the activity list, and the project start date is shown at the end of the sequence list.

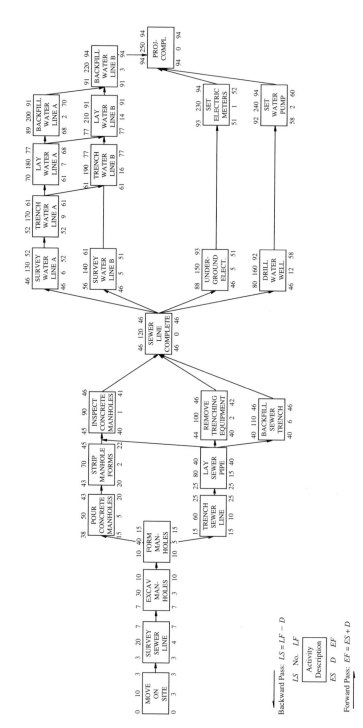

Figure 8-6 CPM diagram for construction phase of sewer and water lines project.

```
                    *********************
                    **  INPUT DETAILS  **
                    *********************

PROJECT:   SEWER & WATER LINE

ACTIVITY LIST:
NUMBER   CODE    DESCRIPTION                    DURATION    COST  ASSIGNED START
-------------------------------------------------------------------------------
    10   5000    MOVE ON SITE                       3      1400.
    20   1100    SURVEY SEWER LINES                 4      2700.
    30   1200    EXCAVATE FOR MANHOLES              3      3500.
    40   1200    INSTALL MANHOLE FORMWORK           5      6000.
    50   1200    PLACE CONCRETE MANHOLES            5      4700.
    60   1300    TRENCH SEWER LINE                 10     12600.
    70   1200    STRIP MANHOLE FORMWORK             2      2100.
    80   1400    LAY SEWER PIPE                    15     11250.
    90   1200    INSPECT MANHOLES                   1       800.
   100   1300    REMOVE TRENCHING EQUIPMENT         2      1400.
   110   1500    BACKFILL SEWER TRENCH              6      3600.
   120   5000    SEWER LINE COMPLETE                0         0.
   130   2110    SURVEY WATER LINE A                6      4000.
   140   2120    SURVEY WATER LINE B                5      3400.
   150   3000    UNDERGROUND ELECTRICAL             5      2500.
   160   4000    DRILL WATER WELL                  12      7000.
   170   2310    TRENCH WATER LINE A                9      8800.
   180   2410    LAY PIPE FOR WATER LINE A          7     16800.
   190   2320    TRENCH WATER LINE B               16     15600.
   200   2510    BACKFILL WATER LINE A              2       900.
   210   2420    LAY PIPE FOR WATER LINE B         14     33600.
   220   2520    BACKFILL WATER LINE B              3      2850.
   230   3000    INSTALL WATER METERS               1       600.
   240   4000    SET WATER PUMP                     2      1400.
   250   5000    PROJECT COMPLETE                   0         0.

SEQUENCE OF ACTIVITIES:
                         FROM         TO
                         ------      ------
                          10          20
                          20          30
                          30          40
                          40          50
                          40          60
                          50          70
                          60          80
                          70          90
                          80          90
                          80         110
                          80         100
                          90         120
                         100         120
                         110         120
                         120         130
                         120         160
                         120         150
                         120         140
                         130         170
                         140         190
                         150         230
                         160         240
                         170         180
                         170         190
                         180         200
                         180         210
                         190         210
                         200         220
                         210         220
                         220         250
                         230         250
                         240         250

                  PROJECT START DATE:   APRIL 1, 2016
                  FIVE-DAY WORK WEEK
                  NO ASSIGNED HOLIDAYS
```

Figure 8-7 Computer input data file for sewer and water lines project.

Figure 8-8 shows the activity schedule report for the project that is typically available from a CPM computer program. Both calendar and work days are shown. Start dates represent beginning of the day while finish dates represent end of the day. The free and total floats are shown for each activity. The letter "C" at the left of an activity denotes it is a critical activity; that is, it has zero total and free float.

```
*****************************
**   ACTIVITY SCHEDULE   **
*****************************
```

PROJECT: SEWER AND WATER LINES   ** Page 1 **
SCHEDULE FOR ALL ACTIVITIES   ACTIVITY SCHEDULE

ACTIVITY NUMBER	ACTIVITY DESCRIPTION	DURA-TION	EARLY START	EARLY FINISH	LATE START	LATE FINISH	TOTAL FLOAT	FREE FLOAT
C 10	MOVE ON SITE	3	1APR2016 1	5APR2016 3	1APR2016 1	5APR2016 3	0	0
C 20	SURVEY SEWER LINES	4	6APR2016 4	11APR2016 7	6APR2016 4	11APR2016 7	0	0
C 30	EXCAVATE FOR MANHOLES	3	12APR2016 8	14APR2016 10	12APR2016 8	14APR2016 10	0	0
C 40	INSTALL MANHOLE FORMWORK	5	15APR2016 11	21APR2016 15	15APR2016 11	21APR2016 15	0	0
50	PLACE CONCRETE MANHOLES	5	22APR2016 16	28APR2016 20	25MAY2016 39	31MAY2016 43	23	0
C 60	TRENCH SEWER LINE	10	22APR2016 16	5MAY2016 25	22APR2016 16	5MAY2016 25	0	0
70	STRIP MANHOLE FORMWORK	2	29APR2016 21	2MAY2016 22	1JUN2016 44	2JUN2016 45	23	18
C 80	LAY SEWER PIPE	15	6MAY2016 26	26MAY2016 40	6MAY2016 26	26MAY2016 40	0	0
90	INSPECT MANHOLES	1	27MAY2016 41	27MAY2016 41	3JUN2016 46	3JUN2016 46	5	5
100	REMOVE TRENCHING EQUIPMENT	2	27MAY2016 41	30JUN2016 42	2JUN2016 45	3JUN2016 46	4	4
C 110	BACKFILL SEWER TRENCH	6	27MAY2016 41	3JUN2016 46	27MAY2016 41	3JUN2016 46	0	0
C 120	SEWER LINE COMPLETE	0	6JUN2016 47	6JUN2016 47	6JUN2016 47	6JUN2016 47	0	0
C 130	SURVEY WATER LINE A	6	6JUN2016 47	13JUN2016 52	6JUN2016 47	13JUN2016 52	0	0
140	SURVEY WATER LINE B	5	6JUN2016 47	10JUN2016 51	20JUN2016 57	24JUN2016 61	10	10
150	UNDERGROUND ELECTRICAL	5	6JUN2016 47	10JUN2016 51	3AUG2016 89	9AUG2016 93	42	0
160	DRILL WATER WELL	12	6JUN2016 47	21JUN2016 58	22JUL2016 81	8AUG2016 92	34	0
230	INSTALL WATER METERS	1	13JUN2016 52	13JUN2016 52	10AUG2016 94	10AUG2016 94	42	42
C 170	TRENCH WATER LINE A	9	14JUN2016 53	24JUN2016 61	14JUN2016 53	24JUN2016 61	0	0
240	SET WATER PUMP	2	22JUN2016 59	23JUN2016 60	9AUG2016 93	10AUG2016 94	34	34
180	LAY PIPE FOR WATER LINE A	7	27JUN2016 62	5JUL2016 68	8JUL2016 71	18JUL2016 77	9	9
C 190	TRENCH WATER LINE B	16	27JUN2016 62	18JUL2016 77	27JUN2016 62	18JUL2016 77	0	0
200	BACKFILL WATER LINE A	2	6JUL2016 69	7JUL2016 70	4AUG2016 90	5AUG2016 91	21	21
C 210	LAY PIPE FOR WATER LINE B	14	19JUL2016 78	5AUG2016 91	19JUL2016 78	5AUG2016 91	0	0
C 220	BACKFILL WATER LINE B	3	8AUG2016 92	10AUG2016 94	8AUG2016 92	10AUG2016 94	0	0
C 250	PROJECT COMPLETE	0	10AUG2016 94	10AUG2016 94	10AUG2016 94	10AUG2016 94	0	0

```
********************************** END OF SCHEDULE **********************************
```

Figure 8-8 Computer-generated activity schedule for sewer and water lines project.

Schedule Coding System

One of the advantages of CPM scheduling by computer methods is the ability to sort specific activities from the complete list of activities for the project. For example, the project manager may only want information about sewer activities, the time required for trenching equipment, or the assignment of the surveying crew. The sorting of these activities can easily be accomplished by a coding system.

Table 8-4 is a simple 4-digit coding system to illustrate sorting capabilities for the sewer and water lines project that is shown in Figure 8-6. All activities related to the sewer line are represented by the number "1" in the first digit. Water line activities are represented by the number "2." The second code digit represents the type of work, such as surveying, forming manholes, trenching, laying pipe, and backfilling. Thus, a 4-digit code is assigned to each activity in the project. For example, the code for activity number 180 is 2410. This code indicates that the activity pertains to water line, laying pipe, and line A. The 4-digit code number for each activity in the sewer and water lines project is shown in the activity list of Figure 8-7.

The coding system provides numerous options for selection of activities by the project manager. For example, all sewer line activities can be sorted from the complete list of project activities by selecting those activities that have a "1" in the first digit. A schedule report for these activities is shown in Figure 8-9. A project manager may also print a bar chart for these activities as shown in Figure 8-10. A coding system provides a means of obtaining many other reports. For example, a sort

TABLE 8-4   Coding System for Sewer and Water Lines Project

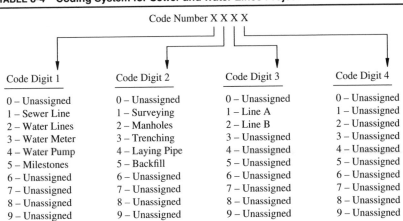

Code Number X X X X

Code Digit 1	Code Digit 2	Code Digit 3	Code Digit 4
0 – Unassigned	0 – Unassigned	0 – Unassigned	0 – Unassigned
1 – Sewer Line	1 – Surveying	1 – Line A	1 – Unassigned
2 – Water Lines	2 – Manholes	2 – Line B	2 – Unassigned
3 – Water Meter	3 – Trenching	3 – Unassigned	3 – Unassigned
4 – Water Pump	4 – Laying Pipe	4 – Unassigned	4 – Unassigned
5 – Milestones	5 – Backfill	5 – Unassigned	5 – Unassigned
6 – Unassigned	6 – Unassigned	6 – Unassigned	6 – Unassigned
7 – Unassigned	7 – Unassigned	7 – Unassigned	7 – Unassigned
8 – Unassigned	8 – Unassigned	8 – Unassigned	8 – Unassigned
9 – Unassigned	9 – Unassigned	9 – Unassigned	9 – Unassigned

```
                          ************************
                          **  ACTIVITY SCHEDULE  **
                          ************************

PROJECT: SEWER AND WATER LINES                                          ** Page 1 **
SCHEDULE FOR SEWER LINE ACTIVITIES ONLY                                 ACTIVITY SCHEDULE

***********************************************************************************************
```

ACTIVITY NUMBER	ACTIVITY DESCRIPTION	DURA-TION	EARLY START	EARLY FINISH	LATE START	LATE FINISH	TOTAL FLOAT	FREE FLOAT
C 20	SURVEY SEWER LINES	4	6APR2016 4	11APR2016 7	6APR2016 4	11APR2016 7	0	0
C 30	EXCAVATE FOR MANHOLES	3	12APR2016 8	14APR2016 10	12APR2016 8	14APR2016 10	0	0
C 40	INSTALL MANHOLE FORMWORK	5	15APR2016 11	21APR2016 15	15APR2016 11	21APR2016 15	0	0
50	PLACE CONCRETE MANHOLES	5	22APR2016 16	28APR2016 20	25MAY2016 39	31MAY2016 43	23	0
C 60	TRENCH SEWER LINE	10	22APR2016 16	5MAY2016 25	22APR2016 16	4MAY2016 25	0	0
70	STRIP MANHOLE FORMWORK	2	29APR2016 21	2MAY2016 22	1JUN2016 44	2JUN2016 45	23	18
C 80	LAY SEWER PIPE	15	6MAY2016 26	26MAY2016 40	6MAY2016 26	26MAY2016 40	0	0
90	INSPECT MANHOLES	1	27MAY2016 41	27MAY2016 41	3JUN2016 46	3JUN2016 46	5	5
100	REMOVE TRENCHING EQUIPMENT	2	27MAY2016 41	30JUN2016 42	2JUN2016 45	3JUN2016 46	4	4
C 110	BACKFILL SEWER TRENCH	6	27MAY2016 41	3JUN2016 46	27MAY2016 41	3JUN2016 46	0	0

```
*********************************     END OF SCHEDULE     *************************************
```

Figure 8-9 Computer printout of sewer line activities only (sort of activities list by code digit #1 equal to one).

```
                                        **********************
                                        **     BAR CHART    **
                                        **********************

PROJECT:  SEWER AND WATER LINES                                                     **  PAGE 1  **
SCHEDULE FOR SEWER LINE ACTIVITIES ONLY                                               BAR CHART

                                            1        10         20         30         40         50         60
ACTIVITY   DESCRIPTION          DURATION  1APR2016 14APR2016 28APR2016 12MAY2016 26MAY2016  9JUN2016 23JUN2016
                                            +.........+.........*.........*.........*.........+.........+
C    20    SURVEY SEWER LINES          4        *XXX
                                            .    .    .    .    .    .    .    .    .    .    .    .    .
C    30    EXCAVATE FOR MANHOLES       3            *XX
                                            .    .    .    .    .    .    .    .    .    .    .    .    .
C    40    INSTALL MANHOLE             5               *XXXX
                                            .    .    .    .    .    .    .    .    .    .    .    .    .
     50    PLACE CONCRETE MANHOLES     5                    XXXXX-----------------*----
                                            .    .    .    .    .    .    .    .    .    .    .    .    .
C    60    TRENCH SEWER LINE          10                    *XXXXXXXXX
                                            .    .    .    .    .    .    .    .    .    .    .    .    .
     70    STRIP MANHOLE FORMWORK      2                              XX---------------*-
                                            .    .    .    .    .    .    .    .    .    .    .    .    .
C    80    LAY SEWER PIPE             15                                   *XXXXXXXXXXXXXX
                                            .    .    .    .    .    .    .    .    .    .    .    .    .
     90    INSPECT MANHOLES            1                                                  X----*
                                            .    .    .    .    .    .    .    .    .    .    .    .    .
    100    REMOVE TRENCHING EQUIPMENT  2                                                  XX--*-
                                            .    .    .    .    .    .    .    .    .    .    .    .    .
C   110    BACKFILL SEWER TRENCH       6                                                  *XXXXX

                                            +.........*.........*.........*.........*.........*.........*
                              WORK DAYS     1        10         20         30         40         50         60
                          CALENDAR DAYS   1APR2016 14APR2016 28APR2016 12MAY2016 26MAY2016  9JUN2016 23JUN2016
```

Figure 8-10 Computer-generated bar chart for sewer line activities only.

of all activities related to trenching and laying pipe can be obtained by selecting activities that have a second code digit number that is greater than "2" and less than "5" (reference Table 8-4).

Cost Distribution

The distribution of costs, with respect to time, must be known to successfully manage a project. In the preceding sections the scheduled early and late starts, and finishes, were calculated based on the duration and sequencing of activities. A cost analysis can also be performed by assigning the cost that is anticipated to complete each activity. The cost of an activity may be distributed over the duration of the activity; however, the activity may be performed over a range of time, starting from the early to late start and ending from the early to late finish.

Because activities can occur over a range of time, a cost analysis must be performed based on activities starting on an early start, late start, and target schedule. The target schedule is the midpoint between the early start and late start. Table 8-5 illustrates the early start cost analysis calculations for the sewer and water lines project shown in Figure 8-6. For each day in the project, the cost per day of each activity that is in progress is summed to obtain the total cost of the project for that day. Cumulative project costs are divided by the total project cost of $147,500 to obtain the percentage cost for each day. The percentage time for each day is calculated by dividing the number of the working days by the total project duration of 94 days. Similar calculations can be performed for activities starting on a late start schedule and target schedule.

Although the calculations for a cost analysis are simple, many are required, as illustrated by the small sewer and water lines project that has only 25 activities and a 94-day project duration. A small microcomputer can perform all the calculations for cost analysis of a project with several hundred activities in less than 2 seconds.

Figure 8-11 is a computer printout of the daily distribution of costs for the calculations illustrated in Table 8-5. A similar analysis can be performed for other resources, such as labor and equipment. For example, a daily distribution of labor-hours, similar to Figure 8-11, can be used to detect periods of time when the need for labor is high or low. The project manager and his or her team can detect this problem early and appropriately adjust the project plan or acquire additional personnel if needed and available.

A tabular format of the distribution of costs on an early start, late start, and target basis is presented in Figure 8-12. The target scheduled costs are average values between the early and late start schedules. The two right-hand columns of Figure 8-12 show the percentage-cost and

TABLE 8-5 **Calculations for Project Costs per Day on an Early Start Basis**
Total Project Duration = 94 Working Days
Total Project Cost = $147,500.00

Day	%-Time	Activities in progress				Project cost/day	Cumulative project cost	% Cost
1	1.06%	Act. 10	$1,400/3	=	$466.67/Day	$466.67	$466.67	0.32%
2	2.12%	" "	"	=	"	"	$933.33	0.63%
3	3.19%	" "	"	=	"	"	$1,400.00	0.95%
4	4.25%	Act. 20	$2,700/4	=	$675.00/Day	$675.00	$2,075.00	1.41%
5	5.32%	" "	"	=	"	"	$2,750.00	1.86%
6	6.38%	" "	"	=	"	"	$3,425.00	2.32%
7	7.45%	" "	"	=	"	"	$4,100.00	2.78%
8	8.51%	Act. 30	$3,500/3	=	$1,167.67/Day	$1,167.67	$5,267.67	3.57%
9	9.57%	" "	"	=	"	"	$6,433.33	4.36%
10	10.63%	" "	"	=	"	"	$7,600.00	5.15%
11	11.70%	Act. 40	$6,000/5	=	$1,200.00/Day	$1,200.00	$8,800.00	5.97%
12	12.77%	" "	"	=	"	"	$10,000.00	6.78%
13	13.83%	" "	"	=	"	"	$11,200.00	7.59%
14	14.89%	" "	"	=	"	"	$12,400.00	8.41%
15	15.96%	" "	"	=	"	"	$13,600.00	9.22%
16	17.02%	Act. 50	$4,700/5	=	$940.00/Day			
		Act. 60	$12,600/10	=	$1,260.00/Day	$2,200.00	$15,800.00	10.71%
17	18.09%	" "	"	=	"	"	$18,000.00	12.20%
18	19.15%	" "	"	=	"	"	$20,200.00	13.69%
19	20.21%	" "	"	=	"	"	$22,400.00	15.19%
20	21.28%	" "	"	=	"	"	$24,600.00	16.68%
21	22.34%	Act. 60	$12,600/10	=	$1,260.00/Day			
		Act. 70	$2,100/2	=	$1,050.00/Day	$2,310.00	$25,910.00	18.24%
⋮	⋮						⋮	⋮
94	100.0%						$147,500.00	100.0%

percentage-time values for the target schedule. As shown in the figure there is a non-linear relationship between the time and cost for a project.

The cumulative cost graph for a project is commonly called the *S-curve*, because it resembles the shape of the letter "S." The early, late, and target cumulative distribution of costs can be superimposed onto one graph to form the envelope of time over which costs may be distributed for the project (reference Figure 8-13). This graph links two of the basic elements of a project, time and cost. The third element, accomplished

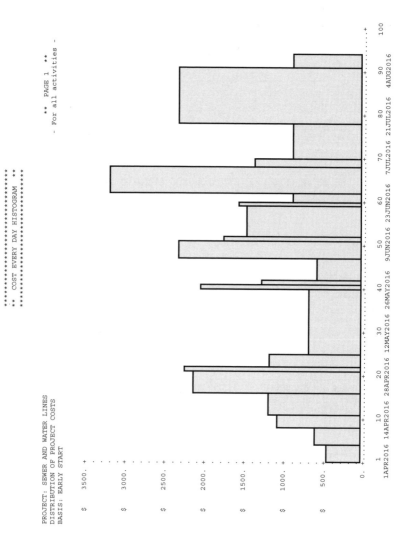

Figure 8-11 Computer printout of daily distribution of costs for sewer and water lines project.

```
                          ********************************
                          **  DAILY COST SCHEDULE  **
                          ********************************
PROJECT: SEWER AND WATER LINES                              ** DAILY COST SCHEDULE FOR ALL ACTIVITIES **
DAILY DISTRIBUTION OF COST FOR ALL ACTIVITIES                   ** EARLY START - LATE START - TARGET **
START DATE: 1 APR 2016 FINISH DATE: 10 AUG 2016
```

WORK DAY	CALENDER DATE	EARLY START COST/DAY	EARLY START CUMULATIVE COST	LATE START COST/DAY	LATE START CUMULATIVE COST	TARGET SCHEDULE COST/DAY	TARGET SCHEDULE CUMULATIVE COST	% TIME	% COST
1	1APR2016	$ 467.	$ 467.	$ 467.	$ 467.	$ 467.	$ 467.	1.1%	0.3%
2	4APR2016	$ 467.	$ 933.	$ 467.	$ 933.	$ 467.	$ 933.	2.1%	0.6%
3	5APR2016	$ 467.	$ 1,400.	$ 467.	$ 1,400.	$ 467.	$ 1,400.	3.2%	0.9%
4	6APR2016	$ 675.	$ 2,075.	$ 675.	$ 2,075.	$ 675.	$ 2,075.	4.3%	1.4%
5	7APR2016	$ 675.	$ 2,750.	$ 675.	$ 2,750.	$ 675.	$ 2,750.	5.3%	1.9%
6	8APR2016	$ 675.	$ 3,425.	$ 675.	$ 3,425.	$ 675.	$ 3,425.	6.4%	2.3%
7	11APR2016	$ 675.	$ 4,100.	$ 675.	$ 4,100.	$ 675.	$ 4,100.	7.4%	2.8%
8	12APR2016	$ 1,167.	$ 5,267.	$ 1,167.	$ 5,267.	$ 1,167.	$ 5,267.	8.5%	3.6%
9	13APR2016	$ 1,167.	$ 6,433.	$ 1,167.	$ 6,433.	$ 1,167.	$ 6,433.	9.6%	4.4%
10	14APR2016	$ 1,167.	$ 7,600.	$ 1,167.	$ 7,600.	$ 1,167.	$ 7,600.	10.6%	5.2%
11	15APR2016	$ 1,200.	$ 8,800.	$ 1,200.	$ 8,800.	$ 1,200.	$ 8,800.	11.7%	6.0%
12	18APR2016	$ 1,200.	$ 10,000.	$ 1,200.	$ 10,000.	$ 1,200.	$ 10,000.	12.8%	6.8%
13	19APR2016	$ 1,200.	$ 11,200.	$ 1,200.	$ 11,200.	$ 1,200.	$ 11,200.	13.8%	7.6%
14	20APR2016	$ 1,200.	$ 12,400.	$ 1,200.	$ 12,400.	$ 1,200.	$ 12,400.	14.9%	8.4%
15	21APR2016	$ 1,200.	$ 13,600.	$ 1,200.	$ 13,600.	$ 1,200.	$ 13,600.	16.0%	9.2%
16	22APR2016	$ 2,200.	$ 15,800.	$ 1,260.	$ 14,860.	$ 1,730.	$ 15,330.	17.0%	10.4%
17	25APR2016	$ 2,200.	$ 18,000.	$ 1,260.	$ 16,120.	$ 1,730.	$ 17,060.	18.1%	11.6%
18	26APR2016	$ 2,200.	$ 20,200.	$ 1,260.	$ 17,380.	$ 1,730.	$ 18,790.	19.1%	12.7%
19	27APR2016	$ 2,200.	$ 22,400.	$ 1,260.	$ 18,640.	$ 1,730.	$ 20,520.	20.2%	13.9%
85	28JUL2016	$ 2,400.	$130,250.	$ 2,983.	$120,767.	$ 2,692.	$125,508.	90.4%	85.1%
86	29JUL2016	$ 2,400.	$132,650.	$ 2,983.	$123,750.	$ 2,692.	$128,200.	91.5%	86.9%
87	1AUG2016	$ 2,400.	$135,050.	$ 2,983.	$126,733.	$ 2,692.	$130,892.	92.6%	88.7%
88	2AUG2016	$ 2,400.	$137,450.	$ 2,983.	$129,717.	$ 2,692.	$133,583.	93.6%	90.6%
89	3AUG2016	$ 2,400.	$139,850.	$ 3,483.	$133,200.	$ 2,942.	$136,525.	94.7%	92.6%
90	4AUG2016	$ 2,400.	$142,450.	$ 3,933.	$137,133.	$ 3,167.	$139,692.	95.7%	94.7%
91	5AUG2016	$ 2,400.	$144,650.	$ 3,933.	$141,067.	$ 3,167.	$142,858.	96.8%	96.9%
92	8AUG2016	$ 950.	$145,600.	$ 2,033.	$143,100.	$ 1,492.	$144,350.	97.9%	97.9%
93	9AUG2016	$ 950.	$146,550.	$ 2,150.	$145,250.	$ 1,550.	$145,900.	98.9%	98.9%
94	10AUG2016	$ 950.	$147,500.	$ 2,250.	$147,500.	$ 1,600.	$147,500.	100.0%	100.0%

```
*********************************** END OF DAILY COST SCHEDULE ***************************************
```

Figure 8-12 Computer printout of daily costs for all activities of sewer and water lines project.

work, must also be linked to time and cost. Chapter 9 discusses linking accomplished work to the S-curve.

The type of reports presented in this section are typical examples of the reports that can be obtained from the many computer software programs that are available. The only input data that a project

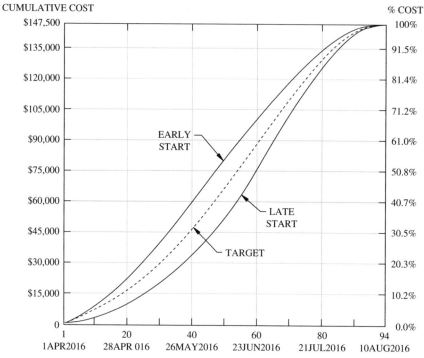

CUMULATIVE COST

% COST

Figure 8-13 Illustrative S-curve for cumulative cost curve on an early start, late start, and target basis.

manager must prepare to obtain the described analyses are shown in Figure 8-7.

Resource Allocations for Design

Efficient utilization of resources is critical to successful management. The primary resource during design is the work-hours of the design team. The project manager depends on the design team to create design alternatives, produce drawings, and write specifications for the proposed project. To properly coordinate all aspects of the design effort, the project manager must ensure the correct expertise is available when needed. Generally, design team members are assigned to the project by their respective home departments. Since each designer is often involved in several projects at the same time, the project manager must develop a resource allocation plan for each project. The plan should then be distributed to each design team member's home department to ensure that each resource will be available when needed.

The project manager can resource load the project plan to include the number of work-hours required for each design discipline. The resource plan is similar to the cost distribution analysis presented earlier in this chapter, except work-hours are used in place of dollars of cost. Thus, the resource plan is simply a histogram of work-hours versus time for each design discipline. The project manager should provide the resource plan for each project to the design team's manager. The design managers can then integrate the resource plans of all active projects into their department's demand for technical expertise. This is necessary to ensure adequate resources are available for the projects when required.

Resource Allocations for Construction

During the construction phase, the primary resources are labor, materials, and equipment. The correct quantity and quality of material must be ordered and delivered to the job-site at the right time to ensure efficiency of labor. Equipment to be installed in the project often requires a long lead time from the fabricator. Thus, the project plan should include material and equipment required by the construction work force.

Labor represents a major cost of construction. The labor force on the job operates construction equipment and installs the materials. A resource allocation plan is required to ensure high efficiency and productivity during construction. The project manager can resource load the project plan to include the number of work-hours required for each craft of construction labor. The resource plan is a histogram of work-hours versus time, similar to the cost distribution analysis presented earlier in the chapter.

The construction plan shows the desired sequence of work. However, to be workable, the plan must also show the distribution of resources, such as the required labor for each craft on the job. The demand for labor should be uniformly distributed for each craft on the project, to prevent irregularities. The resource plan can be used as a tool to ensure a relatively uniform distribution of labor on the project. Figure 8-14 is a simple bar chart for a project, showing each work activity, number of crews, and number of people in the crew for all labor in the project. The lower portion of Figure 8-14 shows irregularity in the distribution of workers per day during the middle of the project.

Figure 8-15 illustrates the same project as Figure 8-14, except Activity F is started one day later and Activity H is started 2 days earlier. The lower portion of Figure 8-15 shows a relative uniform distribution of labor by making these two minor adjustments in the project plan. A similar analysis can be made for each craft of labor to ensure uniform distribution of workers on the project. A resource allocation for a particular craft typically has a flat appearance on the graph, whereas a resource allocation for all crafts typically has a bell-shaped graph as shown in Figure 8-15.

Activity Number	Duration in Days	Numbers of Crews	Workers in Crew	Work Day of Project				
				1 2 3 4 5	6 7 8 9 10	11 12 13 14 15	16 17 18 19 20	21 22 23 24 25
A	4	1	4	4 4 4 4				
B	7	2	3	6 6 6	6 6 6 6			
C	9	1	2	2	2 2 2 2 2	2 2 2		
D	7	2	4	8	8 8 8 8 8	8		
E	3	1	5		5 5 5			
F	9	3	4		12 12	12 12 12 12 12	12 12	
G	4	1	5		5	5 5 5		
H	8	3	6			18 18	18 18 18 18 18	18
I	11	2	4			8 8	8 8 8 8 8	8 8 8 8
J	4	3	2				6	6 6 6
K	6	1	3				3	3 3 3 3 3
		Totals		4 4 10 10 16	16 21 21 33 27	27 19 19 35 35	38 38 26 26 35	35 17 17 11 3

```
Number of Workers per Day

35
                                          38 38
                                          38 38
                                          38 38
35                              35 35  38 38        35 35
                                35 35  38 38        35 35
              33                35 35  38 38        35 35
              33                35 35  38 38        35 35
              33                35 35  38 38        35 35
30            33                35 35  38 38        35 35
              33                35 35  38 38        35 35
              33                35 35  38 38        35 35
              33 27  27         35 35  38 38  26 26 35  35
              33 27  27         35 35  38 38  26 26 35  35
25            33 27  27         35 35  38 38  26 26 35  35
              33 27  27         35 35  38 38  26 26 35  35
              33 27  27         35 35  38 38  26 26 35  35
          21 21 33 27  27       35 35  38 38  26 26 35  35
20        21 21 33 27  27       35 35  38 38  26 26 35  35
          21 21 33 27  27 27 19 19 35 35  38 38 26 26 35  35
          21 21 33 27  27 27 19 19 35 35  38 38 26 26 35  35
          21 21 33 27  27 27 19 19 35 35  38 38 26 26 35  35  17 17
15     16 16 21 21 33 27  27 27 19 19 35 35  38 38 26 26 35  35  17 17
       16 16 21 21 33 27  27 27 19 19 35 35  38 38 26 26 35  35  17 17
       16 16 21 21 33 27  27 27 19 19 35 35  38 38 26 26 35  35  17 17
       16 16 21 21 33 27  27 27 19 19 35 35  38 38 26 26 35  35  17 17
       16 16 21 21 33 27  27 27 19 19 35 35  38 38 26 26 35  35  17 17 11
10  10 10 16 16 21 21 33 27  27 27 19 19 35 35  38 38 26 26 35  35  17 17 11
    10 10 16 16 21 21 33 27  27 27 19 19 35 35  38 38 26 26 35  35  17 17 11
    10 10 16 16 21 21 33 27  27 27 19 19 35 35  38 38 26 26 35  35  17 17 11
    10 10 16 16 21 21 33 27  27 27 19 19 35 35  38 38 26 26 35  35  17 17 11
    10 10 16 16 21 21 33 27  27 27 19 19 35 35  38 38 26 26 35  35  17 17 11
5   10 10 16 16 21 21 33 27  27 27 19 19 35 35  38 38 26 26 35  35  17 17 11
  4 4 10 10 16 16 21 21 33 27  27 27 19 19 35 35  38 38 26 26 35  35  17 17 11
  4 4 10 10 16 16 21 21 33 27  27 27 19 19 35 35  38 38 26 26 35  35  17 17 11 3
  4 4 10 10 16 16 21 21 33 27  27 27 19 19 35 35  38 38 26 26 35  35  17 17 11 3
1 4 4 10 10 16 16 21 21 33 27  27 27 19 19 35 35  38 38 26 26 35  35  17 17 11 3
```

Figure 8-14 Irregular distribution of labor.

Calculations to Verify Schedules and Cost Distributions

Previous sections of this chapter presented the principles of planning, techniques of developing schedules, and methods of cost-loading CPM schedules to obtain the distribution of costs. Planning identifies the

Activity Number	Duration in Days	Number of Crews	Workers in Crew	1 2 3 4 5	6 7 8 9 10	11 12 13 14 15	16 17 18 19 20	21 22 23 24 25
A	4	1	4	4 4 4 4				
B	7	2	3	6 6 6	6 6 6 6			
C	9	1	2	2	2 2 2 2 2	2 2 2		
D	7	2	4	8	8 8 8 8 8	8		
E	3	1	5		5 5 5			
F	9	3	4		12	12 12 12 12 12	12 12 12	
G	4	1	5		5	5 5 5		
H	8	3	6			18 18 18 18	18 18 18 18	
I	11	2	4			8 8	8 8 8 8 8	8 8 8 8
J	4	3	2				6	6 6 6
K	6	1	3				3	3 3 3 3 3
Totals				4 4 10 10 16	16 21 21 21 27	27 37 37 38 38	38 38 38 26 17	17 17 17 11 3

```
Number of Workers
     per Day
                                                38 38  38 38 38
                                             37 37 38 38  38 38 38
                                             37 37 38 38  38 38 38
     35                                      37 37 38 38  38 38 38
                                             37 37 38 38  38 38 38
                                             37 37 38 38  38 38 38
                                             37 37 38 38  38 38 38
                                             37 37 38 38  38 38 38
     30                                      37 37 38 38  38 38 38
                                             37 37 38 38  38 38 38
                                             37 37 38 38  38 38 38
                                          27 27 37 37 38 38  38 38 38
                                          27 27 37 37 38 38  38 38 38 26
     25                                   27 27 37 37 38 38  38 38 38 26
                                          27 27 37 37 38 38  38 38 38 26
                                          27 27 37 37 38 38  38 38 38 26
                                          27 27 37 37 38 38  38 38 38 26
                                    21 21 21 27 27 37 37 38 38  38 38 38 26
     20                             21 21 33 27 27 37 37 38 38  38 38 38 26
                                    21 21 33 27 27 37 37 38 38  38 38 38 26
                                    21 21 33 27 27 37 37 38 38  38 38 38 26
                                    21 21 33 27 27 37 37 38 38  38 38 38 26 17 17 17
                              16 16 21 21 33 27 27 37 37 38 38  38 38 38 26 17 17 17
     15                       16 16 21 21 33 27 27 37 37 38 38  38 38 38 26 17 17 17
                              16 16 21 21 33 27 27 37 37 38 38  38 38 38 26 17 17 17
                              16 16 21 21 33 27 27 37 37 38 38  38 38 38 26 17 17 17
                              16 16 21 21 33 27 27 37 37 38 38  38 38 38 26 17 17 17
                              16 16 21 21 33 27 27 37 37 38 38  38 38 38 26 17 17 17 11
     10                 10 10 16 16 21 21 33 27 27 37 37 38 38  38 38 38 26 17 17 17 11
                       10 10 16 16 21 21 33 27 27 37 37 38 38  38 38 38 26 17 17 17 11
                       10 10 16 16 21 21 33 27 27 37 37 38 38  38 38 38 26 17 17 17 11
                       10 10 16 16 21 21 33 27 27 37 37 38 38  38 38 38 26 17 17 17 11
                       10 10 16 16 21 21 33 27 27 37 37 38 38  38 38 38 26 17 17 17 11
      5                10 10 16 16 21 21 33 27 27 37 37 38 38  38 38 38 26 17 17 17 11
                 4  4 10 10 16 16 21 21 33 27 27 37 37 38 38  38 38 38 26 17 17 17 11
                 4  4 10 10 16 16 21 21 33 27 27 37 37 38 38  38 38 38 26 17 17 17 11 3
                 4  4 10 10 16 16 21 21 33 27 27 37 37 38 38  38 38 38 26 17 17 17 11 3
      1          4  4 10 10 16 16 21 21 33 27 27 37 37 38 38  38 38 38 26 17 17 17 11 3
```

Figure 8-15 Distribution of labor after starting activity F one day later and Activity H two days earlier.

activities and sequencing of activities that must be performed to complete the project. Scheduling establishes the start and finish times of activities based on durations and interdependency of activities. The distribution of costs over the life of a project is obtained by assigning costs to activities from the cost estimate of the project.

The project schedule is crucial to managing a project because it has all the key elements in a project: what needs to be done, when it will be done, and how much it will cost. Thus, a good quality schedule and cost estimate is essential to managing projects. The schedule is the centerpiece of a project work plan.

Computers are used to perform many calculations in a CPM schedule. Computers are good for performing calculations for early/late start and finish times of activities, total and free float times, and cost distributions on an early, late, and target basis. However, computers cannot identify activities and cannot determine the logical sequencing of activities. Only people can identify and sequence work activities.

Before entering data into a computer scheduling software package, a careful review of the list of activities should be performed to ensure work is sufficiently defined to complete the project. A logic diagram should be created and reviewed to ensure activities are sequenced in a realistic order to guide field operations. There should also be a careful check of the duration that is assigned to each activity to ensure the work can be accomplished in a reasonable time. These items can only be accomplished by people, not the computer.

It is easy for people to become complacent and depend too much on the computer. Although the computer can perform calculations in seconds, it is highly likely that errors are created in the process of entering data in the computer. Therefore, it is necessary to check the output from the computer to ensure a quality schedule. Example 8-1 shows hand calculations that are easily performed to check and verify the results of computer analysis of a cost-loaded CPM diagram. For example, it is easy to perform a forward pass on the CPM diagram to calculate the early start and finish of an activity as a check to verify the correct output from the computer.

Example 8-1 A project team has identified the following activities required to complete a project. The duration of each activity is based on the quantity of work, available resources, and productivity of crews. The cost of each activity is based on quantity of work and historical cost data from previously completed projects.

Activity	Duration	Cost	Preceded by	Followed by
A	2 days	$500	None	B, C, and D
B	3 days	$900	A	E
C	4 days	$1,600	A	F
D	5 days	$500	A	G
E	7 days	$1,400	B	H
F	7 days	$1,500	C	I and L
G	8 days	$2,400	D	J and K
H	4 days	$800	E	L
I	2 days	$1,000	F	N

(Continued)

Activity	Duration	Cost	Preceded by	Followed by
J	12 days	$3,600	G	M and O
K	5 days	$2,000	G	P
L	6 days	$1,200	F and H	Q
M	2 days	$900	J	N
N	2 days	$700	I and M	S
O	6 days	$1,800	J	R and T
P	4 days	$1,200	K	T
Q	4 days	$2,000	L	U
R	4 days	$1,600	O	S
S	2 days	$1,400	N and R	V
T	9 days	$1,800	O and P	V
U	2 days	$1,200	Q	V
V	3 days	$300	S, T, and U	None

From the preceding information a CPM precedence diagram can be developed to calculate the schedule for the project. The critical path is shown as double arrows between critical activities.

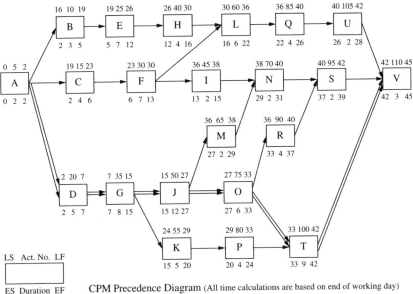

CPM Precedence Diagram (All time calculations are based on end of working day)

From the calculations shown on the CPM diagram the following table is prepared which shows the early/late starts and finishes as well as the total and free float for each activity. All time calculations are based on end of working day. For example, Activity A starts at the end of day 0, which is the beginning of day 1.

The early start is the earliest time an activity can be started. It is governed by the largest early finish of the immediately preceding activity. The early finish is the earliest time an activity can be finished and is equal to the early start plus the duration of the activity. The late start is the latest time an activity can be started without delaying the completion date of the project.

It is calculated as the late finish minus the duration of the activity. The late finish is the latest time an activity can be finished without delaying the completion date of the project. It is governed by the smallest late start of the immediately following activity.

The total float is the amount of time an activity may be delayed without delaying the completion date of the project. It is calculated as the late finish minus the early finish, or as the late start minus the early start of the activity. The free float is the amount of time an activity may be delayed without delaying the early start time of the immediately following activity. The free float of an activity is calculated as the early start of the following activity minus the early finish of the activity. Critical activities are those activities that have a total float and free float equal to zero. The critical path activities are noted as CP in the right hand column of the following table.

No.	Activity	Preceded by	Followed by	D	Total Cost	ES	EF	LS	LF	TF	FF	
5	A	None	B,C,D	2	$500	0	2	0	2	0	0	CP
10	B	A	E	3	$900	2	5	16	19	14	0	
15	C	A	F	4	$1,600	2	6	19	23	17	0	
20	D	A	G	5	$500	2	7	2	7	0	0	CP
25	E	B	H	7	$1,400	5	12	19	26	14	0	
30	F	C	I,L	7	$1,500	6	13	23	30	17	0	
35	G	D	J,K	8	$2,400	7	15	7	15	0	0	CP
40	H	E	L	4	$800	12	16	26	30	14	0	
45	I	F	N	2	$1,000	13	15	36	38	23	14	
50	J	G	M,O	12	$3,600	15	27	15	27	0	0	CP
55	K	G	P	5	$2,000	15	20	24	29	9	0	
60	L	F,H	Q	6	$1,200	16	22	30	36	14	0	
65	M	J	N	2	$900	27	29	36	38	9	0	
70	N	I,M	S	2	$700	29	31	38	40	9	6	
75	O	J	R,T	6	$1,800	27	33	27	33	0	0	CP
80	P	K	T	4	$1,200	20	24	29	33	9	9	
85	Q	L	U	4	$2,000	22	26	36	40	14	0	
90	R	O	S	4	$1,600	33	37	36	40	3	0	
95	S	N,R	V	2	$1,400	37	39	40	42	3	3	
100	T	O,P	V	9	$1,800	33	42	33	42	0	0	CP
105	U	Q	V	2	$1,200	26	28	40	42	14	14	
110	V	S,T,U	None	3	$300	42	45	42	45	0	0	CP

(Note: All time calculations are based on end of working day)

Management is always interested in both time and cost. The cost of each activity is distributed over the duration of the activity. However, some activities may begin on their early start date, whereas other activities may begin on their late start date. Likewise, some activities may finish on their early finish date, whereas other activities may finish on their late finish dates. Therefore, costs may be analyzed on an early start basis, or a late start basis. Following is a cost analysis of the project on an early start and late start basis. The S-curve for the project shows the cost/time relationship with percent costs and percent times.

Early Start Cost Analysis   (Note: Activity F calculated daily cost of $214.29 is rounded to $214 on days 7 through 13. The rounding is adjusted in the early start cumulative cost on Day 13.)

Day	Early start %-time	Early start basis activities in progress	Early start cost/day	Early start cumulative cost	Early start %-cost
1	2.22%	A@$250	$250	$250	0.83%
2	4.44%	"	"	$500	1.65%
3	6.67%	B@$300 + C@$400 + D@$100	$800	$1,300	4.29%
4	8.89%	" " "	"	$2,100	6.93%
5	11.11%	" " "	"	$2,900	9.57%
6	13.33%	C@$400 + D@$100 + E@$200	$700	$3,600	11.88%
7	15.56%	D@$100 + E@$200 + F@$214	$514	$4,114	13.58%
8	17.18%	E@$200 + F@$214 + G@$300	$714	$4,828	15.93%
9	20.00%	" " "	"	$5,542	18.29%
10	22.22%	" " "	"	$6,256	20.65%
11	24.44%	" " "	"	$6,970	23.00%
12	26.67%	" " "	"	$7,684	25.36%
13	28.89%	F@$214 + G@$300 + H@$200	$714	$8,400	27.72%
14	31.11%	G@$300 + H@$200 + I@$500	$1,000	$9,400	31.02%
15	33.33%	" " "	"	$10,400	34.32%
16	35.56%	H@$200 + J@$300 + K@$400	$900	$11,300	37.29%
17	37.38%	J@$300 + K@$400 + L@$200	$900	$12,200	40.26%
18	40.00%	" " "	"	$13,100	43.23%
19	42.22%	" " "	"	$14,000	46.20%
20	44.44%	" " "	"	$14,900	49.18%
21	46.67%	J@$300 + L@$200 + P@$300	$800	$15,700	51.82%
22	48.89%	" " "	"	$16,500	54.46%
23	51.11%	J@$300 + P@$300 + Q@$500	$1,100	$17,600	58.09%
24	53.33%	" " "	"	$18,700	61.72%
25	55.56%	J@$300 + Q@$500	$800	$19,500	64.36%
26	57.78%	" "	"	$20,300	67.00%
27	60.00%	J@$300 + U@$600	$900	$21,200	69.97%
28	62.22%	M@$450 + O@$300 + U@$600	$1,350	$22,550	74.42%
29	64.44%	M@$450 + O@$300	$750	$23,300	76.90%
30	66.67%	N@$350 + O@$300	$650	$23,900	79.04%
31	68.89%	" "	"	$24,600	81.19%
32	71.11%	O@$300	$300	$24,900	82.18%
33	73.33%	"	"	$25,200	83.17%
34	75.56%	R@$400 + T@$200	$600	$25,800	85.15%
35	77.78%	" "	"	$24,600	87.13%
36	80.00%	" "	"	$27,000	89.11%
37	82.22%	" "	"	$27,600	91.09%
38	84.44%	S@$700 + T@$200	$900	$28,500	94.06%
39	86.67%	" "	"	$29,400	97.69%
40	88.89%	T@$200	$200	$29,600	98.35%
41	91.11%	"	"	$29,800	98.35%
42	93.33%	"	"	$30,000	99.01%
43	95.56%	V@$100	$100	$30,100	99.34%
44	97.78%	"	"	$30,200	99.67%
45	100.00%	"	"	$30,300	100.00%

Late Start Cost Analysis   (Note: Activity F calculated daily cost of $214.29 is rounded to $214 on days 24 through 30. The rounding is adjusted in the late start cumulative cost on Day 30.)

Day	Late start %-time	Late start basis activities in progress	Late start cost/day	Late start cumulative cost	Late start %-cost
1	2.22%	A@$250	$250	$250	0.83%
2	4.44%	"	"	$500	1.65%
3	6.67%	D@$100	$100	$600	1.98%
4	8.89%	"	"	$700	2.31%
5	11.11%	"	"	$800	2.64%
6	13.33%	"	"	$900	2.97%
7	15.56%	"	"	$1,000	3.30%
8	17.18%	G@$300	$300	$1,300	4.29%
9	20.00%	"	"	$1,600	5.28%
10	22.22%	"	"	$1,900	6.27%
11	24.44%	"	"	$2,200	7.26%
12	26.67%	"	"	$2,500	8.25%
13	28.89%	"	"	$2,800	9.24%
14	31.11%	"	"	$3,100	10.23%
15	33.33%	"	"	$3,400	11.22%
16	35.56%	J@$300	$300	$3,700	12.21%
17	37.78%	B@$300 + J@$300	$600	$4,300	14.19%
18	40.00%	" "	"	$4,900	16.17%
19	42.22%	" "	"	$5,500	18.15%
20	44.44%	C@$400 + E@$200 + J@$300	$900	$6,400	21.12%
21	46.67%	" " "	"	$7,300	24.09%
22	48.89%	" " "	"	$8,200	27.06%
23	51.11%	" " "	"	$9,100	30.03%
24	53.33%	E@$200 + F@$214 + J@$300	$714	$9,814	32.39%
25	55.56%	E@$200 + F@$214 + J@$300 + K@$400	$1,114	$10,928	36.07%
26	57.78%	" " " "	"	$12,042	39.74%
27	60.00%	F@$214 + H@$200 + J@$300 + K@$400	$1,114	$13,156	43.42%
28	62.22%	F@$214 + H@$200 + K@$400 + O@$300	$1,114	$14,270	47.10%
29	64.44%	" " " "	"	$15,384	50.77%
30	66.67%	F@$214 + H@$200 + O@$300 + P@$300	$1,014	$16,400	54.13%
31	68.89%	L@$200 + O@$300 + P@$300	$800	$17,200	56.77%
32	71.11%	" " "	"	$18,000	59.41%
33	73.33%	" " "	"	$18,800	62.05%
34	75.56%	L@$200 + T@$200	$400	$19,200	63.37%
35	77.78%	" "	"	$19,600	64.69%
36	80.00%	" "	"	$20,000	66.01%
37	82.22%	I@$500 + M@$450 + Q@$500 + R@$400 + T@$200	$2,050	$22,050	72.77%
38	84.44%	" " " " "	"	$24,100	79.54%
39	86.67%	N@$350 + Q@$500 + R@$400 + T@$200	$1,450	$25,550	84.32%
40	88.89%	" " " "	"	$27,000	89.11%
41	91.11%	S@$700 + T@$200 + U@$600	$1,500	$28,500	94.06%
42	93.33%	" " "	"	$30,000	99.01%
43	95.56%	V@$100	$100	$30,100	99.34%
44	97.78%	"		$30,200	99.67%
45	100.00%	"	"	$30,300	100.00%

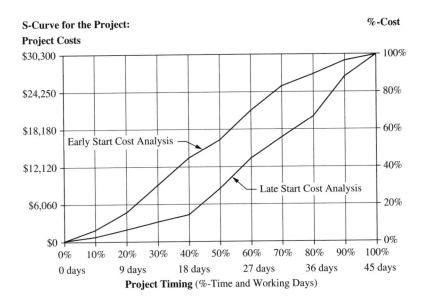

S-Curve for the Project: %-Cost
Project Costs

Early Start Cost Analysis

Late Start Cost Analysis

Project Timing (%-Time and Working Days)

Program Evaluation and Review Technique (PERT)

In the Critical Path Method of scheduling projects, the duration of each
activity is usually defined with a reasonable degree of certainty. For
most projects the type and amount of work is known, which enables the
project manager to establish the approximate duration for each work
activity. For example, the time to produce drawings may be 4 weeks, the
time to perform a soil investigation to gather and test soil samples may
be 2 weeks, or the time to erect forms for concrete may be 3 days. Using
the CPM, the assignment of one duration to each activity provides a
deterministic process for the start and finish dates of each activity and
a single finish date for the entire project.

For some projects it may be difficult to estimate a reasonable single
duration for one or more of the activities in the project schedule. There
may be a range of durations that may apply to a particular activity,
which makes it difficult to select just one duration to assign to the activity. The Program Evaluation and Review Technique (PERT) method of
scheduling uses three durations for each activity and the fundamental
statistics to determine the probability of a project finishing earlier or
later than expected. Although the PERT method is not used extensively
in engineering and construction projects, it provides valuable information for assessing the risks of a schedule slippage of a project.

The PERT method uses an arrow network diagram to show the logical
sequence of activities in a project, whereas the CPM uses a precedence
diagram as discussed in preceding sections of this book. In a PERT
diagram, activities are represented by arrows with circles at each end

of the arrow. The circles are called events that represent an instant in time. The circle at the beginning of the activity represents the start of an activity, and the circle at the end of the arrow represents the finish of the activity.

The major difference between the PERT method and CPM is the estimation of durations of activities. PERT is applicable for projects where there is a high degree of uncertainty about how long any given activity will take to complete, where even the most experienced manager can give only an educated guess of the estimated time, and that guess is subject to a wide margin of error. Using PERT there are three durations that are assigned to each activity:

$$a = \text{optimistic time}$$
$$b = \text{pessimistic time}$$
$$m = \text{most likely time}$$

The optimistic time is the shortest possible time in which the activity could possibly be completed, assuming that everything goes well. There is only a very small chance of completing the activity in less than this time. The pessimistic time is the longest time the activity could ever require, assuming that everything goes poorly. There is only a very small chance of expecting this activity to exceed this time. The most likely time is the time the activity could be accomplished if it could be repeated many times under exactly the same conditions. It is the time that it would take more often than any other time. The most likely time is the time the manager would probably give if asked for a single time estimate. It is important to note that the optimistic time and the pessimistic time may not deviate the same amount from the most likely time. In simple mathematical terms, a and b may not be symmetrical about m.

PERT uses a weighted average of the three times to find the overall project duration. This average is called the expected time, t_e, and is found by the following simple equation:

$$t_e = \frac{a + 4m + b}{6}$$

where t_e is the expected time of an activity.

The above equation is used to calculate a single duration for each activity in the PERT network diagram. Then a forward pass can be performed to calculate the early start and early finish for each activity in the PERT network and a total project duration, similar to the CPM scheduling method. A backward pass can be performed to calculate the late start and late finish for each activity. Total float, free float, and the critical path can be identified. At this stage of a PERT analysis it is identical to the CPM analysis.

The three time estimates of PERT can be used to measure the degree of uncertainty involved in the activity. The measure of the spread of the distribution is called the *standard deviation* and is denoted by the Greek letter s. To determine the probability of the project completing earlier or later than expected using PERT, the variance of each activity along the critical path must be calculated. The variance of an activity is the square of the standard deviation of the activity and can be calculated using the following equation:

$$v = \sigma^2 = \left(\frac{(b-a)}{6}\right)^2$$

where v is the variance of an activity.

Since the duration (t_e) of each of the activities in the arrow network diagram is uncertain, the time of occurrence of each event is also subject to uncertainty. The expected time of an event is denoted as T_E. Although the distribution of uncertainty of individual activities may not necessarily be symmetrical, the distribution for event times is assumed symmetrical because there are numerous activities in the chain ahead of the event. For example, the T_E for the final event in a project has all the critical path activities in the chain that lead up to the final event in the project.

The measure of uncertainty of the final event in a PERT diagram is the standard deviation of the expected time, denoted as σ_{TE}. The σ_{TE} is the square root of the sum of the activities ahead of the event. Thus, the σ_{TE} for the last event in a PERT diagram is the square root of the sum of the variance of all activities along the critical path. The σ_{TE} is calculated with the following equation:

$$\sigma_{TE} = \sqrt{v_{1-2} + v_{2-3} + v_{3-4} + \cdots + v_{i-j}}$$

where σ_{TE} is the standard deviation of the expected time.

The final calculation necessary to perform PERT calculations is the deviation, which is denoted by the symbol z in the following equation:

$$z = \frac{T_S - T_E}{\sigma_{TE}}$$

where z is the deviation.

Table 8-6 provides the value of z for various probabilities of certainty. The term T_E is the expected time of the event, which is calculated from the PERT diagram, and T_S is the scheduled time.

TABLE 8-6   PERT Probability Table

Value of deviation z	Probability of completion
−3.0	0.00
−2.5	0.01
−2.0	0.03
−1.5	0.07
−1.4	0.08
−1.3	0.09
−1.2	0.11
−1.1	0.14
−1.0	0.16
−0.9	0.18
−0.8	0.21
−0.7	0.24
−0.6	0.27
−0.5	0.31
−0.4	0.35
−0.3	0.38
−0.2	0.42
−0.1	0.46
−0.0	0.50
0.1	0.54
0.2	0.58
0.3	0.62
0.4	0.66
0.5	0.69
0.6	0.73
0.7	0.76
0.8	0.79
0.9	0.82
1.0	0.84
1.1	0.86
1.2	0.88
1.3	0.90
1.4	0.92
1.5	0.93
2.0	0.98
2.5	0.99
3.0	1.00

Example 8-2 The following PERT diagram shows the optimistic time (a), the most likely time (t_m), and the pessimistic time (b) at the bottom of each activity. The number shown in parentheses below each activity is the expected time (t_e) for the activity, calculated from the equation, $t_e = (a + 4m + b)/6$. All times denote working days in the project.

The optimistic duration of the project can be calculated by performing a forward pass to calculate the start and finish times using the optimistic duration (smallest duration) of each activity. For the data in this PERT diagram, the optimistic project duration is 48 days, which follows through events 1, 2, 4, 5, 6, 9, 12, 13, 14, and 15.

The pessimistic duration of the project can be calculated by performing a forward pass to calculate the start and finish times using the pessimistic

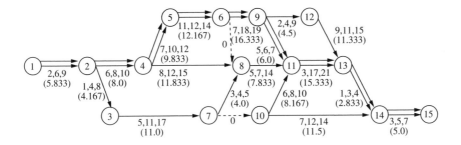

(largest duration) of each activity. For the data in this PERT diagram, the pessimistic project duration is 103 days, which follows through events 1, 2, 4, 5, 6, 9, 11, 13, 14, and 15.

The project duration (T_E) for event 15 can be calculated by performing a forward pass to calculate the start and finish times using the expected duration (t_e) of each activity. For the data in this PERT diagram, the expected time of the last event number 15 is $T_E = 81.3$ days, which follows through events 1, 2, 4, 5, 6, 9, 11, 13, 14, and 15.

In summary, for this PERT diagram the optimistic project duration is 48 days, the pessimistic project duration is 103 days, and the expected project duration is 81.3 days. The probability of completing the project ahead or behind day 81.3 will depend on the σ_{TE} of event 15. The value of σ_{TE} can be calculated as the square root of the sum of the variances of all activities along the critical path. Based on the expected time of each activity, the critical path for this PERT network follows events 1, 2, 4, 5, 6, 9, 11, 13, 14, and 15. The σ_{TE} for the final event, number 15, is the square root of the variances of these events, which can be calculated as follows:

$$\sigma_{TE} = \sqrt{v_{1-2} + v_{2-4} + v_{4-5} + v_{5-6} + v_{6-9} + v_{9-11} + v_{11-13} + v_{13-14} + v_{14-15}}$$

$$= \sqrt{\left(\frac{9-2}{6}\right)^2 + \left(\frac{10-6}{6}\right)^2 + \left(\frac{12-7}{6}\right)^2 + \left(\frac{14-11}{6}\right)^2 + \left(\frac{19-7}{6}\right)^2 + \left(\frac{7-5}{6}\right)^2 + \left(\frac{21-3}{6}\right)^2 + \left(\frac{4-1}{6}\right)^2 + \left(\frac{7-3}{6}\right)^2}$$

$$= \sqrt{1.3611 + 0.4444 + 0.6944 + 0.2500 + 4.0000 + 0.1111 + 9.0000 + 0.2500 + 0.4444}$$

$$= \sqrt{16.5554}$$

$$= 4.07$$

All the calculations necessary to perform an analysis of the project finishing ahead or behind the expected project duration of 81.3 days can be determined from the deviation equation. For example, it may be desired to determine the probability of the project finishing later than expected, that is, later that 81.3 days. The probability of completing the project by the 84th day can be calculated as follows:

$$z = \frac{T_S - T_E}{\sigma_{TE}}$$

$$= \frac{84.0 - 81.3}{4.07}$$

$$= 0.663$$

From Table 8-6, based on a deviation z of 0.663, the probability is 0.75. Thus, there is a 75% probability that the project will be completed by the 84th day.

Other probability scenarios can be calculated. For example, the probability of the project finishing earlier than expected, such as in 78 days, can be calculated as follows:

$$z = \frac{T_S - T_E}{\sigma_{TE}}$$

$$= \frac{78.0 - 81.3}{4.07}$$

$$= -0.811$$

From Table 8-6, based on a deviation z of −0.811, the probability is 0.21. Thus there is only a 21% probability that the project will be completed by the 78th day.

It may be desired to calculate the number of days to complete the project based on a 70% probability. This scenario can be calculated as follows. For a 70% probability the value of z from Table 8-6 is 0.5333:

$$z = \frac{T_S - T_E}{\sigma_{TE}}$$

$$0.533 = \frac{T_S - 81.3}{4.07}$$

$$T_S = 83.5 \text{ days}$$

Thus, there is a 70% probability that the project will be completed by 83.5 days.

The following is a graphical plot of the probability versus the project duration. This curve provides the full range of probabilities for assessing the risks of finishing the project ahead of or behind the expected duration of 81.3 days.

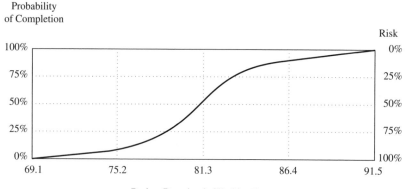

Project Duration in Working Days

This analysis is based on a σ_{TE} for event 15 based on the critical path activities. A full analysis is more complex if all near-critical paths are analyzed.

For example, it is possible to have another path through the PERT diagram that is near critical, which would yield a shorter T_E for event 15 and a higher value for σ_{TE} due to larger variances in the activities in the non-critical path.

Successor/Predecessor Relationships

The CPM diagram is a graphical representation that shows the sequencing of activities in the project. The purpose of the CPM diagram is to model the logical flow of work, showing each activity and one or more activities that follow each activity. In the activity relationships, the preceding activity is called the predecessor activity and the following activity is called the successor activity. In a pure CPM network diagram a finish-to-start (F-S) relationship of each activity is assumed. That is, the preceding activity must be completed before starting the following activity. All the CPM diagrams and calculations presented earlier in this book are based on an F-S relationship of activities to depict a pure CPM logic network diagram.

Most commercially available computer software programs for CPM provide activity relationships other than F-S. These additional activity relationships include start-to-start (S-S), start-to-finish (S-F), and finish-to-finish (F-F). The S-S relationship means the successor activity can start at the same time or later than the predecessor activity. The F-F relationship means the successor activity can finish at the same time as or later than the predecessor activity. The S-F relationship means the predecessor activity must start before the successor activity can finish. There is no practical application of the S-F relationship in the engineering and construction industry. It is presented here just to show all possible successor/predecessor relationships.

Delays in the relationship of one activity to another is defined by lag. *Lag* is the amount of time that an activity follows or is delayed from the start or finish of its predecessor. For example, an F-S relationship with a 2-day lag means the start of the successor activity cannot occur until 2 days after the finish of the predecessor activity. Lag can be assigned to any activity relationship, including F-S, F-F, S-S, and S-F. In addition, the lag can also be a negative number. Lead is the opposite of lag, which is the amount of time that an activity precedes the start or finish of its successor. Thus, the use of successor/predecessor relationships provides many options for modeling the network logic to show constraints between activities and the sequential flow of work in the project.

A model of the F-S with lag relationship is illustrated in Figure 8-16a. An alternative of this model is shown in Figure 8-16b, which shows the same results without using the F-S with lag relationship. Thus, adding one intermediate activity between the successor and predecessor activities clearly produces the same results as the F-S relationship.

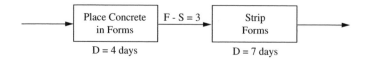

a. Model with Successor/Predecessor with F-S = 3 days of lag

b. Model without Successor/Predecessor Relationships

Figure 8-16 Comparison of CPM diagrams with and without a finish-to-start (F-S) successor/predecessor relationship.

The S-S relationship is used to overlap activities, allowing the successor activity to start before the predecessor activity is completed. To tie the finish of the predecessor activity to the successor activity, the F-F relationship is normally used together with the S-S relationship. Figure 8-17*a* illustrates S-S and F-F with lag relationships between

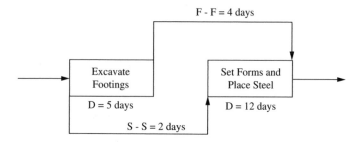

a. Model with Successor/Predecessor S - S = 2 days and F - F = 4 days of lag

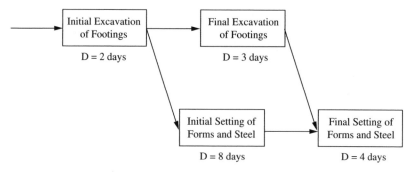

b. Model without Successor/Predecessor Relationships

Figure 8-17 Comparison of CPM diagrams with and without a start-to-start (S-S) successor/predecessor relationship.

two activities. Figure 8-17b is a model of the activities without the S-S and F-F relationships. Thus, for any CPM diagram the S-S and F-F activity relationships can be eliminated by adding additional activities. Although additional activities may appear as a disadvantage, the additional activities provide a clearer understanding of the sequence of work.

As previously stated, there is no practical application of the S-F relationship in the engineering and construction industry. Some computer scheduling software packages do not allow the use of the S-F relationship.

Problems Using Successor/Predecessor Relationships

Successor/predecessor relationships between activities allow overlapping of concurrent activities, thereby reducing the number of activities in CPM precedence diagrams. However, misunderstandings, confusion, and serious problems have been created by assigning successor/predecessor relationships to activities in the project schedule. Too often a good CPM schedule at the beginning of the project becomes a schedule that is incoherent and unmanageable due to successor/predecessor relationships that are assigned to activities, particularly during schedule updates.

In a pure CPM diagram, successor/predecessor activity relationships are not used. During the forward pass, the path that yields the largest early finish of the preceding activities becomes the early start of the following activities. During the backward pass, the path that yields the smallest late start of the following activities is the late finish of the preceding activities. Thus, in a pure CPM diagram there are no questions regarding the start and finish times of activities.

When successor/predecessor activity relationships are introduced into a CPM diagram, the start, finish, and float times may be governed by the successor and predecessor relationships, which makes the calculations more complex. Using successor/predecessor activity relationships with positive or negative lag provides numerous opportunities to model the network logic, but it can also create numerous opportunities for creating errors, confusion, and misunderstandings in the project schedule.

Figure 8-18 illustrates problems that can occur in calculating the start and finish dates when modeling multiple successor/predecessor activity relationships. Each successor/predecessor path must be evaluated to determine start and finish times. A forward pass for the activities in Figure 8-18 shows Activity B can start as early as 2 days or as late as 7 days after the start of Activity A, depending on the path chosen. Also, depending on the successor/predecessor path chosen, the early finish for Activity B is 9, 12, 14, or 19 days. Similarly, a backward pass will result in multiple late starts and late finishes for these activities. All these multiple calculations can cause confusion and a lack of confidence in the project schedule.

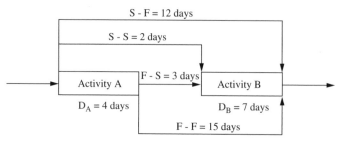

Figure 8-18 Example of multiple successor/predecessor activity relationships.

Figure 8-19 is a CPM diagram with successor/predecessor relationships of construction work related to foundation construction work. Figure 8-20 is a pure CPM diagram, without successor/predecessor activity relationships, for the same construction work. These figures are presented to illustrate the complexity of successor/predecessor calculations. As shown in Figure 8-19, the early start of 10 days for Activity 30 is calculated by subtracting its duration from the latest early finish time. The largest early finish of Activity 30 is 16 days, which is governed by the F-F relationship between Activities 20 and 30.

When multiple S-S and F-F relationships are used to relate the predecessor activity to its successor activities, the calculations of total and free float are very complex and require a close examination of the relationships. Using multiple relationships, it is possible to show activities that are critical, when they are really not critical.

A review of Figure 8-19 indicates that all the activities are critical, although that is not the case. Figure 8-20 is a pure CPM diagram for the same work shown in Figure 8-19. Figure 8-20 eliminates the S-S and F-F relationships by splitting activities to show the actual sequencing of work. As shown in Figure 8-20 much of the work is actually non-critical, compared to that of Figure 8-19. For example, the network of Figure 8-20 shows

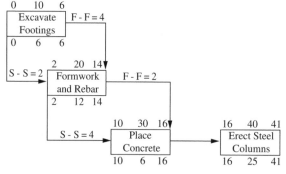

Figure 8-19 Successor/predecessor activity relationships. (note all activities appear critical.)

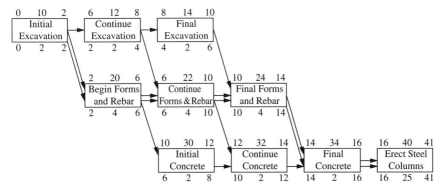

Figure 8-20 Pure CPM diagram without successor/predecessor activity relationships. (Double arrows denote path of critical activities: 10, 20, 22, 24, 34, and 40.)

all formwork as critical, but most of the excavation work and placement of concrete is non-critical. A pure CPM diagram as illustrated in Figure 8-20 provides an accurate description of the work without raising questions regarding start/finish times, total/free float, or critical path activities. Thus, a pure CPM diagram is recommended, rather than modeling with successor/predecessor activity relationships.

Unfortunately, there is no industry standard to define how to finalize the early/late starts and finishes using successor/predecessors relationships with lags. Each computer software developer defines the algorithm used by his or her software to determine start and finish dates using successor/predecessor activity relationships. Since different software packages handle the calculations differently, it is possible to get different results in the schedule, depending on the software used, which creates problems in understanding and interpreting project schedules. Therefore, take extreme caution when using successor/predecessor activity relationships because serious problems can arise. To prevent these problems, a pure CPM diagram can be developed without successor/predecessor relationships by simply adding additional activities. Although additional activities may appear as a disadvantage, adding additional activities provides a clear understanding of the sequence of work, thus preventing confusion and misunderstandings of the project schedule.

The following examples are presented to illustrate calculations based on each of the successor/predecessor activity relationships, F-S, S-S, F-F, and S-F. These examples show how each path must be carefully investigated to determine the start, finish, and float times. When the calculations are performed by computer software, it is possible to get project schedules that sometimes do not make logical sense. Thus, it is important to understand the calculations and be able to check the results of computer outputs to ensure the results are presented as intended by the person who prepared the schedule as well as by the persons who will be interpreting and using the schedule.

Example 8-3: Finish-to-Start Activity Relationships Calculate the start, finish, and float for each activity in the following CPM precedence network diagram. Activity C and E have an F-S relationship with a 5-day negative lag. The 5-day negative lag is equivalent to a 5-day lead, which means the finish of Activity C must occur no later than 5 days after the start of Activity E, given by the relationship: $EF_C \leq ES_E + 5$. The use of negative lags is not recommended because they lead to much confusion with respect to the meaning of the relationship and the schedule as discussed in the following paragraphs. The following diagram and accompanying table show the early and late start, early and late finish, and the total and free float for each activity.

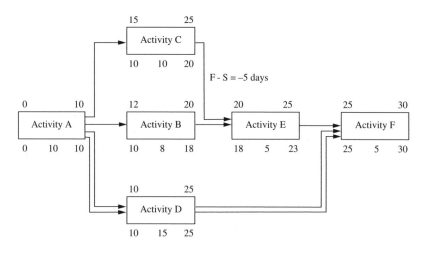

LS		LF
	Activity	
ES	D	EF

For Activity C, the calculations for total and free float are:

$$TF_C = LF_C - EF_C$$
$$= 25 - 20$$
$$= 5 \text{ days}$$

$$FF_C = ES_E - EF_C - Lag$$
$$= 18 - 20 - (-5)$$
$$= 3 \text{ days}$$

Activity	Duration	Early start	Early finish	Late start	Late finish	Total float	Free float
A*	10	0	10	0	10	0	0
B	8	10	18	12	20	2	0
C	10	10	20	15	25	5	3
D*	15	10	25	10	25	0	0
E	5	18	23	20	25	2	2
F*	5	25	30	25	30	0	0

*Denotes critical activities.

Based on the forward and backward pass calculations, the critical path is through Activities A, D, and F. Note the impact on the schedule and possible confusion due to the F-S relationship with 5 days of negative lag between Activity C and E. The EFC is 20 days (governed by the equation $EF_C = ES_C + D = 10 + 10 = 20$). The EFC is not governed by the equation $EF_C \leq$

$ES_E + 5 = 18 + 5 = 23$ days, because the ES_E is governed by 18 from the EF_B. Also, a review of the diagram shows the preceding activity $EF_C = 20$, which is greater than the following activity $ES_E = 18$, which is illogical. A computer schedule that shows the ES of a following activity to be earlier than the EF of a preceding activity will make no sense to workers in the field and will cause them to lose confidence in the schedule.

Similar confusion exists for the free float of Activity C. The FF_C is governed by the equation $FF_C = ES_E - EF_C - Lag = 18 - 20 - (-5) = 3$ days. However, a review of the diagram shows the $EF_C = 20$, which is greater than the $ES_E = 18$. Thus, the use of F-S relationships with negative lag is not recommended.

Example 8-4: Finish-to-Finish Activity Relationships Calculate the start, finish, and float for each activity in the following CPM precedence network diagram. Activity C and E have an F-F relationship with a lag of 7 days. The following diagram and accompanying table show the early and late start, early and late finish, and the total and free float for each activity.

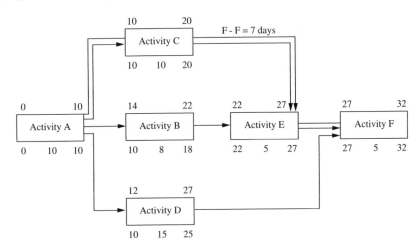

For Activity C, the calculations for total and free float are:

$$TF_C = LF_C - EF_C$$
$$= 20 - 20$$
$$= 0 \text{ days}$$

$$FF_C = LF_E - LF_C - Lag$$
$$= 27 - 20 - 7$$
$$= 0 \text{ days}$$

Activity	Duration	Early start	Early finish	Late start	Late finish	Total float	Free float
A*	10	0	10	0	10	0	0
B	8	10	18	14	22	4	4
C*	10	10	20	10	20	0	0
D	15	10	25	12	27	2	2
E*	5	22	27	22	27	0	0
F*	5	27	32	27	32	0	0

*Denotes critical activities.

Based on the forward and backward pass calculations, the critical path is through Activities A, C, E, and F. Note that the free float of Activity C is governed by the F-F relationship.

Example 8-5: Start-to-Start Activity Relationships Calculate the start, finish, and float for each activity in the following CPM precedence network diagram. Activity C and E have an S-S relationship with a lag of 10 days. The following diagram and accompanying table show the early and late start, early and late finish, and the total and free float for each activity.

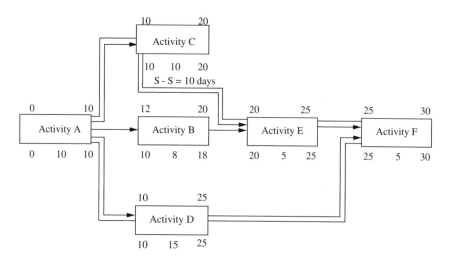

For Activity C, the calculations for total and free float are:

$$TF_C = LF_C - EF_C$$
$$= 20 - 20$$
$$= 0 \text{ days}$$

$$FF_C = ES_E - ES_C - Lag$$
$$= 20 - 10 - 10$$
$$= 0 \text{ days}$$

Activity	Duration	Early start	Early finish	Late start	Late finish	Total float	Free float
A*	10	0	10	0	10	0	0
B	8	10	18	12	20	2	2
C*	10	10	20	10	20	0	0
D*	15	10	25	10	25	0	0
E*	5	20	25	20	25	0	0
F*	5	25	30	25	30	0	0

*Denotes critical activities.

Based on the forward and backward pass calculations, the critical path is through Activities A, C, D, E, and F. Note the free float of Activity C is governed by the S-S relationship between Activity C and E.

Example 8-6: Start-to-Finish Activity Relationships   Although the start-to-finish relationship is seldom used, this example is presented to illustrate the calculations for an S-F relationship with a lag of 7 days between Activity C and E. The following diagram and accompanying table show the early and late start, early and late finish, and the total and free float for each activity.

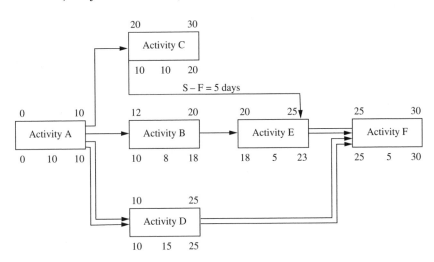

LS ⬚ LF
⬚ Activity ⬚
ES   D   EF

For Activity C, the calculations for total and free float are:

$$TF_C = LF_C - EF_C$$
$$= 30 - 20$$
$$= 10 \text{ days}$$

$$FF_C = EF_E - ES_C - Lag$$
$$= 23 - 10 - 5$$
$$= 8 \text{ days}$$

Activity	Duration	Early start	Early finish	Late start	Late finish	Total float	Free float
A*	10	0	10	0	10	0	0
B	8	10	18	12	20	2	0
C	10	10	20	20	30	10	8
D*	15	10	25	10	25	0	0
E	5	18	23	20	25	2	2
F*	5	25	30	25	30	0	0

*Denotes critical activities.

Based on the forward and backward pass calculations, the critical path is through Activities A, D, and F.

References

1. Antill, J. M. and Woodhead, R. W., *Critical Path Methods in Construction Practice,* 3rd ed., Wiley, New York, NY, 1982.
2. Armstrong-Writhe, A. T., *Critical Path Method,* Longman Group Ltd., London, 1969.

3. Burman, P. J., *Precedence Networks for Project Planning and Control*, McGraw-Hill, Inc., London, 1972.
4. Callahan, M. T., Quackenbush, D. G., and Rowings, J. E., *Construction Project Scheduling*, McGraw-Hill, Inc., New York, NY, 1992.
5. Duffy, G. A., Oberlender, G. D., and Jeong, H. S., "Linear Scheduling Model with Varying Production Rates," *Journal of Construction Engineering and Management*, ASCE, Reston, VA, Vol. 137, No. 8, August 2011.
6. Harris, R. B., *Precedence and Arrow Networking Techniques for Construction*, Wiley, New York, NY, 1978.
7. Hinze, J. W., *Construction Planning and Scheduling*, Prentice-Hall, Inc., 4th ed. Upper Saddle River, NJ, 2011.
8. Moder, J. J., Phillips, C. J., and Davis, E. D., *Project Management with CPM, PERT, and Precedence Diagraming*, 3rd ed., Blitz Publising Company, Middleton, WI, 1995.
9. Paulson, B. C., Jr., "Man-Computer Concepts for Planning and Scheduling," *Journal of Construction Division*, ASCE, Reston, VA, Vol. 100, No. 3, September, 1974.
10. Plotnick, F. L., and O'Brien, J. J., *CPM in Construction Management, 7th ed.*, McGraw-Hill, New York, NY, 2010.
11. Popescu, C. M., and Charoenngam, C., *Project Planning, Scheduling, and Control in Construction: An Encyclopedia of Terms and Applications*, John Wiley & Sons, New York, NY, 1995.
12. Priluck, H. M. and Hourihan, P. R., *Practical CPM for Construction*, Robert S. Means, Co., Duxbury, MA, 1968.
13. Spencer, G. R. and Rahbar, F. F., *Automation of the Scheduling Analysis Process*, Transactions of the American Association of Cost Engineers, San Diego, CA, October, 1989.
14. Stevens, J. D., *Techniques for Construction Network Scheduling*, McGraw-Hill, Inc., New York, NY, 1990.
15. Wiest, J. D. and Levy, F. K., *A Management Guide to PERT/CPM*, Prentice-Hall, Englewood Cliffs, NJ, 1969.
16. Willis, E. M., *Scheduling Construction Projects*, Wiley, New York, NY, 1986.

9

Tracking Work

Control Systems

Effective project management requires planning, measuring, evaluating, forecasting, and controlling all aspects of a project: quality and quantity of work, costs, and schedules. An all encompassing project plan must be defined before starting a project; otherwise there is no basis for control. Project tracking cannot be accomplished without a well-defined work plan, budget, and schedule as discussed in the previous chapters of this book.

The project plan must be developed with input from people who will be performing the work, and it must be communicated to all participants. The tasks, costs, and schedules of the project plan establish the benchmarks and check-points that are necessary for comparing actual accomplishments to planned accomplishments, so the progress of a project can be measured, evaluated, and controlled.

At the end of any reporting period (N), a project is expected to have achieved an amount of work (X) with a level of quality (Q) at a predicted cost (C). The objective of project control is to measure the actual values of these variables and determine if the project is meeting the targets of the work plan, and to make any necessary adjustments to meet project objectives. Project control is difficult because it involves a quantitative and qualitative evaluation of a project that is in a continuous state of change.

To be effective, a project control system must be simple to administer and easily understood by all participants in a project. Control systems tend to fall into two categories; they are either so complex that no one can interpret the results that are obtained, or they are too limited because they apply to only costs or schedules rather than integrating costs, schedules, and work accomplished. A control system must be developed so information can be routinely collected, verified, evaluated, and communicated to all participants in a project; so it will serve as a tool for project improvement rather than reporting flaws that irritate people.

Since the introduction of small personal computers in the early 1980s, the automation of the concept of an integrated project control system has become widely discussed. Many papers have been written that describe different, but similar, approaches to integrated project control systems. Common among the approaches is development of a well-defined work breakdown structure (WBS) as a starting point in the system. The smallest unit in the WBS is a work package, which defines the work in sufficient detail so it can be measured, budgeted, scheduled, and controlled.

The Critical Path Method (CPM) is used to develop the overall project schedule from the WBS by integrating and sequencing the work in accordance with the work packages. A coding system is designed that identifies each component of the WBS so information from the WBS can be related to the project control system. To control costs the WBS is linked to the cost breakdown structure (CBS) by the code of accounts. Likewise, the WBS is linked to the organizational breakdown structure (OBS) to coordinate personnel to keep the project on schedule. A coding system allows sorting of information to produce a variety of reports that are subsets of the entire project.

This general concept of project control was presented by the Department of Energy for federal and energy projects. Since that time, several modifications have been suggested to simplify the process of transferring information from the WBS to the CPM, linking the WBS and OBS to the coding system, and measurement of work accomplished.

Linking the WBS and CPM

The work packages of the WBS provide the information necessary to develop a CPM logic network diagram. With a well-defined detailed WBS a single work package often becomes one activity on the diagram. However, sometimes it is necessary to combine several work packages into a single activity or to develop a single work package into several activities. The process of developing the CPM diagram requires good judgement with extensive involvement of key participants in the project. Although the level of detail should be kept to a minimum, all activities that may influence the project completion date must be included in the diagram.

A CPM diagram for project scheduling and control can be classified as one of three types: design, construction, or engineering /procurement/construction (EPC). For each, the WBS defines the project framework for planning, scheduling, and control of the work. The level of detail of the CPM diagram depends upon the completeness of the WBS.

The products of design are production drawings and specifications. As discussed in Chapter 8, a bar chart is frequently preferred for scheduling individual design activities. However, for effective scheduling control of the whole project a composite of the individual bar charts must be integrated into a CPM diagram that shows the interrelationship and sequencing of related work. Thus, a CPM diagram for design is often a summary level schedule. CII Publication 6-1, *Project Control for Engineering,* provides a good description of current industry practice of scheduling and control of design.

CPM logic diagrams have been successfully used for scheduling and control of construction for many years. A detailed WBS can be developed as a cooperative effort of the estimating, project control, and field operations management personnel. The estimate must be prepared so that costs, durations, and resources can be assigned to work packages in the WBS. The work packages then become activities on the CPM diagram. The purchase and delivery of long lead-time material must also be included in the diagram. In addition, the work performed by subcontractors on the job must be integrated with other work to form a complete integrated CPM diagram. For large projects, a separate CPM diagram can be prepared for each individual area of the project; then a master CPM diagram can be developed that links the individual area diagrams.

The CPM diagram for an EPC project must interface the design work packages with the procurement and construction activities. It is often best to develop separate individual working schedules for design, procurement, and construction. Then link the individual schedules into a summary EPC schedule that integrates the total system. It is imperative to sequence all related activities that may influence the completion date of the project.

To illustrate the linking of the WBS to the CPM, the WBS shown in Figure 9-1 is used to develop the CPM diagram in Figure 9-2. This EPC project is an expansion of the design project presented in Chapter 8, to now include the procurement and construction activities. It is a service facility for maintenance operations and consists of site-work, on-site utilities, an employee's office building, and a maintenance building. To handle this project, the contracting strategy is to use in-house personnel to design the on-site utilities, site-work, and the maintenance building. A separate contract is assigned for design of the employee's building. The maintenance building is denoted as Building A and the office building as Building B on the WBS and CPM.

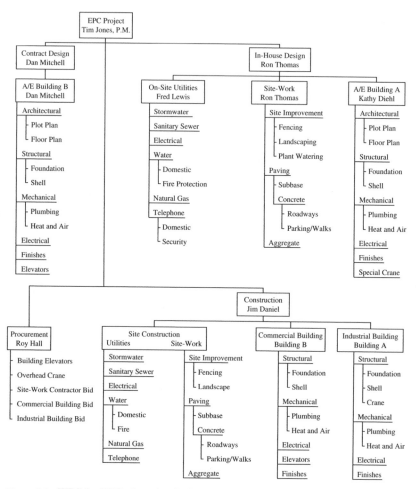

Figure 9-1 WBS for EPC of service facility project.

The contracting strategy for construction is to assign one contract to a heavy construction contractor to build all on-site utilities and site-work activities. Two building contractors will be used, one for the office building and the other for the industrial maintenance building. The construction activities are limited on the EPC schedule but will be expanded into more detail by each construction contractor as a part of his or her contractual requirements.

As shown in Figure 9-2, the appropriate design activities are directly linked to the procurement activities for materials and equipment. For example, the design of the overhead crane for Building A is followed by procurement and then construction. Likewise, design of the elevator for Building B is linked to the respective procurement and construction activities.

Coding System for Project Reports

A coding system can be developed that identifies each component of a project to allow the sorting of information in order to produce a variety of reports for project monitoring and control. A code number can be assigned to each work item that identifies a variety of information, such as phase of project, type of work, responsible person, or facility of which the work item is a part. Table 9-1 is an illustrative example of a simple 4-digit coding system for the project shown in Figure 9-2.

Table 9-1 can be used to assign a code number that is unique to each activity in the CPM diagram of Figure 9-2 to link the WBS to the OBS and CPM. For example, Activity 95 (Design of Foundations and Structure for Building A) is assigned the code number of 1735. This code identifies the activity as the in-house structural engineering for Building A involving structural design that is the responsibility of team member Kathy Diehl. Table 9-2 is a list of the code numbers assigned to each activity in the project.

Using the coding system of Table 9-1, a schedule report can be obtained for all structural work by selecting activities that have a 3 in the third digit of their code number. Similarly, all activities related to Building B can be obtained by selecting activities that have an 8 in the second digit of their code number.

Multiple sorting of code numbers enables a project manager to obtain various levels of reports for project control. This can be accomplished even with a simple 4-digit coding system as just described. For example, a report for all structural work for Buildings A and B can be obtained by sorting activity code numbers that have a code digit 3 equal to 3, and

TABLE 9-1   Coding System for the Project in Figure 9-2

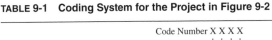

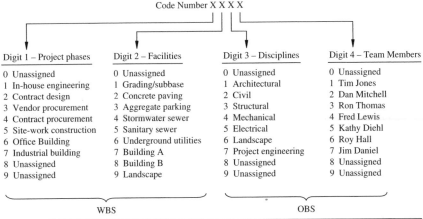

Digit 1 – Project phases	Digit 2 – Facilities	Digit 3 – Disciplines	Digit 4 – Team Members
0 Unassigned	0 Unassigned	0 Unassigned	0 Unassigned
1 In-house engineering	1 Grading/subbase	1 Architectural	1 Tim Jones
2 Contract design	2 Concrete paving	2 Civil	2 Dan Mitchell
3 Vendor procurement	3 Aggregate parking	3 Structural	3 Ron Thomas
4 Contract procurement	4 Stormwater sewer	4 Mechanical	4 Fred Lewis
5 Site-work construction	5 Sanitary sewer	5 Electrical	5 Kathy Diehl
6 Office Building	6 Underground utilities	6 Landscape	6 Roy Hall
7 Industrial building	7 Building A	7 Project engineering	7 Jim Daniel
8 Unassigned	8 Building B	8 Unassigned	8 Unassigned
9 Unassigned	9 Landscape	9 Unassigned	9 Unassigned

WBS   OBS

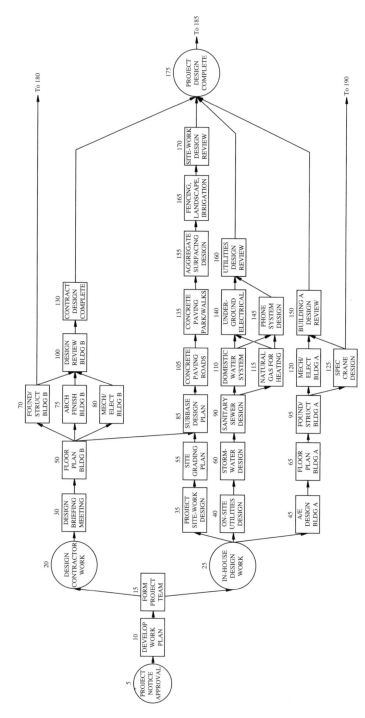

Figure 9-2 CPM diagram for EPC project. (Continued on next page.)

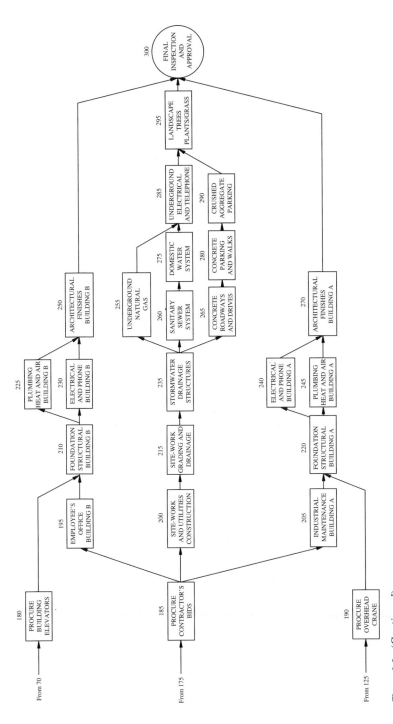

Figure 9-2 *(Continued)*

TABLE 9-2   EPC Project Activity List—Example EPC Maintenance Facility Project

No.	Code	Activity description	Duration	Cost	Team member
5	0071	Project notice approval	3	500.	Tim Jones
10	0071	Develop work plan	7	12000.	Tim Jones
15	0071	Form project team	5	850.	Tim Jones
20	2872	Design contractor's work	2	3000.	Dan Mitchell
25	1073	In-house design work	3	1500.	Ron Thomas
30	2872	Design briefing meeting	1	1200.	Dan Mitchell
35	1073	Project site-work design	1	1400.	Ron Thomas
40	1624	On-site utilities design	1	1200.	Fred Lewis
45	1715	A/E design building A	1	1500.	Kathy Diehl
50	2812	Floor plan building B	10	9900.	Dan Mitchell
55	1123	Site grading plan	12	14000.	Ron Thomas
60	1424	Stormwater design	10	2000.	Fred Lewis
65	1715	Floor plan building A	15	26000.	Kathy Diehl
70	2832	Found/struct building B	45	31200.	Dan Mitchell
75	2812	Arch finishes building B	30	49500.	Dan Mitchell
80	2842	Mech/elect building B	45	37300.	Dan Mitchell
85	1123	Subbase design plan	5	4000.	Ron Thomas
90	1524	Sanitary sewer design	10	12000.	Fred Lewis
95	1735	Found/struct building A	30	92700.	Kathy Diehl
100	2871	Design review building B	10	8000.	Tim Jones
105	1223	Concrete paving roads	20	12000.	Ron Thomas
110	1624	Domestic water system	7	9000.	Fred Lewis
115	1624	Natural gas system	8	6000.	Fred Lewis
120	1745	Mech/elect building A	30	22200.	Kathy Diehl
125	1735	Special overhead crane	11	10800.	Kathy Diehl
130	2872	Contract design complete	1	1000.	Dan Mitchell
135	1223	Concrete paving parking/walks	10	7000.	Ron Thomas
140	1654	Underground electrical	14	12000.	Fred Lewis
145	1654	Underground telephone system	4	3000.	Fred Lewis
150	1771	Building a design review	3	5000.	Tim Jones
155	1323	Aggregate surfacing design	8	6000.	Ron Thomas
160	1677	Utilities design review	1	1100.	Tim Jones
165	1963	Fencing/landscape/irrigation	14	28000.	Ron Thomas
170	1071	Site-work design review	5	7000.	Tim Jones
175	0071	Project design complete	1	1000.	Tim Jones
180	3876	Procure building elevators	25	95000.	Roy Hall
185	4076	Procure contractors' bids	20	7000.	Roy Hall
190	3776	Procure overhead crane	40	55000.	Roy Hall
195	6887	Employee's office building B	3	1000.	Jim Daniel
200	5087	Site-work/utilities construction	4	1500.	Jim Daniel
205	7787	Industrial/maintenance bldg A	2	1400.	Jim Daniel
210	6882	Found/struct building B	45	195000.	Dan Mitchell
215	5083	Site-work/grading/drainage	18	85000.	Ron Thomas
220	7785	Found/struct building A	110	390000.	Kathy Diehl
225	6882	Plumbing, heat, & air bldg B	75	285000.	Dan Mitchell
230	6882	Electrical/phone building B	60	215000.	Dan Mitchell
235	5484	Stormwater/drainage structures	15	22000.	Fred Lewis
240	7785	Electrical/ phone building A	65	167000.	Kathy Diehl

(Continued)

TABLE 9-2   EPC Project Activity List—Example EPC Maintenance Facility Project (Continued)

No.	Code	Activity description	Duration	Cost	Team member
245	7785	Plumbing, heat, & air bldg A	85	192000.	Kathy Diehl
250	6882	Arch finishes building B	50	260000.	Dan Mitchell
255	5684	Underground natural gas	5	10500.	Fred Lewis
260	5584	Sanitary sewer system	21	33200.	Fred Lewis
265	5283	Concrete paving roads & drives	60	185000.	Ron Thomas
270	7785	Arch finishes building A	30	175000.	Kathy Diehl
275	5684	Domestic water system	7	13200.	Fred Lewis
280	5283	Concrete parking & walkways	15	35000.	Ron Thomas
285	5684	Underground elect & phone	14	47000.	Fred Lewis
290	5383	Crushed aggregate parking	40	76000.	Ron Thomas
295	5983	Landscape trees/plants/grass	20	62000.	Ron Thomas
300	9977	Final inspection & approval	3	3500.	Tim Jones

code digit 2 greater than 6 and less than 9. Thus, many types of sorts can be obtained by selecting combinations of code digits that are greater than, equal to, or less than a number.

A coding system can also be designed using letters, rather than numbers. Using letters there are 26 characters that can be designated (one for each letter in the alphabet), for each location in the code, instead of 10 numbers that can be designated for each digit in a pure numeric system. However, for computer applications it is more difficult to sort letters on a greater than, equal to, and less than basis, than it is for a numeric coding system. A combination alpha-numeric coding system can be designed using numbers in locations of the code that require sorting multiple capabilities and letters in locations where sorting is not required.

Control Schedules for Time and Cost

The CPM network diagram shows the sequencing of activities that represent the work packages identified by the WBS. The expected time and cost that is required to perform each activity can be obtained from the work packages of the WBS in order to establish the parameters for control of cost and time. For each activity of the EPC project shown in Figure 9-2, Table 9-2 provides the cost and duration along with the team member who is responsible for the activity. These costs and durations are directly related to the WBS, and their expenditure is directly related to the work that is produced. To measure the progress of the project, the actual costs and durations are compared to these control costs and durations.

The total cost for the project includes the direct costs (from the previous paragraph) plus indirect costs, contingency reserve, and profit. A CBS for the project includes all these costs. However, only the direct costs that are tied to the WBS are used for project control purposes to manage the accomplishment of work. Indirect costs include support personnel, equipment, and supplies that are not directly chargeable to the project. The cost of insurance, bonds, general office overhead, etc., are also excluded from the project control system for monitoring costs and managing work because these items are fixed at the beginning of the project and they are independent of the work accomplished. The management of these costs is generally a function of the accounting department, because the project manager and his or her team usually do not have control of these costs. These costs are typically distributed over a specified period of time and will expand or contract with the schedule.

A good description of the relationship between engineering CBS and WBS is provided in CII Publication 6-1, *Project Control for Engineering*. Figure 9-3 is an example from the publication that illustrates a total engineering budget matrix. The CBS includes all elements in the budget matrix that have been given a dollar amount. The total dollar value of all the elements is the total engineering budget. The WBS for the project consists of budget items from the CBS for tasks that produce deliverables: design calculations, drawings, and specifications.

For the example of Figure 9-3, the functions that are chosen for work control are shaded in the matrix and are design and drawings, specifications, procurement support, and field support. The detailed WBS would be expanded from these budget items into areas, systems, and subsystems that define the total project. For example, the deliverable to be produced by the electrical would be a drawing list that included all drawings for electrical work. The number of work-hours (WH) for each drawing and the number of WH of calculations that are required to produce the drawings would represent the budget.

The schedule for the work is the total time to produce the final drawings, including the overlap of design calculations and design drafting. As discussed in Chapter 8 most engineers prefer a bar chart for scheduling individual design tasks. However, for project control the individual bar charts must be developed into activities on the CPM diagram for the total project schedule. The start and finish of each activity of the CPM engineering design schedule is a composite of all tasks of the work package. The following illustrates the evaluation of

Engineering Budget Matrix

ACTIVITY OR COST ELEMENT / FUNCTION	DESIGN AND DRAWINGS	SPECS	PROCUREMENT SUPPORT	FIELD SUPPORT	SUPERVISION AND CONTROL	TRAVEL	SUPPLIES AND SERVICES
MANAGEMENT					WH & $		
PROCUREMENT			WH & $	WH & $		$	$
DISCIPLINES — CIVIL	WH & $	WH & $	WH & $	WH & $	WH & $		
ELECTRICAL	WH & $	WH & $	WH & $	WH & $	WH & $		
ETC.	WH & $	WH & $	WH & $	WH & $	WH & $		

Figure 9-3 Engineering cost breakdown structure and work breakdown structure.
(*Source:* Construction Industry Institute, Publication No. 6-1.)

overlapping tasks of the work package to determine the duration of an activity on the CPM diagram.

Tasks of Work Package	Duration
Project Engineering	7 days
Electrical Engineering	19 days
CADD Operator	8 days
Total Budgeted Days =	34 days

Project Engineering
Electrical Engineering
CADD Operator

27 days duration
for CPM scheduling
control diagram

CII Publication 6-4, *Contractor Planning for Fixed-Price Construction,* provides a good description of the relationship of the CBS and WBS for construction. Figure 9-4 is an example from the publication that shows the WBS is the direct cost portion of the CBS. The WBS includes work that is budgeted, scheduled, and controlled. The estimate should be prepared in the same organizational format as the WBS. The quantity takeoff from the plans and specifications is used as the basis for

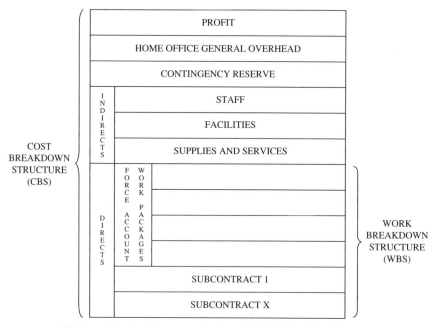

Figure 9-4   WBS vs. CBS. (*Source:* Construction Industry Institute, Publication No. 6-4.)

direct labor, material, and equipment costs. The cost estimate should also consider the method of construction and the sequencing of work for development of the project schedule.

The general superintendent who will be responsible for the project on the construction job-site must be involved in developing the detailed construction schedule. However, during the early stages of project development it is often necessary to develop a construction schedule before the construction contractor has been identified. For this type of situation, the initial CPM diagram for construction must be developed without excessive constraints in the sequencing of activities. The CPM schedule should show the sequencing of major areas of the project and identify the general flow of work. Then, prior to construction, a detailed CPM diagram can be developed by the construction personnel who will actually perform the work.

A construction contractor usually performs some work with direct-hire (force account) personnel and contracts portions of the work to one or more subcontractors. Since many subcontractors do not have an elaborate project control system, the assignment of work to a subcontractor should be a work package that has a scope of work, budget, and schedule that is defined in sufficient detail so the subcontractor's responsibilities and duties are clearly understood. The subcontract work package must be compatible with the WBS; otherwise there is no basis for control.

Milestone dates that are required for the start and completion of each subcontractor's work should be clearly defined, including any hold in work that may be necessary in order to schedule the work of other subcontractors. Each subcontractor is an independent company and not an employee of the general contractor. However, each subcontractor's work must be included in the total project schedule since the work of any one contractor usually affects the work of other contractors on the job, which can impact the completion date of the project.

The unsuccessful procurement of material is a common source of delays during construction. A procurement plan must be included in the project schedule to guide the purchase of contractor furnished materials. Although the contractor generally obtains most of the material as a part of his or her construction contract, many projects require the procurement of special material and equipment that is unique to the project. Also, the owner may procure equipment or bulk materials that will be installed by the construction contractor. The project schedule should identify and sequence all activities that can impact the delivery of special equipment and material.

The previous sections presented the list of activities, costs, durations, and coding system for the EPC project shown in Figure 9-2. The preparation of this data provides the base for a project monitoring and control system. For example, to evaluate the engineering design phase of the project an S-curve (see Figure 9-5) can be produced that shows the

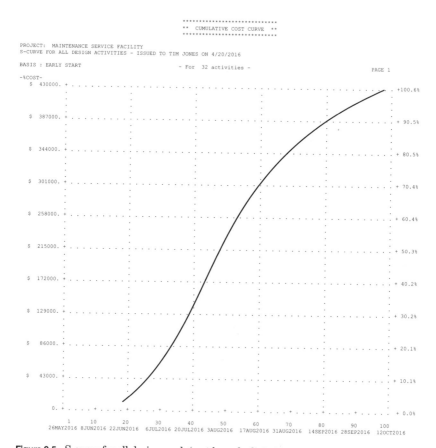

```
                            ********************************
                            **  CUMULATIVE COST CURVE  **
                            ********************************
PROJECT:  MAINTENANCE SERVICE FACILITY
S-CURVE FOR ALL DESIGN ACTIVITIES - ISSUED TO TIM JONES ON 4/20/2016

BASIS : EARLY START                      - For   32 activities -                              PAGE 1
-%COST-
    $  430000. + . . . . . . . . . . . . . . . . . . . . . . . . . . . . . . . . . . . . . . . .  . +100.6%

    $  387000. + . . . . . . . . . . . . . . . . . . . . . . . . . . . . . . . . . . . . . + 90.5%

    $  344000. + . . . . . . . . . . . . . . . . . . . . . . . . . . . . . . . . . + 80.5%

    $  301000. + . . . . . . . . . . . . . . . . . . . . . . . . . . . + 70.4%

    $  258000. + . . . . . . . . . . . . . . . . . . . . . . . . . + 60.4%

    $  215000. + . . . . . . . . . . . . . . . . . . . . . . . + 50.3%

    $  172000. + . . . . . . . . . . . . . . . . . . . . . + 40.2%

    $  129000. + . . . . . . . . . . . . . . . . . . . + 30.2%

    $   86000. + . . . . . . . . . . . . . . . . . + 20.1%

    $   43000. + . . . . . . . . . . . . . . . + 10.1%

        0. + . . . . + . . . . + . . . . + . . . . + . . . . + . . . . + . . . . + . . . . + . . . . + . . . . + + 0.0%
                1      10       20       30       40       50       60       70       80       90      100
             26MAY2016 8JUN2016 22JUN2016  6JUL2016 20JUL2016 3AUG2016  17AUG2016 31AUG2016  14SEP2016 28SEP2016  12OCT2016
```

Figure 9-5   S-curve for all design work (sort by code digit 1 greater than 0 and less than 3).

distribution of costs for all design work. This curve is obtained using the coding system of Table 9-1, by selecting all activities in which the first digit of their code is greater than zero and less than 3.

Additional reports can be computer generated as discussed in Chapter 8. For example, Figure 9-6 shows the graph of daily cost for all in-house engineering design. It can be obtained by selecting all activities that have a 1 in the first digit of their code number. A similar curve can be obtained for work-hours which would show the personnel requirements needed to perform the work.

Table 9-3 is a partial listing of the project control schedule for all activities in the project. Only the first and last portion of the total schedule is shown to illustrate the type of information that can be obtained for all the 60 activities in the total project. Critical activities are those activities that have zero float time and are denoted by the letter "C" at

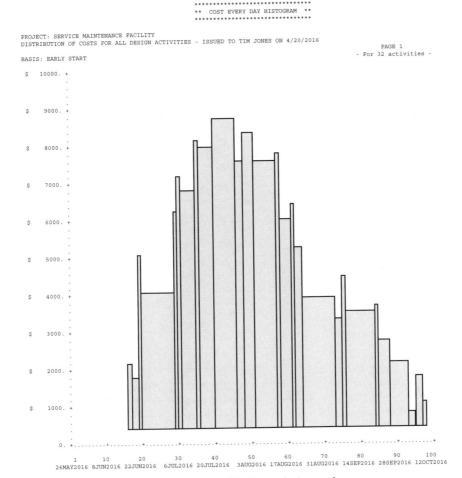

Figure 9-6 Distribution of daily costs for all in-house design work.

the left of the activity number. Major milestone events are also noted on the schedule. This schedule is a summary-level schedule report that integrates engineering, procurement, and construction.

Table 9-4 is a monthly cost schedule for the entire project. The distribution of costs are shown on an early start, late start, and target basis. The percentage-time and percentage-cost distribution are shown in the two rightmost columns. There is a non-linear distribution between costs and time. For example, at the end of the 6th month, which represents 33.0% of the time of completion, only 17.9% of the costs are anticipated to be expended. However, at the end of the 11th month, 64.1% of the time, 54.3% of the costs are expected to be expended.

TABLE 9-3   Schedule of All Activities

```
****************************
**   ACTIVITY SCHEDULE   **
****************************
```

PROJECT: SERVICE MAINTENANCE FACILITY
SCHEDULE FOR ALL ACTIVITIES - ISSUED TO TIM JONES ON 4/20/2016

** Page 1 **
ACTIVITY SCHEDULE

ACTIVITY NUMBER	ACTIVITY DESCRIPTION	DURA-TION	EARLY START	EARLY FINISH	LATE START	LATE FINISH	TOTAL FLOAT	FREE FLOAT
C 5	PROJECT NOTICE APPROVAL	3	26MAY2016 1	30MAY2016 3	26MAY2016 1	30MAY016 3	0	0
C 10	DEVELOP WORK PLAN	7	31MAY2016 4	8JUN2016 10	31MAY2016 4	8JUN2016 10	0	0
C 15	FORM PROJECT TEAM	5	9JUN2016 11	15JUN2016 15	9JUN2016 11	15JUN2016 15	0	0
20	DESIGN CONTRACTOR'S WORK	2	16JUN2016 16	17JUN2016 17	27JUN2016 23	28JUN2016 24	7	0
C 25	IN-HOUSE DESIGN WORK	3	16JUN2016 16	20APR2016 18	16JUN2016 16	20JUN2016 18	0	0
30	DESIGN BRIEFING MEETING	1	20JUN2016 18	20JUN2016 18	29JUN2016 25	29JUN2016 25	7	0
35	PROJECT SITE-WORK DESIGN	1	21JUN2016 19	21JUN2016 19	27JUN2016 23	27JUN2016 23	4	0
40	ON-SITE UTILITIES DESIGN	1	21JUN2016 19	21JUN2016 19	9AUG2016 54	9AUG2016 54	35	0
C 45	A/E DESIGN BUILDING A	1	21JUN2016 19	21JUN2016 19	21JUN2016 19	21JUN2016 19	0	0
50	FLOOR PLAN BUILDING B	10	21JUN2016 19	4JUL2016 28	30JUN2016 26	13JUL2016 35	7	3
55	SITE GRADING PLAN	12	22JUN2016 20	7JUL2016 31	28JUN2016 24	13JUL2016 35	4	0
60	STORM-WATER DESIGN	10	22JUN2016 20	5JUL2016 29	10AUG2016 55	23AUG2016 64	35	0
260	SANITARY SEWER SYSTEM	21	29DEC2016 156	26JAN2017 176	27JUN2017 284	25JUL2017 304	128	0
265	CONCRETE PAVING ROADS & DRIVES	60	29DEC2016 156	22MAR2017 215	16MAR2017 211	7JUN2017 270	55	0
225	PLUMBING, HEAT & AIR BLDG B	75	13JAN2017 167	27APR2017 241	30MAR2017 221	12JUL2017 295	54	0
230	ELECTRICAL/PHONE BUILDING B	60	13JAN2017 167	6APR2017 226	20APR2017 236	12JUL2017 295	69	15
275	DOMESTIC WATER SYSTEM	7	27JAN2017 177	6FEB2017 183	26JUL2017 305	3AUG2017 311	128	0
285	UNDERGROUND ELECT & PHONE	14	7FEB2017 184	24FEB2017 197	4AUG2017 312	23AUG2017 325	128	73
280	CONCRETE PARKING AND WALKWAYS	15	23MAR2017 216	12APR2017 230	8JUN2017 271	28JUN2017 285	55	0
240	ELECTRICAL/PHONE BUILDING A	65	13APR2017 231	12JUL2017 295	11MAY2017 251	9AUG2017 315	20	20
C 245	PLUMBING, HEAT & AIR BLDG A	85	13APR2017 231	9AUG2017 315	13APR2017 231	9AUG2017 315	0	0
290	CRUSHED AGGREGATE PARKING	40	13APR2017 231	7JUN2017 270	29JUN2017 286	23AUG2017 325	55	0
250	ARCH FINISHES BUILDING B	50	28APR2017 242	6JUL2017 291	13JUL2017 296	20SEP2017 345	54	54
295	LANDSCAPE TREES/PLANTS/GRASS	20	8JUN2017 271	5JUL2017 290	24AUG2017 326	20SEP2017 345	55	55
C 270	ARCH FINISHES BUILDING A	30	10AUG2017 316	20SEP2017 345	10AUG2017 316	20SEP2017 345	0	0
C 300	FINAL INSPECTION & APPROVAL	3	21SEP2017 346	25SEP2017 348	21SEP2017 346	25SEP2017 348	0	0

```
********************************   END OF SCHEDULE   ********************************
```

TABLE 9-4   Monthly Cost Distribution for All Activities in the Project

```
********************************
**  MONTHLY COST SCHEDULE  **
********************************
```

PROJECT: SERVICE MAINTENANCE FACILITY
SCHEDULE FOR ALL ACTIVITIES - ISSUED TO TIM JONES ON 4/20/2016
START DATE: 26 May 2016   FINISH DATE: 25 Sep 2017

MONTHLY COST SCHEDULE
- For all activities -

NO.	MONTH YEAR	EARLY START COST/MON	EARLY START CUMULATIVE COST	LATE START COST/MON	LATE START CUMULATIVE COST	TARGET SCHEDULE COST/MON	TARGET SCHEDULE CUMULATIVE COST	%TIME	%COST
1	MAY 2016	$ 3929.	$ 3929.	$ 3929.	$ 3929.	$ 3929.	$ 3929.	1.4%	.1%
2	JUN 2016	$ 48841.	$ 52770.	$ 34645.	$ 38573.	$ 41743.	$ 45672.	7.5%	1.5%
3	JUL 2016	$ 177261.	$ 230031.	$ 101204.	$ 139778.	$ 139233.	$ 184904.	14.1%	6.1%
4	AUG 2016	$ 130583.	$ 360614.	$ 130838.	$ 270616.	$ 130711.	$ 315615.	20.1%	10.3%
5	SEP 2016	$ 166931.	$ 527545.	$ 160525.	$ 431141.	$ 163728.	$ 479343.	26.4%	15.7%
6	OCT 2016	$ 69255.	$ 596800.	$ 63784.	$ 494925.	$ 66519.	$ 545863.	33.0%	17.9%
7	NOV 2016	$ 180187.	$ 776987.	$ 62507.	$ 557432.	$ 121347.	$ 667210.	38.8%	21.9%
8	DEC 2016	$ 247116.	$ 1024103.	$ 111945.	$ 669377.	$ 179531.	$ 846740.	45.4%	27.7%
9	JAN 2017	$ 324067.	$ 1348170.	$ 180933.	$ 850311.	$ 252500.	$ 1099240.	51.7%	36.0%
10	FEB 2017	$ 332900.	$ 1681070.	$ 235742.	$ 1086053.	$ 284321.	$ 1383561.	57.5%	45.3%
11	MAR 2017	$ 316279.	$ 1997348.	$ 234362.	$ 1320415.	$ 275320.	$ 1658882.	64.1%	54.3%
12	APR 2017	$ 242022.	$ 2239371.	$ 245967.	$ 1566382.	$ 243995.	$ 1902877.	70.4%	62.3%
13	MAY 2017	$ 250489.	$ 2489860.	$ 308343.	$ 1874725.	$ 279416.	$ 2182293.	76.4%	71.5%
14	JUN 2017	$ 284017.	$ 2773877.	$ 329589.	$ 2204314.	$ 306803.	$ 2489096.	82.8%	81.5%
15	JUL 2017	$ 89479.	$ 2863356.	$ 322710.	$ 2527024.	$ 206094.	$ 2695190.	89.1%	88.3%
16	AUG 2017	$ 110461.	$ 2973817.	$ 338893.	$ 2865917.	$ 224677.	$ 2919867.	95.4%	95.6%
17	SEP 2017	$ 79333.	$ 3053150.	$ 187233.	$ 3053150.	$ 133283.	$ 3053150.	100.0%	100.0%

```
*************** END OF MONTHLY COST SCHEDULE ***************
```

Relationships between Time and Work

Measurement of design work is difficult because design is a creative process that involves ideas, calculations, evaluation of alternatives, and other tasks that are not physically measurable quantities. Considerable time and cost can be expended in performing these tasks before end results such as drawings, specifications, reports, etc., which are measurable quantities of work, are ever seen.

The measurement of design is further complicated because of the diversity of work. For example, all the design calculations may be complete, half of the drawings may be produced, and yet only one-fourth of the specifications may be written. For this situation it is difficult to determine how much work has been accomplished because the work that is produced does not have a common unit of measure. Because of this, a percentage of completion is commonly used as a unit of measure of design work. The criteria for determining the percent complete for measuring work must be developed and confirmed in writing with the project team members prior to commencing design. This provides a common basis for the monthly evaluation of progress.

A weighting multiplier can be assigned to design tasks to define the magnitude of effort that is required to achieve each task. The sum of the weighting factors is 1.0, which represents 100% of the total design effort. The determination of each weight should be a joint effort between the project manager and the designer who is responsible for performing the work. This should be done before starting work.

A significant amount of overlap work is necessary during the design process. For example, initial drafting normally starts before the design calculations are finished. Likewise, final calculations are often not complete before the final production drawings are started. The project manager and his or her team members can jointly define the overlap of related work to show the timing of tasks throughout the duration of a project. Table 9-5 is an illustrative example that lists work items, weight multiplier, and estimated timing for each task required in a design effort. Further division of weight multipliers within each category may also be necessary.

The information in Table 9-5 is provided for illustrative purposes only. Because each project is unique, it is necessary to define the weight multipliers that are appropriate for each individual project. The timing of design work depends upon the availability of personnel. This information can be compiled from a summary of the individual design work packages.

To manage the overall design effort a work/time curve can be developed from the information in Table 9-5, see Figure 9-7. The upper portion of Figure 9-7 is a series of graphs that represent each design task individually, arranged in the order of occurrence. The slope of each graph is the ratio of the weight multiplier to the time required for the work to be performed. The lower portion of Figure 9-7 is the work/time curve for the entire design effort and is obtained by a composite, superposition of the individual graphs. This curve represents the planned accomplishment of work and serves as a basis of control for comparison of actual work accomplished. It can be superimposed onto the time/cost S-curve that was discussed in the preceding section. This forms the overall integrated cost/schedule/work curve that is discussed later in subsequent sections of this chapter.

TABLE 9-5   Illustrative Weight Multipliers For Design Work

Design work	Weight multiplier	Project timing
Review backup material	0.05	0% to 10%
Design calculations	0.10	10% to 25%
Initial drafting	0.25	15% to 45%
Final calculations	0.20	35% to 60%
Production drawings	0.30	50% to 90%
Drawing approval	0.10	90% to 100%
	1.00	

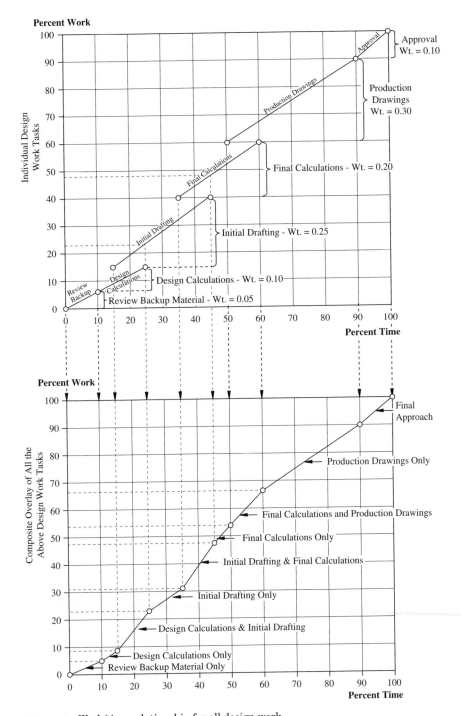

Figure 9-7 Work/time relationship for all design work.

TABLE 9-6   Illustrative Weight Multiplier for Construction

Facility	Weight multiplier	Project timing
Site-work	0.25	0% to 35%
Concrete building	0.40	15% to 75%
Metal building	0.35	65% to 100%
	1.00	

Construction involves many types of different work that have different units of measure such as cubic yards, square feet, pounds, or each. Thus, it is convenient to use percentage as a unit of measure for management and control of overall construction.

The procedure used for measurement of design can also be applied to construction. For example, a project may consist of three major facilities; site-work, a concrete office building, and a preengineered metal building. A weight multiplier can be applied to each of the three major facilities, along with their planned sequence of occurrence (reference Table 9-6).

Table 9-6 only lists three major parts of the project. A more accurate definition of the planned work can be achieved by dividing each of the major facilities into smaller components. For example, site-work can be divided into grading, drainage, paving, landscaping, etc. Likewise, each of the two buildings can be divided into smaller components. Regardless of the level of detail of the project, the sum of all the weight multipliers would be equal to 1.0, to represent 100% of the project. The weight and timing of each major facility is established by key participants of the project team, before starting construction, to serve as a basis of control during construction.

Figure 9-8 shows the composite of the planned accomplishment of work. The upper portion of Figure 9-8 is a graphical display of the overlap and sequence of each of the three major facilities. The lower portion is the integrated work/time curve for the entire project and is obtained by a composite superposition of the three individual graphs of each major facility.

The procedure is a top level summary of all facilities in the total project. This same procedure can be used for each facility, or parts of a facility, depending upon the complexity of the project and the level of control that is desired by the project manager.

At the lowest level it may be possible to use a unit of measure of work that can be physically measured at the job-site, rather than using a percentage. For example, "wire pulling" can be easily measured in linear feet or "concrete piers" can be measured in cubic yards. However, some precautions must be used when physical quantities are used to represent work in place of a percentage. To illustrate, the construction of concrete piers involves drilling, setting steel, and placement of concrete. For a project that has 18 piers, a progress report may show all piers

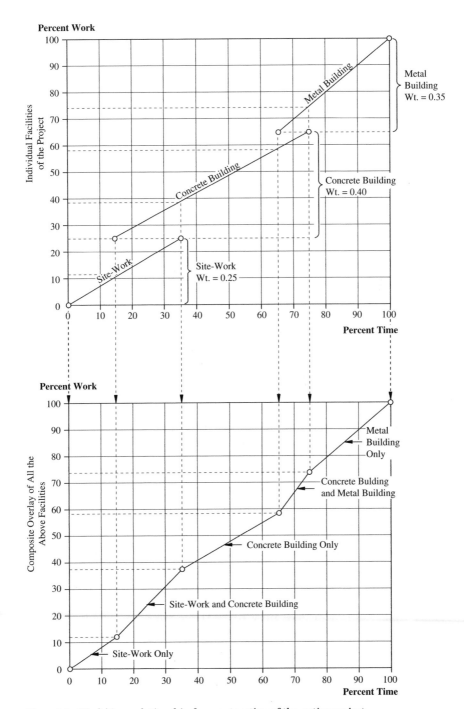

Figure 9-8 Work/time relationship for construction of the entire project.

drilled, steel set in 9 piers, and concrete placed in 3 piers. If cubic yards is the only measure of control, only 3 of the 18 piers would be reported as complete, which would not include the drilling and steel work that is accomplished. For this situation, a weight multiplier system must be developed to account for each task that is required to construct the piers. As previously discussed the sum of all weights must equal 1.0 to represent 100% of the work.

Integrated Cost/Schedule/Work

Experienced project managers are familiar with the problems of using only partial information, such as only costs or time to track the status of a project. To illustrate, half of a project budget may be expended by the midpoint of the scheduled duration, but only 20% of the work may be accomplished. A monitoring of only time and cost would indicate the project is going well; however, upon completion of the project there would likely be a cost overrun and a delay in schedule because the measurement of work was not included in the project control system. Thus, a project manager must develop an integrated cost/schedule/work system which provides meaningful feedback during the project rather than afterward. The status of a project can then be determined and corrective actions taken when corrections can be made at the least cost.

The preceding sections presented the relationships of cost/time and work/time for project control. However, evaluating these relationships separately does not provide an accurate status of a project. A cost/schedule/work graph can be prepared that shows the integrated relationship of the three basic components of a project: scope (work), budget (cost), and schedule (time). Figure 9-9 is a graph that links costs on the left-hand ordinate, time on the abscissa, and work on the right-hand ordinate. The upper curve is simply the cost/time S-curve that has been discussed in previous sections of this book. The lower work/time curve shows the relationship between work and time throughout the duration of the project. Thus, the graph is simply a composite overlay of the information previously presented.

The unit for costs is dollars and the unit for schedule is days, which are easily determined units of measure for any type of project. The unit of measure for work is represented as a percentage which provides a common base for all parts of the project. As previously discussed, a project may have three types of buildings: concrete, steel, and wood frame. An appropriate unit of measure for work would be cubic yards for the concrete building, pounds of steel for the steel building, and board feet of lumber for the wood frame building. However, since it is not possible to add cubic yards to pounds or board feet, percent is a dimensionless unit of measure that can conveniently be used to represent work.

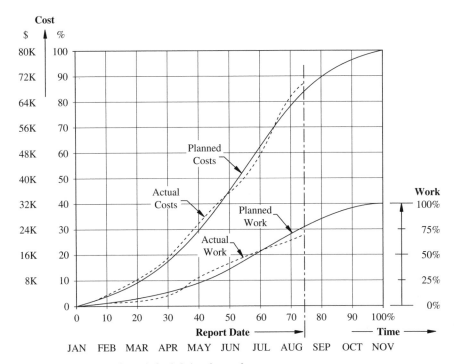

Figure 9-9 Integrated cost/schedule/work graph.

Although percentage provides a common unit of measure for work, a multiplier is necessary to define the distribution of work for each part of the project. For example, the project with three buildings of the previous paragraph may use a multiplier of 45% for the concrete structure, 35% for the steel structure, and 20% for the wood structure which when combined, represents 100% for the total project. Factors that may be considered for determining the multiplier include work-hours, costs, and/or the time to complete the work. The multiplier that is selected involves both a quantitative and qualitative evaluation of the project, based upon good judgement, and should be determined as a joint effort by key participants of the project team.

The actual cost and work accomplished can be superimposed onto the curves to compare with the planned cost and work in order to determine the status of the project; see Figure 9-10a. For this example the actual accomplished work curve is below the planned work curve, which shows a schedule slippage. For the same reporting period there is a cost over-run as noted on the upper curve of the graph. Thus, there is a schedule slippage and cost overrun for the reporting period. Other scenarios are possible as illustrated in Figure 9-10b to 9-10d.

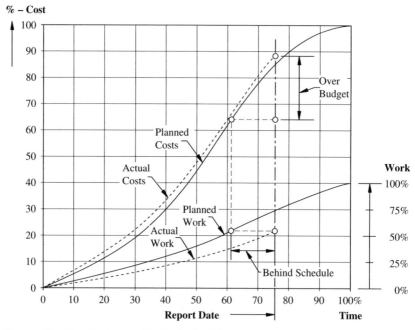

Figure 9-10a Over budget and behind schedule.

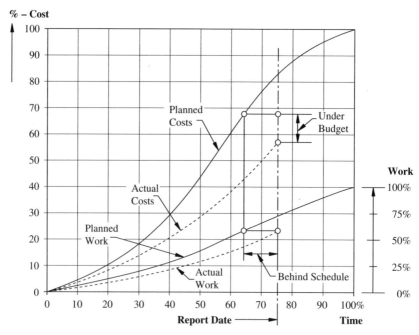

Figure 9-10b Under budget and behind schedule.

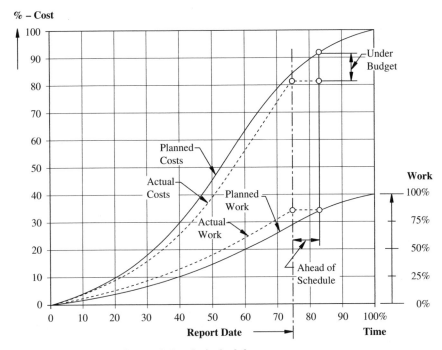

Figure 9-10c Under budget and ahead of schedule.

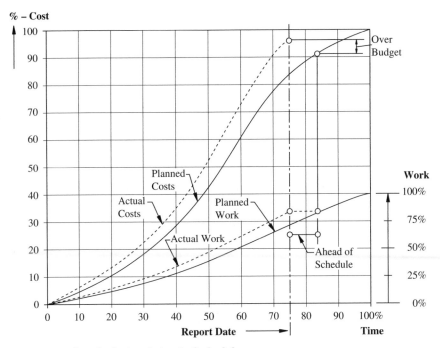

Figure 9-10d Over budget and ahead of schedule.

An integrated cost/schedule/work graph provides a good summary level report for the status of the overall project. Lower-level graphs, that is, the concrete building only, can be prepared to evaluate the status of a part of the total project. Thus, multiple graphs can be developed, depending upon the complexity of the project and the level of control desired by the project manager.

The integrated cost/schedule/work method is especially useful for tracking design work because design work is typically measured in percent complete. For example, a design review may be held at 25% design complete. For architectural design work the design may represent 40% of the design effort and 60% of the work may represent development of drawings, reference Tables 7-1 and 9-5. Also, the cost of design is sometimes measured in design-hours or CADD-hours.

Example 9-1   The following data represent the information that has been jointly compiled by the various disciplines of the design team for a project. These data represent the planned cost and work anticipated during the design phase. Prepare an integrated planned cost/time/work graph for the project.

Expected time	Anticipated cost	Planned work
0%	0%	0%
10%	5%	5%
20%	15%	10%
30%	20%	20%
40%	30%	30%
50%	45%	40%
60%	65%	50%
70%	80%	70%
80%	90%	80%
90%	95%	90%
100%	100%	100%

Evaluate the status report for the project for the following scenarios, assuming the status report is issued at 70% of the project duration. For each scenario, determine if the project is behind or ahead of schedule. Is there a cost overrun or underrun, and if so, by how much? A graphical evaluation of the four scenarios is shown below, followed by a summary table.

Scenario	Status report at 70% of the project duration
A	Reported cost to date is 55% and work accomplished is reported as 40%
B	Reported cost to date is 75% and work accomplished is reported as 50%
C	Reported cost to date is 95% and work accomplished is reported as 70%
D	Reported cost to date is 85% and work accomplished is reported as 80%

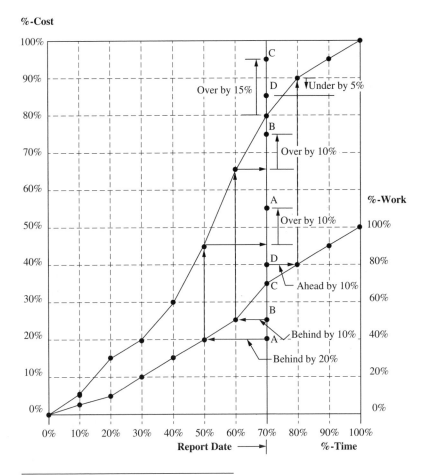

Scenario	Schedule	Cost
A	Behind by 20%	Over by 10%
B	Behind by 10%	Over by 10%
C	On Schedule	Over by 15%
D	Ahead by 10%	Under by 5%

Percent Complete Matrix Method

A very simple technique for determining the overall status of a project is the percent complete matrix method. It can be used for any size project and only requires a minimal amount of information that is readily available in the work packages. The overall status can be measured as a percent complete matrix, based upon the budget for each work package in a project. The budget can be measured as any one of three variables: cost, work-hours, or physical quantity of work. To describe this method, only cost is used in the following paragraphs; however, the method can be just as easily applied to work-hours or physical quantities of work. There may be times when a project manager may wish to use all three measures to determine the status of a project.

The percent complete matrix method requires only two input variables for each work package: estimated cost and percent complete. A spreadsheet can be developed on a small microcomputer that contains six pieces of information from the work package for a given area (see Figure 9-11). Formulas in the spreadsheet calculate the "percent unit" and "percent project" based on the "estimated cost" that is entered for each work package. Likewise, formulas in the spreadsheet calculate the "cost to date" and "percent complete project" based on entry of the second input variable, "percent complete."

Figure 9-12 is a spreadsheet that illustrates the percent complete matrix method for a project that consists of five buildings. Each building is defined by four summary level components: foundation/structural, mechanical/electrical, finishes/furnishings, and special equipment. The "total cost" and "percent of project" for each building are shown in the right-hand two columns of the spreadsheet. The total cost of all buildings in the project represents the total project budget, $240,000. Formulas in the spreadsheet calculate the "total cost" of each component in the project. For example, $81,000 represents the budget costs for all structural work.

The summary level components of each building contain the six pieces of information from the work package discussed in Figure 9-11. For example, the work packages for Building A structural is $15,000, which represents 30% of Building A and 6% of the total $240,000 project budget.

Work Package	
Estimated Cost*	Percent Unit
Percent Complete*	Percent Project
Cost to Date	% - Complete Project
Spreadsheet Formulas	
Input from Estimate*	Estimated Cost/ Total Unit Cost
Variable Input by User*	Estimated Cost/ Total Project Cost
Percent Complete × Estimated Cost	Cost to Date/ Total Project Cost

*Required Input Data by User

Figure 9-11 Work package information and spreadsheet formulas for %-complete matrix.

Figure 9-12 Percent complete matrix for project.

FACILITY	FOUNDATION/STRUCTURAL						MECHANICAL/ELECTRICAL						FINISHES/FURNISHINGS						SPECIAL/EQUIPMENT						TOTAL COST	PERCENT TOTAL
	ESTIMATED COST	PERCENT COMPLETE	COST TO DATE	PERCENT UNIT	PERCENT PROJECT	%-COMPLETE TOTAL	ESTIMATED COST	PERCENT COMPLETE	COST TO DATE	PERCENT UNIT	PERCENT PROJECT	%-COMPLETE TOTAL	ESTIMATED COST	PERCENT COMPLETE	COST TO DATE	PERCENT UNIT	PERCENT PROJECT	%-COMPLETE TOTAL	ESTIMATED COST	PERCENT COMPLETE	COST TO DATE	PERCENT UNIT	PERCENT PROJECT	%-COMPLETE TOTAL		
BUILDING A	$15000.	70.%	$10500.	30.%	6.%	4.%	$8000.	35.%	$2800.	16.%	3.%	1.%	$10000.	0.%	0.	20.%	4.%	0.%	$17000.	0.%	0.	34.%	7.%	0.%	$50000.	21.%
BUILDING B	$25000.	10.%	$2500.	28.%	10.%	1.%	$9000.	0.%	$0.	10.%	4.%	0.%	$23000.	0.%	0.	26.%	10.%	0.%	$33000.	0.%	0.	37.%	14.%	0.%	$90000.	38.%
BUILDING C	$8000.	100.%	$8000.	40.%	3.%	3.%	$3000.	80.%	$2400.	15.%	1.%	1.%	$4000.	0.%	0.	20.%	2.%	0.%	$5000.	0.%	0.	25.%	2.%	0.%	$20000.	8.%
BUILDING D	$2000.	100.%	$2000.	20.%	1.%	1.%	$1000.	100.%	$1000.	1.%	0.%	0.%	$	0.%	0.	0.%	0.%	0.%	$7000.	0.%	0.	70.%	3.%	0.%	$10000.	4.%
BUILDING E	$31000.	5.%	$1550.	44.%	13.%	1.%	$18000.	0.%	0.	26.%	8.%	0.%	$21000.	0.%	0.	30.%	9.%	0.%	$	0.%	0.	0.%	0.%	0.%	$70000.	29.%
SUM EST. & %-PROJ	$81000.			34.%			$39000.			16.%			$58000.			24.%			$62000.			26.%			$240000.	100.%
SUM COST TO DATE & SUM %-COMPLETE	$24550.			10.%			$6200.			3.%			0.			0.%			0.			0.%			$30750.	13.%

245

Upon entry in the spreadsheet of 70% complete for Building A, the formulas automatically calculate the $10,500 "cost to date" and the 4%, which represents the percent complete for the total project. The only entry required to obtain the project status is the cell of "%-complete," the entry highlighted by an underline in the spreadsheet illustrated in Figure 9-12.

As entries are made for each work package in the matrix, the total values for each component in the project are calculated at the bottom of the spreadsheet. For example, all the "cost to date" values in the vertical column under the heading of structural work sums to $24,550, which represents 10% of the total project. Similarly, formulas in the spreadsheet sum all the total cost values of each component in the bottom horizontal row to calculate the total project cost to date of $30,750, which represents 13% complete for the entire project. This percentage can be used for the evaluation of the integrated cost/time/work graphs that were discussed in the preceding section.

Progress Measurement of Design

The primary objectives of a progress measurement method during design are to (1) support project management in establishing a realistic plan for execution of the project, (2) provide the project manager and client with a consistent analysis of project performance, and (3) provide an early warning system to indentify deviations from the project plan and scope growth.

The drawing list and specification list are typical deliverables of the design effort. However, for some projects there may be additional deliverables. For example, deliverables for projects in the process industry typically include an equipment list and instrument list. Sometimes procurement of long lead-time equipment or procurement of special materials are included in the design contract. For this situation a list of procurement activities becomes a part of deliverables of the design effort.

The percentage completion of a single activity can usually be determined by one of four systems: Units Completed, Incremental Milestone, Start/Finish Percentages, or Ratio. The Units Completed method may be suitable for writing specifications, provided that each part of the specifications can be considered as having equal effort of work. The measurement of work can be determined as a percentage that is calculated by dividing the number of specifications completed by the total number of specifications that are to be produced.

The Incremental Milestone method is appropriate for measurement of production drawings or procurement activities that consist of easily recognized milestones. The following percentages are typical examples of measurement of production for development of production drawings and procurement.

Production drawings	
Start drafting	0%
Drawn, not checked	20%
Complete for office check	30%
To owner for approval	70%
First issue	95%
Final issue	100%
Procurement	
Bidders list developed	5%
Inquiry documents complete	10%
Bids analyzed	20%
Contract awarded	25%
Vendor drawings submitted	45%
Vendor drawings approved	50%
Equipment shipped	90%
Equipment received	100%

The Start/Finish Percentage method is applicable to those activities that lack readily definable intermediate milestones, or the effort and time required is difficult to estimate. For these tasks 20% to 50% is given when the activity is started and 100% when finished. A percentage is assigned for starting to account for the long time between the start and finish when no credit is being given. This method is appropriate for work such as planning, designing, model building, and studies. It can also be used for specification writing.

The Ratio method is applicable to tasks such as project management or project control that are involved throughout the duration of the project. Such tasks have no particular end product and are estimated and budgeted on a bulk allocation basis, rather than on some measure of production. It can also be used on tasks that are appropriate for the start/finish method. The percent complete at any point in time is found by dividing hours (or costs) spent to date by the current estimate of hours (or costs) at completion.

Percent complete may be determined by one of the above described methods. Earned-value techniques can be used to summarize overall work status. The earned value of any one item being controlled is

Earned work-hours = (Budgeted work-hours) × (Percent complete)

Budgeted work-hours equal original budget plus approved changes.

Overall percent complete of the project or of a work package can be calculated as

$$\text{Percent complete} = \frac{\text{Sum of earned work-hours of tasks included}}{\text{Sum of budgeted work-hours of tasks included}}$$

Trends can be tracked through various indices, such as the Cost Performance Index (CPI) and the Schedule Performance Index (SPI). The CPI provides a comparison of the number of work-hours being spent on work tasks to the hours budgeted and is an indicator of productivity. The equation for CPI is

$$CPI = \frac{\text{Sum of earned work-hours of tasks included}}{\text{Sum of actual work-hours of tasks included}}$$

For the above to be a true indicator of productivity, only those tasks for which budgets have been established should be included in summations. If for some reason there is no budget for an item and people are working on that item, the project manager should prepare a change order for out of scope work. All actual work-hours need to be properly reported to establish accurate historical records for subsequent projects of a similar nature.

The SPI relates the amount of work performed to the amount scheduled to a point in time. The equation for SPI is

$$SPI = \frac{\text{Sum of earned work-hours to date}}{\text{Sum of scheduled work-hours to date}}$$

Scheduled work-hours in the above equation are summarized from the task schedules.

In both the CPI and SPI equations, an index of 1.0 or greater is favorable. Trends can be noted by plotting both "this period" and "cumulative" CPI and SPI values on a graph. While the SPI for the total project, or for a work package, is somewhat of an indicator of schedule performance, it only compares volume of work performed to volume of work scheduled. There can be an SPI in excess of 1.0 that can still be in danger of not meeting milestones and final completion dates if managers are expending effort on non-critical activities at the expense of critical activities. The SPI does not show that work is being completed in the proper sequence. Thus, as part of schedule control, the schedules of all included tasks in each work package must be regularly examined so that any items behind schedule can be identified and corrective action taken to bring them back on schedule.

Measurement of Construction Work

A construction project requires completion of numerous tasks, beginning with initial clearing and site-work through the final punch list and cleanup. Throughout the duration of a project there must be a systematic reporting of the progress of work for each part of the project. The following paragraphs describe six methods for measuring progress during construction: Units Completed, Incremental Milestone, Start/

Finish, Supervisor Opinion, Cost Ratio, and Weighted Units. The system that is selected depends upon the nature and complexity of the project and the desired level of control by the project manager. Each of the six methods may be used on a given project.

The U.S. Departments of Defense and Energy have established what is known as the Cost and Schedule Control Systems Criteria (C/SCSC) for control of selected federal projects. While intended primarily for high-value, cost-reimbursable research and development projects, it may be applied to selected construction projects. Various methods for measuring the status of a project that have become common usage in the construction industry are described in C/SCSC.

The Units Completed method of measuring progress during construction is applicable to tasks that are repetitive and require a uniform effort. Generally, the task is the lowest level of control so only one unit of work is necessary to define the work. To illustrate, the percent complete for installation of wire is determined as a percentage, by dividing the number of feet installed by the total number of feet that is required to be installed.

The Incremental Milestone method is applicable to tasks that include subtasks that must be handled in sequence. For example, the installation of a major vessel in an industrial facility may include the sequential tasks as shown below. Completion of any subtask is considered a milestone, which represents a certain percentage of the total installation. The percentage may be established based upon the estimated work-hours to accomplish the work.

Received and inspected	15%
Setting complete	35%
Alignment complete	50%
Internals installed	75%
Testing complete	90%
Accepted by owner	100%

The Start/Finish method is applicable to tasks that do not have well-defined intermediate milestones or for which the time required is difficult to estimate. For example, the alignment of a piece of equipment may take a few hours to a few days, depending upon the situation. The workers may know when the work will start and when it is finished, but never know the percentage completion in between. For this method, an arbitrary percent complete is assigned at the start of the task and a 100% complete is assigned when it is finished. A starting percentage of 20% to 30% might be assigned for tasks that require a long duration, while 0% may be assigned for tasks with a short duration.

The Supervisor Opinion method is a subjective approach that may be used for minor tasks, such as construction support facilities, where development of a more discrete method cannot be used.

The Cost Ratio method is applicable for administrative tasks, such as project management, quality assurance, contract administration, or project control. These tasks involve a long period of time or are continuous throughout the duration of the project. Generally these tasks are estimated and budgeted as lump-sum dollars and work-hours, rather than measurable quantities of production work. For this method the percent complete can be calculated by the following equation.

$$\text{Percent complete} = \frac{\text{Actual cost or work-hours to date}}{\text{Forecast at completion}}$$

The Weighted Units method is applicable for tasks that involve major efforts of work that occur over a long duration of time. Generally the work requires several overlapping subtasks that each have a different unit of measurement of work. This method is illustrated by the structural steel example shown in Figure 9-13. A common unit of measure for steel is tons. A weight is assigned to each subtask to represent the estimated level of effort. Work-hours is usually a good measure of the required level of effort. As quantities of work are completed for each subtask, these quantities are converted into equivalent tons and percent complete is calculated as shown in Figure 9-13.

As presented in the previous paragraphs, there are many ways of measuring progress of work for each task in a project. After determining

Wt.	Subtask	U/M	Quan total	Equiv steel ton	Quantity to-date	Earned tons*
0.02	Run foundation bolts	each	200	10.4	200	10.4
0.02	Shim	%	100	10.4	100	10.4
0.05	Shakeout	%	100	26.0	100	26.0
0.06	Columns	each	84	31.2	74	27.5
0.10	Beams	each	859	52.0	0	0.0
0.11	Cross-braces	each	837	57.2	0	0.0
0.20	Girts & sag rods	bay	38	104.0	0	0.0
0.09	Plumb & align	%	100	46.8	5	2.3
0.30	Connections	each	2977	156.0	74	3.9
0.05	Punch list	%	100	26.0	0	0.0
1.00	STEEL	TON		520.0		80.5

$$*\text{Earned Tons to Date} = \frac{(\text{Quantity to date})\,(\text{Relative Weight})}{(\text{Total Quantity})}$$

$$\text{Percent Complete} = \frac{80.5 \text{ tons}}{520 \text{ tons}} = 15.5\%$$

Figure 9-13 Illustrative example of weighted-units method for measurement of construction work. (*Source:* Construction Industry Institute, Publication No. 6-5.)

the progress of the different work tasks, the next step is to develop a method that combines the different work tasks to determine the overall percent complete for the project. The earned-value system can be used to define overall percent complete for the entire project. Earned value can be linked to the project budget, which is expressed as work-hours or dollars. For a single account, the earned value can be calculated by the following equation.

Earned value = (Percent complete) × (Budget for that account)

A budgeted amount is "earned" as a task is completed, up to the total amount in that account. For example, an account may be budgeted at $10,000 and 60 work-hours. If the account is reported as 25% complete, as measured by one of the previously described methods, then the earned value is defined as $2,500 and 15 work-hours. Thus, progress in all accounts can be reduced to earned work-hours and dollars, which provides a method for summarizing multiple accounts and calculating overall progress for the total project. The equation for determining the overall project percent complete is given below.

$$\text{Percent complete} = \frac{\text{(Earned work-hours/dollar all accounts)}}{\text{(Budgeted work-hours/dollars all accounts)}}$$

The concepts discussed above provide a system for determining the percent complete of a single work task or combinations of tasks. This provides the basis for analyzing the results to determine how well work is proceeding, as compared to what was planned. The earned-value system provides a method for evaluating performance of the project.

Previous paragraphs have only discussed budgeted and earned work-hours and dollars. Actual work-hours and dollars must also be included to evaluate the performance evaluation of the project. The following definitions are provided to describe the procedure for evaluation of cost and schedule performance.

- Budgeted work-hours or dollars to date represent what is planned to do. The C/SCSC defines this as Budgeted Cost for Work Scheduled (BCWS).

- Earned work-hours or dollars to date represent what was done. The C/SCSC defines this as Budgeted Cost for Work Performed (BCWP).

- Actual work-hours or dollars to date represent what has been paid. The C/SCSC defines this as Actual Cost of Work Performed (ACWP).

Performance against schedule is a comparison of what was planned against what was done, that is, a comparison of budgeted and earned work-hours. If the budgeted work-hours are less than the earned work-hours, it means more was done than planned and the project is ahead of schedule. The opposite would indicate the project is behind schedule.

Performance against budget is measured by comparing what was done to what was paid. This compares earned work-hours to actual work-hours or cost. If more was paid than is done, then the project would have a cost overrun of the budget. The following variance and index equations can be used to calculate these values.

$$\text{Scheduled variance (SV)} = \text{(Earned work-hours or dollars)}$$
$$- \text{(Budgeted work-hours or dollars)}$$
$$\text{SV} = \text{BCWP} - \text{BCWS}$$

$$\text{Schedule Performance Index (SPI)} = \frac{\text{(Earned work-hours or dollars to date)}}{\text{(Budgeted work-hours or dollars to date)}}$$
$$\text{SPI} = \frac{\text{BCWP}}{\text{BCWS}}$$

$$\text{Cost variance (CV)} = \text{(Earned work-hours or dollars)}$$
$$- \text{(Actual work-hours or dollars)}$$
$$\text{CV} = \text{BCWP} - \text{ACWP}$$

$$\text{Cost Performance Index (CPI)} = \frac{\text{(Earned work-hours or dollars to date)}}{\text{(Actual work-hours or dollars to date)}}$$
$$\text{CPI} = \frac{\text{BCWP}}{\text{ACWP}}$$

A positive variance and an index of 1.0 or greater is a favorable performance.

Project Measurement and Control

The purpose of the project plan is to successfully control the project to ensure completion within budget and schedule constraints. The S-curve, which plots cost (or work-hours) against time, provides a measure of schedule performance. Project elements of labor, materials, and equipment are usually evaluated in common units of dollars. Schedule performance indicates the rate at which these project elements are occurring as budgeted expenses. Performance measurements are based on earned-value concepts and S-curve analysis. The earned-value concept provides a quantitative measure of the scheduled budget value of work compared to the actual budgeted value of the work accomplished.

Measuring the progress of a project supports management in establishing a realistic plan for execution of a project and provides the project manager and client with a consistent analysis of project performance. Progress measurement also provides an early warning system to identify deviations from the project plan and scope growth. To control engineering work a drawing list, specification list, equipment list, instrument list, progress S-curve, and histogram of work-hours are used to determine the project status and for continued planning. The project manager can use progress curves and work-hour histograms for different levels of the WBS.

Earned-Value System

The earned-value system is used to monitor the progress of work and compare accomplished work with planned work. The BCWS is the amount of money that was planned, or budgeted, at each time period in the project. It is determined by cost loading the CPM diagram to determine the distribution of cost in accordance with the project plan. The S-curve for a project represents the BCWS. The ACWP is the actual amount of money that has been spent at any point in time during the project. It is determined from accounting records or the responsible party that keeps records of actual expenditure of money. The BCWP is the amount of money earned based on the work that has been completed. It is determined by multiplying the percent completed by the budgeted amount for the work. Below is a summary of the terms used in an earned-value analysis.

Earned-value analysis:

$$BCWS = \text{Budgeted cost of work scheduled} \quad \text{(Planned)}$$
$$ACWP = \text{Actual cost of work performed} \quad \text{(Actual)}$$
$$BCWP = \text{Budgeted cost of work performed} \quad \text{(Earned)}$$

Variances:

$$CV = BCWP - ACWP \quad \text{(Cost variance = Earned - Actual)}$$
$$SV = BCWP - BCWS \quad \text{(Schedule variance = Earned - Planned)}$$

Indices:

$$CPI = \frac{BCWP}{ACWP} \quad \left(\text{Cost Performance Index} = \frac{\text{Earned}}{\text{Actual}} \right)$$
$$SPI = \frac{BCWP}{BCWS} \quad \left(\text{Schedule Performance Index} = \frac{\text{Earned}}{\text{Planned}} \right)$$

Forecasting:

$$BAC = \text{Original project estimate} \quad \text{(Budget at completion)}$$

$$ETC = \frac{BAC - BCWP}{CPI} \quad \text{(Estimate to complete)}$$

$$EAC = (ACWP + ETC) \quad \text{(Estimate at completion)}$$

Ratios are used in the earned-value system to predict the cost to complete a project. The CPI is used to predict the magnitude of a possible cost overrun or underrun. It adjusts the budget based on past performance. The SPI is used to predict the magnitude of a possible time advance or delay. It adjusts the schedule based on past performance. The following examples illustrate the calculations in an earned-value analysis. The examples provide a progressive analysis of the project with an interpretation of the results based on the earned-value system for assessing project performance.

Example 9-2 The EPC project shown in the Appendix has an original planned schedule of 17 months with a budget of $3,053,150, excluding contingency (see Figure A-3). Figure A-10 in the Appendix shows the monthly distribution of costs for all activities of the project. The costs are shown on an early start, late start, and target schedule basis. The target schedule is used for planned BCWS cost in this earned-value analysis. Below are update reports for the first 8 months of the project.

Status report for month	Planned (BCWS)	Earned (BCWP)	Actual (ACWP)
1	$3,929	$3,870	$3,968
2	$45,672	$42,932	$44,721
3	$184,904	$162,715	$191,429
4	$315,615	$309,303	$351,481
5	$479,343	$508,103	$523,818
6	$545,863	$556,780	$535,365
7	$667,210	$713,915	$661,032
8	$846,740	$812,870	$719,354

Perform an earned-value analysis of the project. For each of the 8 months, calculate the variances (CV and SV), the performance indices (CPI and SPI), the estimate to complete (ETC), and the estimate at completion (EAC). After performing an earned-value analysis for each of the 8 months, prepare a graph to show trends in the project performance. Prepare the graph with the SPI in the horizontal direction and the CPI in the vertical direction.

Analysis of Month #1:

The status report after Month #1 into the project includes the following information:

$$\text{Planned, BCWS} = \$3,929$$
$$\text{Earned, BCWP} = \$3,870$$
$$\text{Actual, ACWP} = \$3,968$$
$$\text{Budget at Completion, BAC} = \$3,053,150$$

Cost and Schedule Deviations:

$$\text{Cost Variance} = \text{BCWP} - \text{ACWP}$$
$$= \$3,870 - \$3,968$$
$$= -\$98$$

> A negative value of CV represents a cost overrun. Based on the status report the actual cost is greater than earned by $98.

$$\text{Schedule Variance} = \text{BCWP} - \text{BCWS}$$
$$= \$3,870 - \$3,929$$
$$= -\$59$$

> A negative value of SV represents a schedule slippage. The project is behind the planned schedule.

Cost and Schedule Performance:

$$\text{Cost Performance Index, CPI} = \frac{\text{BCWP}}{\text{ACWP}}$$
$$= \frac{\$3,870}{\$3,968}$$
$$= 0.97$$

> The CPI is less than one, which indicates a poor cost performance. The earned value is less than the actual costs.

$$\text{Schedule Performance Index, SPI} = \frac{\text{BCWP}}{\text{BCWS}}$$
$$= \frac{\$3,870}{\$3,929}$$
$$= 0.98$$

> The SPI is less than one, which indicates the schedule performance is worse than planned. The project is behind schedule.

Forecasting Cost at Completion:

$$\text{Estimate to Complete, ETC} = \frac{(\text{BAC} - \text{BCWP})}{\text{CPI}}$$

$$= \frac{(\$3,053,150 - \$3,870)}{0.97}$$

$$= \$3,143,588$$

Based on the analysis of the status report, the remaining cost to complete the project is $3,143,588.

$$\text{Estimate at Completion, EAC} = \text{ACWP} + \text{ETC}$$

$$= 3,968 + \$3,143,588$$

$$= \$3,147,556$$

Based on analysis of the status report, the estimated cost of the project at completion is $3,147,556, which is $94,406 over the original budget of $3,053,150.

Analysis of Month #2:

The status report after Month #2 into the project includes the following information:

Planned, BCWS = $45,672

Earned, BCWP = $42,932

Actual, ACWP = $44,721

Budget at Completion, BAC = $3,053,150

Cost and Schedule Deviations:

$$\text{Cost Variance} = \text{BCWP} - \text{ACWP}$$

$$= \$42,932 - \$44,721$$

$$= -\$1,789$$

A negative value of CV represents a cost overrun. Based on the status report, the actual cost is greater than earned by $1,789.

$$\text{Schedule Variance} = \text{BCWP} - \text{BCWS}$$

$$= \$42,932 - \$45,672$$

$$= -\$2,740$$

A negative value of SV represents a schedule slippage. The project is behind the planned schedule.

Cost and Schedule Performance:

$$\text{Cost Performance Index, CPI} = \frac{\text{BCWP}}{\text{ACWP}}$$
$$= \frac{\$42,932}{\$44,721}$$
$$= 0.96$$

The CPI is less than one, which indicates a poor cost performance. The earned value is less than the actual costs.

$$\text{Schedule Performance Index, SPI} = \frac{\text{BCWP}}{\text{BCWS}}$$
$$= \frac{\$42,932}{\$45,672}$$
$$= 0.94$$

The SPI is less than one, which indicates the schedule performance is worse than planned. The project is behind schedule.

Forecasting Cost at Completion:

$$\text{Estimate to Complete, ETC} = \frac{(\text{BAC} - \text{BCWP})}{\text{CPI}}$$
$$= \frac{(\$3,053,150 - \$42,932)}{0.96}$$
$$= \$3,135,644$$

Based on the analysis of the status report, the remaining cost to complete the project is $3,135,644.

$$\text{Estimate at Completion, EAC} = \text{ACWP} + \text{ETC}$$
$$= \$44,721 + \$3,135,644$$
$$= \$3,180,365$$

Based on analysis of the status report, the estimated cost of the project at completion is $3,180,365, which is $127,215 over the original budget of $3,053,150.

Analysis of Month #3:

The status report after Month #3 into the project includes the following information:

Planned, BCWS = $184,904

Earned, BCWP = $162,715

Actual, ACWP = $191,429

Budget at Completion, BAC = $3,053,150

Cost and Schedule Deviations:

$$\text{Cost Variance} = \text{BCWP} - \text{ACWP}$$
$$= \$162{,}715 - \$191{,}429$$
$$= -\$28{,}714$$

A negative value of CV represents a cost overrun. Based on the status report, the actual cost is greater than earned by \$28,714.

$$\text{Schedule Variance} = \text{BCWP} - \text{BCWS}$$
$$= \$162{,}715 - \$184{,}904$$
$$= -\$22{,}189$$

A negative value of SV represents a schedule slippage. The project is behind the planned schedule.

Cost and Schedule Performance:

$$\text{Cost Performance Index, CPI} = \frac{\text{BCWP}}{\text{ACWP}}$$
$$= \frac{\$162{,}715}{\$191{,}429}$$
$$= 0.85$$

The CPI is less than one, which indicates a poor cost performance. The earned value is less than the actual costs.

$$\text{Schedule Performance Index, SPI} = \frac{\text{BCWP}}{\text{BCWS}}$$
$$= \frac{\$162{,}715}{\$184{,}904}$$
$$= 0.88$$

The SPI is less than one, which indicates the schedule performance is worse than planned. The project is behind schedule.

Forecasting Cost at Completion:

$$\text{Estimate to Complete, ETC} = \frac{(\text{BAC} - \text{BCWP})}{\text{CPI}}$$
$$= \frac{(\$3{,}053{,}150 - \$162{,}715)}{0.85}$$
$$= \$3{,}400{,}512$$

Based on the analysis of the status report, the remaining cost to complete the project is $3,400,512.

Estimate at Completion, EAC = ACWP + ETC

$$= \$191{,}429 + \$3{,}400{,}512$$

$$= \$3{,}591{,}941$$

Based on analysis of the status report, the estimated cost of the project at completion is $3,591,941, which is $538,791 over the original budget of $3,053,150.

Analysis of Month #4:

The status report after Month #4 into the project includes the following information:

Planned, BCWS = $315,615

Earned, BCWP = $309,303

Actual, ACWP = $351,481

Budget at Completion, BAC = $3,053,150

Cost and Schedule Deviations:

Cost Variance = BCWP – ACWP

$$= \$315{,}615 - \$351{,}481$$

$$= -\$35{,}866$$

A negative value of CV represents a cost overrun. Based on the status report, the actual cost is greater than earned by $35,866.

Schedule Variance = BCWP – BCWS

$$= \$309{,}303 - \$315{,}615$$

$$= -\$6{,}312$$

A negative value of SV represents a schedule slippage. The project is behind the planned schedule.

Cost and Schedule Performance:

$$\text{Cost Performance Index, CPI} = \frac{\text{BCWP}}{\text{ACWP}}$$

$$= \frac{\$309{,}303}{\$351{,}481}$$

$$= 0.88$$

The CPI is less than one, which indicates a poor cost performance. The earned value is less than the actual costs.

$$\text{Schedule Performance Index, SPI} = \frac{\text{BCWP}}{\text{BCWS}}$$

$$= \frac{\$309,303}{\$315,615}$$

$$= 0.98$$

The SPI is less than one, which indicates the schedule performance is worse than planned. The project is behind schedule.

Forecasting Cost at Completion:

$$\text{Estimate to Complete, ETC} = \frac{(\text{BAC} - \text{BCWP})}{\text{CPI}}$$

$$= \frac{(\$3,053,150 - \$309,303)}{0.88}$$

$$= \$3,118,008$$

Based on the analysis of the status report, the remaining cost to complete the project is $3,118,008.

$$\text{Estimate at Completion, EAC} = \text{ACWP} + \text{ETC}$$

$$= \$351,181 + \$3,118,008$$

$$= \$3,469,489$$

Based on analysis of the status report, the estimated cost of the project at completion is $3,469,489, which is $416,339 over the original budget of $3,053,150.

Analysis of Month #5:

The status report after Month #5 into the project includes the following information:

Planned, BCWS = $479,343

Earned, BCWP = $508,103

Actual, ACWP = $523,818

Budget at Completion, BAC = $3,053,150

Cost and Schedule Deviations:

$$\text{Cost Variance} = \text{BCWP} - \text{ACWP}$$

$$= \$508,103 - \$523,818$$

$$= -\$15,715$$

A negative value of CV represents a cost overrun. Based on the status report, the actual cost is greater than earned by $15,715.

$$\text{Schedule Variance} = \text{BCWP} - \text{BCWS}$$
$$= \$508,103 - \$479,343$$
$$= +\$28,760$$

A positive value of SV represents a good schedule performance. The project is ahead of the planned schedule.

Cost and Schedule Performance:

$$\text{Cost Performance Index, CPI} = \frac{\text{BCWP}}{\text{ACWP}}$$
$$= \frac{\$508,103}{\$523,818}$$
$$= 0.97$$

The CPI is less than one, which indicates a poor cost performance. The earned value is less than the actual costs.

$$\text{Schedule Performance Index, SPI} = \frac{\text{BCWP}}{\text{BCWS}}$$
$$= \frac{\$508,103}{\$479,343}$$
$$= 1.06$$

The SPI is greater than one, which indicates the schedule performance is better than planned. The project is ahead of schedule.

Forecasting Cost at Completion:

$$\text{Estimate to Complete, ETC} = \frac{(\text{BAC} - \text{BCWP})}{\text{CPI}}$$
$$= \frac{(\$3,053,150 - \$508,103)}{0.97}$$
$$= \$2,623,760$$

Based on the analysis of the status report, the remaining cost to complete the project is $2,623,760.

$$\text{Estimate at Completion, EAC} = \text{ACWP} + \text{ETC}$$
$$= \$523,818 + \$2,623,760$$
$$= \$3,147,578$$

Based on analysis of the status report, the estimated cost of the project at completion is $3,147,578, which is $94,428 over the original budget of $3,053,150.

Analysis of Month #6:

The status report after Month #6 into the project includes the following information:

$$\text{Planned, BCWS} = \$545{,}863$$
$$\text{Earned, BCWP} = \$556{,}780$$
$$\text{Actual, ACWP} = \$535{,}365$$
$$\text{Budget at Completion, BAC} = \$3{,}053{,}150$$

Cost and Schedule Deviations:

$$\text{Cost Variance} = \text{BCWP} - \text{ACWP}$$
$$= \$556{,}780 - \$535{,}365$$
$$= +\$21{,}415$$

A positive value of CV represents a good cost performance. Based on the status report, the actual cost is less than earned by $21,415.

$$\text{Schedule Variance} = \text{BCWP} - \text{BCWS}$$
$$= \$556{,}780 - \$545{,}863$$
$$= +\$10{,}917$$

A positive value of SV represents a good schedule performance. The project is ahead of the planned schedule.

Cost and Schedule Performance:

$$\text{Cost Performance Index, CPI} = \frac{\text{BCWP}}{\text{ACWP}}$$
$$= \frac{\$556{,}780}{\$535{,}365}$$
$$= 1.04$$

The CPI is greater than one, which indicates a good cost performance. The earned value is greater than the actual costs.

$$\text{Schedule Performance Index, SPI} = \frac{\text{BCWP}}{\text{BCWS}}$$
$$= \frac{\$556{,}780}{\$545{,}863}$$
$$= 1.02$$

The SPI is greater than one, which indicates the schedule performance is better than planned. The project is ahead of schedule.

Forecasting Cost at Completion:

$$\text{Estimate to Complete, ETC} = \frac{(\text{BAC} - \text{BCWP})}{\text{CPI}}$$

$$= \frac{(\$3,053,150 - \$556,780)}{1.04}$$

$$= \$2,400,356$$

Based on the analysis of the status report, the remaining cost to complete the project is $2,400,356.

$$\text{Estimate at Completion, EAC} = \text{ACWP} + \text{ETC}$$

$$= \$535,365 + \$2,400,356$$

$$= \$2,935,721$$

Based on analysis of the status report, the estimated cost of the project at completion is $2,935,721, which is $117,429 under the original budget of $3,053,150.

Analysis of Month #7:

The status report after Month #7 into the project includes the following information:

> Planned, BCWS = $667,210
> Earned, BCWP = $713,915
> Actual, ACWP = $661,032
> Budget at Completion, BAC = $3,053,150

Cost and Schedule Deviations:

$$\text{Cost Variance} = \text{BCWP} - \text{ACWP}$$

$$= \$713,915 - \$661,032$$

$$= +\$52,883$$

A positive value of CV represents a good cost performance. Based on the status report, the actual cost is less than earned by $52,883.

$$\text{Schedule Variance} = \text{BCWP} - \text{BCWS}$$

$$= \$713,915 - \$667,210$$

$$= +\$46,705$$

A positive value of SV represents a good schedule performance. The project is ahead of the planned schedule.

Cost and Schedule Performance:

$$\text{Cost Performance Index, CPI} = \frac{\text{BCWP}}{\text{ACWP}}$$
$$= \frac{\$713{,}915}{\$661{,}032}$$
$$= 1.08$$

The CPI is greater than one, which indicates a good cost performance. The earned value is greater than the actual costs.

$$\text{Schedule Performance Index, SPI} = \frac{\text{BCWP}}{\text{BCWS}}$$
$$= \frac{\$713{,}915}{\$667{,}210}$$
$$= 1.07$$

The SPI is greater than one, which indicates the schedule performance is better than planned. The project is ahead of schedule.

Forecasting Cost at Completion:

$$\text{Estimate to Complete, ETC} = \frac{(\text{BAC} - \text{BCWP})}{\text{CPI}}$$
$$= \frac{(\$3{,}053{,}150 - \$713{,}915)}{1.08}$$
$$= \$2{,}165{,}958$$

Based on the analysis of the status report, the remaining cost to complete the project is $2,165,958.

$$\text{Estimate at Completion, EAC} = \text{ACWP} + \text{ETC}$$
$$= \$661{,}032 + \$2{,}165{,}958$$
$$= \$2{,}826{,}990$$

Based on analysis of the status report, the estimated cost of the project at completion is $2,826,990, which is $226,160 under the original budget of $3,053,150.

Analysis of Month #8:

The status report after Month #8 into the project includes the following information:

Planned, BCWS = $846,740

Earned, BCWP = $812,870

Actual, ACWP = $719,354

Budget at Completion, BAC = $3,053,150

Cost and Schedule Deviations:

$$\text{Cost Variance} = \text{BCWP} - \text{ACWP}$$
$$= \$812,870 - \$719,354$$
$$= +\$93,516$$

A positive value of CV represents a good cost performance. Based on the status report, the actual cost is less than earned by \$93,516.

$$\text{Schedule Variance} = \text{BCWP} - \text{BCWS}$$
$$= \$812,870 - \$846,740$$
$$= -\$33,870$$

A negative value of SV represents a schedule slippage. The project is behind the planned schedule.

Cost and Schedule Performance:

$$\text{Cost Performance Index, CPI} = \frac{\text{BCWP}}{\text{ACWP}}$$
$$= \frac{\$812,870}{\$719,354}$$
$$= 1.13$$

The CPI is greater than one, which indicates a good cost performance. The earned value is greater than the actual costs.

$$\text{Schedule Performance Index, SPI} = \frac{\text{BCWP}}{\text{BCWS}}$$
$$= \frac{\$812,870}{\$846,740}$$
$$= 0.96$$

The SPI is less than one, which indicates a poor schedule performance. The project is behind schedule.

Forecasting Cost at Completion:

$$\text{Estimate to Complete, ETC} = \frac{(\text{BAC} - \text{BCWP})}{\text{CPI}}$$
$$= \frac{(\$3,053,150 - \$812,870)}{1.13}$$
$$= \$1,982,549$$

Based on the analysis of the status report, the remaining cost to complete the project is \$1,982,549.

$$\text{Estimate at Completion, EAC} = \text{ACWP} + \text{ETC}$$
$$= \$719,354 + \$1,982,549$$
$$= \$2,701,903$$

Based on analysis of the status report, the estimated cost of the project at completion is $2,701,903, which is $351,247 under the original budget of $3,053,150. Following is a project performance chart of the earned value analysis for the 8 months.

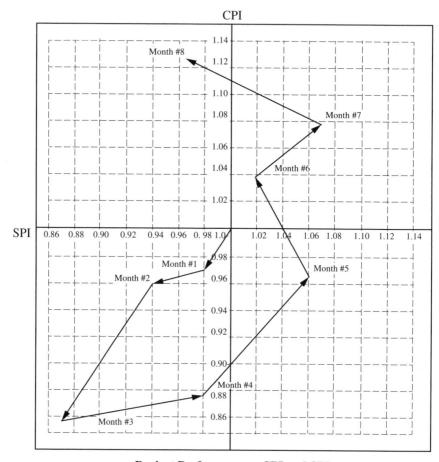

Project Performance - CPI and SPI

Example 9-3   Provide an earned-value analysis to evaluate the progress of the sewer and water lines project shown in Figure 8-6. The original budget is $147,500 and the project is scheduled to be completed in 94 working days. A status report after 10 working days into the project includes the following information:

Activity 10, 100% complete as scheduled, actual cost = $1,500

Activity 20, 100% complete as scheduled, actual cost = $2,200

Activity 30, 100% complete as scheduled, actual cost = $4,000

BCWP = $7,600

$$ACWP = \$7,700$$
$$BCWS = \$7,600$$
$$BAC = \$147,500$$

The BCWS and BCWP shown above are the same values as shown on the 10th working day in Figure 8-12 because Activities 10, 20, and 30 were all completed according to the original planned schedule.

Cost and Schedule Deviations:

$$\text{Cost variance, CV} = BCWP - ACWP$$
$$= \$7,600 - \$7,700$$
$$= -\$100$$

A negative value of CV represents a cost overrun. Based on the status report the actual cost is greater than earned by $100.

$$\text{Schedule variance, SV} = BCWP - BCWS$$
$$= \$7,600 - \$7,600$$
$$= 0$$

Since the SV is zero, the project is progressing as planned. The project is not ahead of or behind the planned schedule.

Cost and Schedule Performance:

$$\text{Cost Performance Index, CPI} = \frac{BCWP}{ACWP}$$
$$= \frac{\$7,600}{\$7,700}$$
$$= 0.987$$

The CPI is less than 1.0, which indicates a poor cost performance. The earned value is less than the actual costs.

$$\text{Schedule Performance Index, SPI} = \frac{BCWP}{BCWS}$$
$$= \frac{\$7,600}{\$7,600}$$
$$= 1.0$$

The SPI is equal to 1.0, which indicates the schedule performance is progressing precisely as planned.

Forecasting Cost at Completion:

$$\text{Estimate to complete, ETC} = \frac{BAC - BCWP}{CPI}$$
$$= \frac{\$147,500 - \$7,600}{0.987}$$
$$= \$141,743$$

Based on the analysis of the status report, the remaining cost to complete the project is $141,743.

$$\text{Estimate at completion, EAC} = \text{ACWP} + \text{ETC}$$
$$= \$7,700 + \$141,743$$
$$= \$149,443$$

Based on the analysis of the status report, the estimated cost of the project at completion is $149,443, which is $1,943 over the original budget of $147,500.

Example 9-4   Provide an earned-value analysis to evaluate the progress of the sewer and water lines project shown in Figure 8-6. The original budget is $147,500 and the project is scheduled to be completed in 94 working days. A status report after 20 working days into the project includes the following information:

Previous period analysis:

BCWP = $7,600; ACWP = $7,700; BCWS = $7,600; BAC = $147,500

Current status report:

Activity 40, 100% complete, 2 days delayed, actual cost = $8,000
Activity 50, 60% complete, 2 days delayed, cost to date = $3,200
Activity 60, 50% complete, started out of sequence, cost to date = $6,300
BCWP = $22,720
ACWP = $25,200
BCWS = $22,400
BAC = $147,500

The BCWS shown above is different than the value shown on the 20th working day in Figure 8-12 because Activities 40, 50, and 60 have been completed at different times than the original planned schedule. The BCWS of $22,400 shown above is based on an updated schedule for the project. Also, the BCWP is based on the updated schedule.

Cost and Schedule Deviations:

$$\text{Cost variance, CV} = \text{BCWP} - \text{ACWP}$$
$$= \$22,720 - \$25,200$$
$$= -\$2,480$$

A negative value of CV represents a cost overrun. Based on the status report, the actual cost is greater than earned by $2,480.

$$\text{Schedule variance, SV} = \text{BCWP} - \text{BCWS}$$
$$= \$22,720 - \$22,400$$
$$= \$320$$

The SV is $320 because the earned value is greater than planned, indicating the project is ahead of schedule.

Cost and Schedule Performance:

$$\text{Cost Performance Index, CPI} = \frac{\text{BCWP}}{\text{ACWP}}$$
$$= \frac{\$22,720}{\$25,200}$$
$$= 0.90$$

The CPI is less than 1.0, which indicates a poor cost performance. The earned value is less than the actual costs for this reporting period.

$$\text{Schedule Performance Index, SPI} = \frac{\text{BCWP}}{\text{BCWS}}$$
$$= \frac{\$22,720}{\$22,400}$$
$$= 1.01$$

The SPI is greater than 1.0, which indicates the schedule performance is progressing better than planned. The project is ahead of schedule, which is slightly better than the status report given in Example 9-3.

Forecasting Cost at Completion:

$$\text{Estimate to complete, ETC} = \frac{\text{BAC} - \text{BCWP}}{\text{CPI}}$$
$$= \frac{\$147,500 - \$22,720}{0.90}$$
$$= \$138,644$$

Based on the analysis of the status report, the remaining cost to complete the project is $138,644.

$$\text{Estimate at completion, EAC} = \text{ACWP} + \text{ETC}$$
$$= \$25,200 + \$138,644$$
$$= \$163,844$$

Based on the analysis of the status report, the estimated cost of the project at completion is $163,844, which is $16,344 over the original budget of $147,500. The cost performance is significantly worse compared to the previous status report given in Example 9-3.

Example 9-5   Provide an earned-value analysis to evaluate the progress of the sewer and water lines project shown in Figure 8-6. The original budget is $147,500 and the project is scheduled to be completed in 94 working days. A status report after 25 working days into the project includes the following information:

Previous period analysis:

BCWP = \$22,720; ACWP = \$25,200; BCWS = \$22,400; BAC = \$147,500

Current status report:

Activity 50, 100% complete, one more day delay, final actual cost = \$6,200
Activity 60, 100% complete, no additional delays, final actual cost = \$12,600
Activity 70, 100% complete, no delays, actual cost = \$2,100
BCWP = \$33,000
ACWP = \$36,600 (Note: Includes final actual costs of Activities 10–70.)
BCWS = \$31,740
BAC = \$147,500

Cost and Schedule Deviations:

$$\text{Cost variance, CV} = \text{BCWP} - \text{ACWP}$$
$$= \$33,000 - \$36,600$$
$$= -\$3,600$$

A negative value of CV represents a cost overrun. Based on the status report, the actual cost is greater than earned by \$3,600.

$$\text{Schedule variance, SV} = \text{BCWP} - \text{BCWS}$$
$$= \$33,000 - \$31,740$$
$$= \$1,260$$

The SV is \$1,260 because the earned value is greater than planned, indicating the project is ahead of schedule.

Cost and Schedule Performance:

$$\text{Cost Performance Index, CPI} = \frac{\text{BCWP}}{\text{ACWP}}$$
$$= \frac{\$33,000}{\$36,600}$$
$$= 0.90$$

The CPI is less than 1.0, which indicates a poor cost performance. The earned value is less than the actual costs for this reporting period.

$$\text{Schedule Performance Index, SPI} = \frac{\text{BCWP}}{\text{BCWS}}$$
$$= \frac{\$33,000}{\$31,740}$$
$$= 1.04$$

The SPI is greater than 1.0, which indicates the schedule performance is progressing better than planned. The project is ahead of schedule, which is better than the previous status report given in Example 9-4.

Forecasting Cost at Completion:

$$\text{Estimate to complete, ETC} = \frac{\text{BAC} - \text{BCWP}}{\text{CPI}}$$

$$= \frac{\$147{,}500 - \$33{,}000}{0.90}$$

$$= \$127{,}222$$

Based on the analysis of the status report, the remaining cost to complete the project is $127,222.

$$\text{Estimate at completion, EAC} = \text{ACWP} + \text{ETC}$$

$$= \$36{,}600 + \$127{,}222$$

$$= \$163{,}822$$

Based on the analysis of the status report, the estimated cost of the project at completion is $163,822, which is $16,322 over the original budget of $147,500. The schedule performance is about the same compared to the previous status report given in Example 9-4.

Example 9-6 Provide an earned-value analysis to evaluate the progress of the sewer and water lines project shown in Figure 8-6. The original budget is $147,500 and the project is scheduled to be completed in 94 working days. A status report after 70 working days into the project includes the following information:
Previous period analysis:

BCWP = $33,000; ACWP = $36,600; BCWS = $31,740; BAC = $147,500

Current status report:

> Activity 80, 100% complete, actual cost = $14,000
> Activity 90, 100% complete, actual cost = $800
> Activity 100, 100% complete, actual cost = $1,400
> Activity 110, 100% complete, actual cost = $5,000
> Activity 130, 100% complete, actual cost = $4,500
> Activity 140, 20% complete, actual cost = $680
> Activity 160, 30% complete, actual cost = $2,100
> BCWP = $56,830
> ACWP = $65,080
> BCWS = $102,800
> BAC = $147,500

Cost and Schedule Deviations:

$$\text{Cost variance, CV} = \text{BCWP} - \text{ACWP}$$

$$= \$56{,}830 - \$65{,}080$$

$$= -\$8{,}250$$

A negative value of CV represents a cost overrun. Based on the status report, the actual cost is greater than earned by $8,250.

$$\text{Schedule variance, SV} = \text{BCWP} - \text{BCWS}$$
$$= \$56,830 - \$102,800$$
$$= -\$45,970$$

The SV is a negative value of $45,970 because the earned value is greater than planned, indicating the project is substantially behind schedule.

Cost and Schedule Performance:

$$\text{Cost Performance Index, CPI} = \frac{\text{BCWP}}{\text{ACWP}}$$
$$= \frac{\$56,830}{\$65,080}$$
$$= 0.87$$

The CPI is less than 1.0, which indicates a poor cost performance. The earned value is less than the actual costs for this reporting period.

$$\text{Schedule Performance Index, SPI} = \frac{\text{BCWP}}{\text{BCWS}}$$
$$= \frac{\$56,830}{\$102,800}$$
$$= 0.55$$

The SPI is less than 1.0, which indicates the schedule performance is significantly worse than planned. The project is behind schedule, which indicates the project will be completed significantly later than planned.

Forecasting Cost at Completion:

$$\text{Estimate to complete, ETC} = \frac{\text{BAC} - \text{BCWP}}{\text{CPI}}$$
$$= \frac{\$147,500 - \$56,830}{0.87}$$
$$= \$104,218$$

Based on the analysis of the status report, the remaining cost to complete the project is $104,218.

$$\text{Estimate at completion, EAC} = \text{ACWP} + \text{ETC}$$
$$= \$65,080 + \$104,218$$
$$= \$169,298$$

Based on the analysis of the status report, the estimated cost of the project at completion is $169,298, which is $21,798 over the original budget of $147,500. The project is projected to have a significant cost overrun.

Monitoring Project Performance

The CPI and SPI provide a quantity measurement of the progress of a project. Values of CPI or SPI that are greater than 1.0 indicate good project performance, whereas values of CPI or SPI that are less than 1.0 indicate poor project performance. Values of these indices can be plotted on a graph as shown in Figure 9-14 to evaluate project performance during routine reporting periods.

As shown in Figure 9-14, a project begins with a CPI and SPI equal to 1.0. Projects are dynamic in nature due to events that can impact the cost and the schedule. Therefore, it is expected that these performance indices will deviate from this initial starting point of zero as the project progresses from week to week. The project manager should not overreact to slight changes in the CPI and SPI. For example, the first 2 weeks of the project show small changes in the performance indices, which might be expected during the early progress of the project.

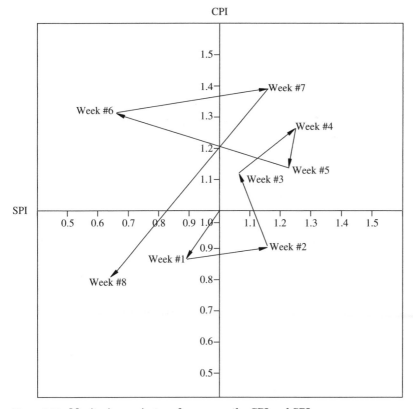

Figure 9-14 Monitoring project performance—the CPI and SPI.

Although minor deviations are expected, major deviations from one report period to the next should alert the project manager to investigate the reason for the significant change in the project performance. For example, the report between weeks 7 and 8 shows a dramatic change toward poor project performance. This type of report merits the attention of the project manager to solicit team members to identify what has happened in the project and the reason for this dramatic change in project performance.

Concern should also be raised if the trend of the CPI and SPI is progressing toward poor performance. For example, from week 2 to week 4, the project performance trend is toward good. However, significant changes in the project begin to appear during weeks 6, 7, and 8.

Thus, the project manager must be continually alert to changes in the project. The CPI and SPI provide the information needed by the project manager to monitor the progress of work to ensure the project is moving in the right direction toward successful completion.

Interpretation of Performance Indices

The graph shown in Figure 9-14 is a valuable tool for the project manager to monitor project performance based on progress reports. It provides a measure of how well the planned cost compares to actual expenditures and work accomplished. As deviations occur between the values, several interpretations can be made.

Values of SPI greater than 1.0 indicate the project is ahead of schedule. The project may be progressing faster than expected if the original production rates were predicted too low or the actual working conditions are better than originally anticipated. There may be more staffing on the project than anticipated, which would also show the work as progressing ahead of schedule. Values of SPI less than 1.0 indicate the project is behind schedule. The project schedule may slip due to weather delays, understaffing, or disorganized work.

Values of CPI greater than 1.0 indicate good project cost performance. Good cost performance may be reported if actual productivity is better than planned or if the measured percent of completed work is too high. Values of CPI less than 1.0 indicate poor cost performance, which may be a result of poorer than planned productivity or due to an underestimate of the measured percent of completed work.

The project manager should seek information from team members to provide a better interpretation of the meaning of project performance indices. Figures 9-15 and 9-16 provide additional interpretations of CPI and SPI data.

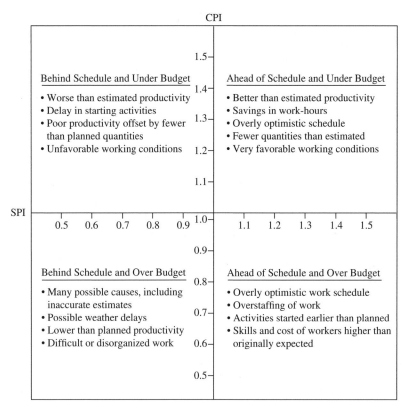

Figure 9-15 Interpretations of cost performance index (CPI) and schedule performance index (SPI).

If CPI is greater than 1.0:
        Overestimated amount of effort required.
        Measurement of percent complete is too optimistic.
        Better than planned productivity.
If CPI is less than 1.0:
        Underestimate amount of effort required.
        Measurement of percent complete is too pessimistic.
        Poorer than planned productivity.
If SPI is greater than 1.0:
        Started work prior to or out of sequence to logical constraints.
        Better than planned productivity resulting in activities completing early.
        Working on critical activities.
If SPI is less than 1.0:
        Understaffing or underequipping activities.
        Delayed in starting of activities.
        Work is more difficult than planned.
        Improper sequencing of activities.
        Work is disorganized.

Figure 9-16 Various interpretations of CPI and SPI.

Analysis Tree of Total Float (TF) and Schedule Performance Index (SPI)

The total float is a measure of slack time in the project, and the SPI is a measure of schedule performance. In a pure CPM precedence diagram, activities with zero total float are identified as critical, and these activities form the critical path in the project schedule. However, using successor/ predecessor activity constraints can cause a total float of greater than zero for activities on the critical path. Also, total float values other than zero are often the result of updates in the schedule when actual starts and finishes are introduced.

Understanding of the schedule can be further complicated when activities are given a specific calendar start or finish date, which can lead to a total float of greater or less than zero for activities on the critical path. For example, the scheduler may specify a time certain for an activity, such as specifying the activity to start or finish on a specific date, to start or finish not earlier than a specific date, or to start or finish not later than a specific date.

For any project, the SPI may also be greater than, less than, or equal to 1.0. Values greater than 1.0 indicate good performance and values less than 1.0 indicate poor performance. Thus, the project manager's analysis of project performance is complicated due to various total float and schedule performance indices. Figure 9-17 is an analysis tree for assessment of total float with respect to SPI.

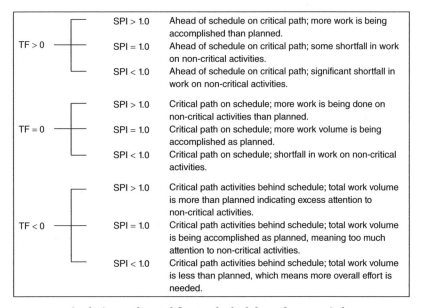

Figure 9-17 Analysis tree for total float and schedule performance index.

Causes of Cost/Schedule Variances

The earned-value system identifies the magnitude of cost and schedule deviations from the original project plan. However, it does not identify the cause of the problem. The project manager and his or her team must assess each status report to identify the reason the project is not progressing as planned. The problems can be the result of numerous situations.

The original cost estimate is the BAC in the earned-value analysis. Therefore, if the original cost estimate for the project is incorrect, then all progress measurements during execution of the project would be measured against an incorrect budget. The system of recording costs charged against the job must be consistent to provide realistic comparisons from one reporting period to another.

Also, the method of measuring work completed must also be consistently applied from one reporting period to another; otherwise the predicted status of the job will vary widely. Each project must be assessed based on the unique circumstances and conditions that apply to the project in order to use the earned-value system to manage the project. Following is a partial list of items that can cause the cost or schedule to vary from the original project plan.

Estimating errors
Technical problems
Design errors
Test data problems
Constructability
Equipment problems
Management problems
Scope control (change orders)
Personnel skill level
Resource availability
Organization structure
Economic/inflation
Delayed material deliveries
Delayed equipment deliveries
Poor production rates
Subcontractor interference and delays
Acts of God (weather, fire, flood, etc.)
Accidents during construction

Trend Analysis and Forecasting

The preceding sections presented methods for measurement of planned and actual costs, schedules, and accomplishment of work. To be effective, a project control system must routinely collect and record the

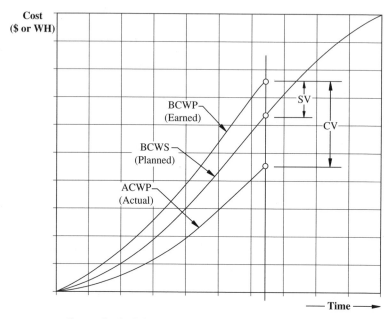

Figure 9-18   Cost and schedule variance graph.

information from the start of the project. At each reporting period the actual status can be compared to the planned status, so that necessary corrective actions can be taken. As the information is accumulated, a trend analysis can be performed to evaluate the productivity and the variances in costs and schedules. Figure 9-18 is a graphical display of the schedule, cost, and time variance at a particular progress reporting period.

A visual display of schedule status is best presented on a bar chart representation. A sample of an excellent format for summary level reporting to management is shown in Figure 9-19.

A project manager is always interested in knowing how the actual productivity compares to the productivity that was used in planning the project and estimating the budget. Although there is no industry standard for calculating productivity, the following equation is a common expression of productivity.

$$\text{Productivity index} = \frac{\text{Estimated unit rate}}{\text{Actual unit rate}}$$

This equation is valid for calculating productivity for tasks that have measurable units of work. A productivity value of 1.0 or greater is favorable, whereas a value of less than 1.0 is unfavorable. The equation can

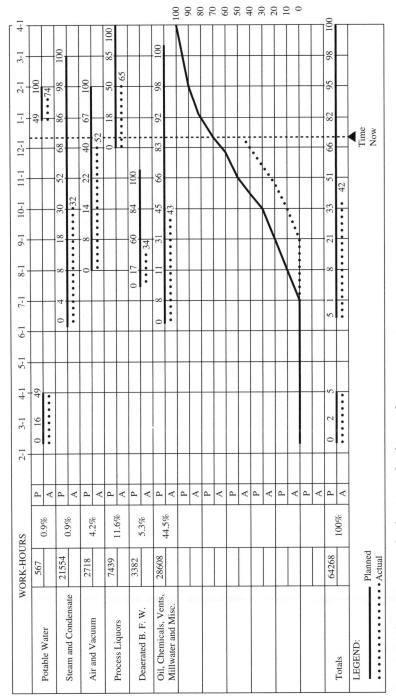

Figure 9-19 Illustrative example of summary level report for management. (*Source:* Construction Industry Institute, Publication No. 6-5.)

279

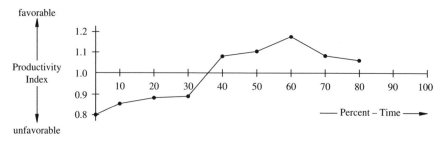

Figure 9-20 Craft productivity profile.

be used to calculate the productivity of tasks at each reporting period. A productivity profile can then be produced for key crafts by plotting productivity against time as illustrated in Figure 9-20. Percent complete can be used in lieu of time for the abscissa of the graph.

As indicated in Figure 9-20, productivity varies over the life of a project. Minor variations are normal and should be expected due to the nature of project work. Significant variations justify the attention of management to identify the source of the problem and assist in any necessary corrective action. The problem may be due to low skill or morale of workers, insufficient staff, late delivery of material, unavailable tools or equipment, inadequate instruction, inclement weather conditions, technical difficulties, or poor field supervision. Since the productivity index includes an estimated unit rate, the source of the problem may be due to poor initial estimates. Trends in the productivity index provide an effective tool for project management.

Almost everyone involved in a project is concerned with costs. As presented in previous sections, the work plan includes a distribution of costs that are anticipated throughout the duration of a project. At each reporting period the actual costs to date can be compared to the planned costs for that period in time. The cost ratio is the ratio of estimated costs to actual costs as defined below.

$$\text{Cost ratio} = \frac{\text{Estimated cost}}{\text{Actual cost}}$$

This equation can be used to calculate the cost ratio at each reporting period and plotted against time to show cost trends for a project (see Figure 9-21). A cost ratio of 1.0 or higher is favorable, whereas a value of less than 1.0 is unfavorable. Cost ratios can be calculated for tasks, grouping of tasks, or for the entire project. A similar analysis can be performed for the schedule.

As a project progresses, the cost ratio trend can be used as an indicator to forecast the probable total cost at completion. Likewise, a schedule variance trend can be developed as an indicator to forecast the

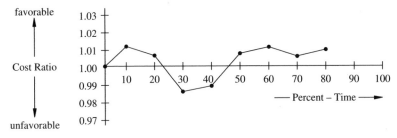

Figure 9-21   Cost ratio.

probable completion date of a project. The cost and schedule forecasting can be accomplished by using one of the various curve fitting techniques. Then, by extrapolating the data to produce a best fit curve, the trend may be extended to the future date.

Work Status System

The Cost and Schedule Control Systems Criteria (C/SCSC) described earlier is an effective method for managing the cost and schedule of a project. It requires a well-defined WBS and a detailed project schedule. It is only effective if the cost and schedule data are reported in a timely manner. For some engineering design, the total time to complete a project is short; therefore, the project may be substantially complete before cost data is reported. For some project managers, the C/SCSC method appears complicated and difficult to use.

An alternate method that is preferred by some project managers for tracking engineering design work is a work status system that tracks costs and work-hours. Figure 9-22 is a cost status report and Figure 9-23 is a work-hour status report that illustrates the work status system method for project tracking. For this example, the project is divided into three general categories: direct engineering, indirect engineering, and expenses. The direct engineering category is subdivided into the disciplines that are required to accomplish the work: architectural, civil, electrical, mechanical, and structural. Likewise, major work that is required for indirect engineering and expenses are subdivided as shown in Figures 9-22 and 9-23.

The cost and work-hour data are entered in a computer spreadsheet program to calculate the status of the project. The titles at the top of the columns identify the entries in the spreadsheet. The first column (A) of data represents the original approved project budget, the second column (B) is the approved change orders, and the third column (C) represents the growth in budget that is required to complete the work. Column (D) is the calculated total budget at the completion of the

EXAMPLE CLIENT
EXAMPLE BOILER REPLACEMENT
Example Project Location
Example Client Project Numbers
Cost Status Report
As of July 20, 2016

Description	A Original Budget Cost	B Approved by Change Order Cost	C Growth Cost	D = A + B + C Budget at Completion Cost	E Cumulative to Last Period Cost	F This Period Cost	G = E + F Cost to Date Cost	H Estimated at Completion Cost	I = G/D Percent Complete %
DIRECT ENGINEERING									
Architectural	$9,799		$2,900	$12,699	$0	$0	$0	$12,699	0.0%
Civil	$4,055		$3,674	$7,729	$217	$1,366	$1,583	$7,729	20.5%
Electrical	$17,203		$2,929	$20,132	$1,327	$529	$1,856	$20,132	9.2%
Mechanical	$44,330		$1,324	$45,654	$7,207	$2,999	$10,206	$45,654	22.4%
Structural	$11,278		$14,086	$25,364	$159	$0	$159	$25,364	0.6%
Subtotal	$86,665	$0	$24,913	$111,578	$8,910	$4,894	$13,804	$111,578	12.4%
INDIRECT ENGINEERING									
Project Manager	$23,412	$12,606		$36,018	$5,794	$1,921	$7,715	$36,018	21.4%
Scheduler	$0	$18,552		$18,552	$3,667	$1,245	$4,912	$18,552	26.5%
Clerical	$0	$3,530		$3,530	$529	$462	$991	$3,530	28.1%
Subtotal	$23,412	$34,688	$0	$58,100	$9,990	$3,628	$13,618	$58,100	23.4%
EXPENSES									
CADD Expense	$9,100			$9,100		$80	$80	$9,100	0.9%
Travel	$1,200			$1,200		$33	$33	$1,200	2.8%
Postage	$300			$300		$0	$0	$300	0.0%
Reproduction	$400			$400		$0	$0	$400	0.0%
Subtotal	$11,000	$0	$0	$11,000	$0	$113	$113	$11,000	1.0%
TOTAL	$121,077	$34,688	$24,913	$180,678	$18,900	$8,635	$27,535	$180,678	15.2%

Figure 9-22 Cost status report.

EXAMPLE CLIENT
EXAMPLE BOILER REPLACEMENT
Example Project Location
Example Client's Project Numbers
Work-Hour Status Report
As of July 20, 2016

Description	A Original Budget Work-Hours	B Approved by Change Order Work-Hours	C Growth Work-Hours	D = A + B + C Budget at Completion Work-Hours	E Cumulative to Last Period Work-Hours	F This Period Work-Hours	G = E + F Hours to Date Work-Hours	H Estimate at Completion Work-Hours	I = G/D Percent Complete %
DIRECT ENGINEERING									
Architectural	240		104	344	0.0	0.0	0.0	344.0	0.0%
Civil	100		94	194	4.0	35.0	39.0	194.0	20.1%
Electrical	324		70	394	25.0	13.0	38.0	394.0	9.6%
Mechanical	828		42	870	135.0	53.0	188.0	870.0	21.6%
Structural	200		352	552	3.0	0.0	3.0	552.0	0.5%
Subtotal	1,692	0	662	2,354	167	101	268	2,354	11.4%
INDIRECT ENGINEERING									
Project Manager	390	210		600	120.0	32.0	152.0	600.0	25.3%
Scheduler	0	400		400	79.0	26.5	105.5	400.0	26.4%
Clerical	0	150		150	24.0	22.0	46.0	150.0	30.7%
Subtotal	390	760	0	1,150	223	81	304	1,150	26.4%
EXPENSES									
CADD Machine Hours	961			961			0.0	961.0	0.0%
Subtotal	961	0	0	961	0	8	0	961	0.0%
TOTAL	3043	760	662	4465	390	189.5	571.5	4465	12.8%

Figure 9-23 Work-hour status report.

project and is calculated in the spreadsheet as the sum of the data in the first three columns.

Columns (E), (F), and (G) represent the cumulative, current, and to-date expenditure of budget as defined by the titles at the top of the columns. The percent complete for the project is calculated by dividing the "cost to date" by the "budget at completion."

It is necessary to report the costs and work-hours separately because there are different wage rates for each classification of project personnel. In addition, there are different wage rates within each discipline. Therefore, the status of a project cannot be accurately determined by reporting only work-hours or only costs. Both must be reported and analyzed.

References

1. "Cost and Schedule Control Systems Criteria for Contract Performance," DOE/MA-0087, U.S. Department of Energy, Washington, D.C., 1980.
2. Diekmann, J. E. and Thresh, K. B., *Project Control for Design Engineering,* Source Document No. 12, Construction Industry Institute, Austin, TX, May 1986.
3. Eldin, N. N., "Measurement of Work Progress: Quantitative Technique," *Journal of Construction Engineering and Management,* ASCE, Reston, VA, Vol. 115, No. 3, September 1991.
4. Elmore, R. L. and Sullivan, D. C., "Project Control Through Work Packaging Concepts," *Transactions of the American Association of Cost Engineers,* Morgantown, WV, 1976.
5. Fleming, Q. W. and Koppelman, J. M., *Earned Value Project Management,* Project Management Institute, Newtown Square, PA, 1996.
6. Geigrich, D. B. and Schlunt, R. J., *Progress Schedule Tracking System,* PMI Symposium, Project Management Institute, Newtown Square, PA, 1984.
7. Halpin, D. W., Escalona, A. L., and Szmiurlo, P. M., *Work Packaging for Project Control,* Source Document No. 28, Construction Industry Institute, Austin, TX, August 1987.
8. Harris, F. W., *Advanced Project Management,* Wiley, New York, NY, 1981.
9. Hetland, P. W., "Toward Ultimate Control of Megaprojects in the North Sea," *Global Project Management Handbook,* McGraw-Hill, New York, NY, 1994.
10. Huot, J. C., "Integration of Cost and Time with the Work Breakdown Structure," *Transactions of the American Association of Cost Engineers,* Morgantown, WV, 1979.
11. Mueller, F. W., *Integrated Cost and Schedule Control for Construction Projects,* Van Nostrand Reinhold Company, New York, NY, 1986.
12. Neil, J. N., *Construction Cost Estimating for Project Control,* Prentice-Hall, Englewood Cliffs, New Jersey, 1982.
13. Neil, J. N., *Project Control for Engineering,* Publication No. 6-1, Construction Industry Institute, Austin, TX, July 1986.
14. Neil, J. N., *Project Control for Construction,* Publication No. 6-5, Construction Industry Institute, Austin, TX, September 1987.
15. Neil, J. N., "A System for Integrated Project Management," *Proceedings Conference on Current Practice of Cost Estimating and Cost Control,* ASCE, Reston, VA, 1983.
16. Oberlender, G. D., "Real-Time Project Tracking, Excellence in the Constructed Project," *Proceedings of Construction Congress I,* ASCE, Reston, VA, 1989.
17. Rasdorf, W. J. and Abudayyeh, O. Y., "Cost and Schedule Control Integration: Issues and Needs," *Journal of Construction Engineering and Management,* ASCE, Reston, VA, Vol. 117 No. 3, September 1991.
18. Riggs, L. S., *Cost and Schedule Control in Industrial Construction,* Source Document No. 24, Construction Industry Institute, Austin, TX, December 1986.
19. Stevens, W. M., "Cost Control: Integrated Cost/Schedule Performance," *Journal of Management in Engineering,* ASCE, Reston, VA, Vol. 2, No. 2, June 1986.

10

Design Coordination

Design Work Plan

To effectively coordinate the design process, the design team leader must develop a work plan. The design work plan should be developed during the design proposal preparation as discussed in Chapter 7. As presented in Chapter 6, any work plan must include the scope, budget, and schedule for performing the work. The design work plan becomes the basis for integrating and interfacing the work to be performed by the various design disciplines. The plan also forms the basis for monitoring scope, cost, and schedules of the design effort.

For design, the scope of work defines the design deliverables, the drawings and specifications, for each discipline in the design effort. The design cost is usually measured in work-hours, rather than dollars of cost. Typically, a milestone bar chart is used to schedule the design work packages. However, for large projects a CPM schedule is recommended.

After reviewing all backup material, the design team leader develops a work breakdown structure (WBS) that defines the various work packages required to generate the design deliverables, the drawings and specifications. Then, the number of design-hours is assigned to each work package for each discipline, including architectural, civil, structural, electrical, and mechanical. Then, a milestone bar chart is developed for the design work packages. The milestone bar chart is cost loaded as presented in Chapter 8 by assigning the number of work-hours to each task in the bar chart. Output from the cost loaded bar chart is usually produced on a weekly basis for monitoring and coordinating the project during design. Then, a simple earned-value analysis as presented in Chapter 9 can be performed for each week during the design process. By multiplying the percent complete times the budgeted design-hours for each task, the earned value can be determined.

The earned value can be compared to actual design-hours billed to the job and the planned design-hours to measure the performance of the design process.

Common Problems in Managing Design

Managing the design phase of a project is difficult because the work begins with limited information, only the information that was approved in the design proposal. There is pressure to start work immediately, although the scope is often not well defined, the budget is not tied to the scope, and the schedule only consists of a start and end date with possibly a few intermediate milestone dates. All these issues lead to problems for the project manager in charge of design. Common complaints of design project managers include

- Project scopes are not well defined
- Too much growth from the beginning to the end of projects
- Project budgets are not tied to scope
- Inconsistency of workload, either too busy or nothing to do
- Trying to handle too many projects at the same time
- Shifting priorities from one project to another
- Poor communications, misunderstandings, and too much rework
- The company is not organized to handle projects
- No system exists to handle projects
- No overall, understandable work plan

Producing Contract Documents

The introduction of the computer for evaluating designs and producing contract documents provides opportunities for evaluating numerous design alternatives that were prohibitive using traditional hand methods of the past. However, the excessive use of computers can lead to overdesigning, overwriting, and overdrafting. Examples are the process engineer or hydraulics engineer wanting to run one more simulation, or the structural engineer wanting one more computer run for another load check. Other examples include the CADD operator wanting to see how a drawing will look if another change is made, or the specification writer continuing to cut and paste a new section in the contract documents. Sometimes oversimulating, overdesigning, or overwriting can lead to errors in the contract documents.

The lead designers must develop a system of monitoring the design effort to ensure the work is progressing without excessive billable hours, but is still producing adequately defined plans and specifications for the contractors to execute the work during construction. This will reduce construction contractor complaints of pretty drawings that are full of errors and lack constructability.

Managing Scope Growth during Design

There is a tendency for some designers to make changes during design to please the client without regard to the impact of the change on the project's cost and schedule. Changes can be cataloged as either project development or scope growth. Project development relates to changes that are needed to accommodate the scope as currently defined. Scope growth relates to changes that alter the project's original scope, the scope that was approved before starting the design process.

For any design effort there must be a process to control scope growth. Both the owner and the engineer must be committed to scope and change control. The owner must be serious about freezing project scope after the conceptual design stage. Every proposed change must be subjected to a formal review and approval process that considers cost and schedule implications, plus consequential effects on other activities. The authority to approve changes during design must be limited.

The owner and engineer should agree on a change management philosophy and plan. For example, under what conditions will changes be considered: If it won't work? If there are legal implications? If there are environmental impacts? When changes are proposed, there must be an answer to such questions as: Does the change add value and is the change necessary? A "no-later-than" date for freezing scope should be agreed to by the owner and the engineer.

Although freezing project scope after the conceptual design stage is the best practice for successful project management, it should be recognized that on some projects the owner must operate in a very competitive market to produce a product. In a competitive market environment the owner may want the flexibility to modify the project scope during design, and even during construction, to best suit the functionality of the completed project. For this type of situation the work of the design team is more complicated and a special effort must be made to keep the owner informed of the full impact of scope changes. The cost of engineering and construction for the scope change must be evaluated against the future financial benefits to the owner, including revenues, operations, and maintenance of the completed facility.

The budget for the project should include a contingency for changes in scope and a management reserve for scope growth. Any proposed change must be transmitted to all discipline managers or others whose work may be affected by the change. These managers then should determine and report the cost and schedule impacts if the change is implemented. Work must not proceed on the change until the above impacts have been reviewed and approved by the owner's representative, the engineering manager, and the project manager. Once a change has been approved, it is important to communicate the reasons for the change to those affected by the change.

Managing Small Projects

It is common for a project manager of small projects to have simultaneous responsibility for multiple projects. The problem is not the management of any one project. The problem is the project manager's ability to give each project the attention needed, which complicates schedule and resource control. For small projects, the project manager must share resources with other project managers because minimal staff is available. This means that the few individuals that are assigned to the project must take responsibility for multiple functions. Time management is crucial. The project manager often finds himself or herself either waiting for information or trying to address pressing needs of several projects at the same time.

The relatively short duration of small projects provides insufficient time for detailed planning or correction of problems at the time the work is being performed. The learning curve for personnel is generally still climbing by the time the project is completed.

To manage multiple small projects, the project manager must develop a single master schedule that contains all the projects under his or her responsibility. This will assist in reducing situations when two or more support personnel are needed at the same time. The problem is scheduling the multiple tasks of the project manager, rather than scheduling the tasks on only one of the projects.

There are more project managers managing small projects than large projects. Many engineering companies create organizational units dedicated to project management of small projects. A special attitude and attention is needed by upper management to recognize the problems associated with management of small projects. Table 10-1 provides attributes of project team personnel who work well in managing small projects.

Project Team Meetings

Design is a creative process that involves diverse areas of expertise and numerous decisions that have major impacts on a project. The work of

TABLE 10-1   Attributes of Small Projects Team Personnel

1. Have a "can do" attitude
2. Prefer a hands-on approach to work
3. Dislike bureaucracy
4. Are decision makers
5. Need little or no supervision
6. Have a value system to make the customer satisfied
7. Are good communicators
8. Prefer to talk out problems
9. Know when to stop an activity when things are going wrong
10. Has the personality and people skills to coerce people to be responsive to his or her needs
11. Has the ability to navigate through the various departments of his or her company to get things done

each designer often influences the work of one or more other designers. A difficult task in design coordination is interfacing related work to ensure compatibility of the whole project. Generally, the problem is not finding design people who know how to do the work, it is interfacing the work of all designers. This can only be accomplished through effective communications at regularly scheduled team meetings.

Team meetings should be held weekly throughout the duration of a project. These meetings are necessary to keep the team acting as a unit and to ensure a continuous exchange of information. A typical project involves numerous conflicts. The best way to resolve these conflicts in a timely manner is by input from all those who are affected. This can only be achieved through open discussions and compromise.

The project manager is the leader of all team meetings; however, he or she should not dominate discussions. Often a team member may be assigned the role of leading discussions to resolve a problem that is related to his or her particular area of expertise. Project managers need to use their own judgement and develop the skillful art of knowing when to lead and when to let others lead.

An agenda should be prepared to direct project team meetings to ensure that important items are addressed and to conclude the meeting in the shortest time. The agenda should include a list, in chronological order, of the items to be discussed, including work completed, work in progress, work scheduled, and special problems. Each attendee should participate in team meetings. The project manager should prepare and distribute minutes of the meeting to all participants.

The project manager must ensure that meetings are productive. Meetings are inevitable and important, but they can be a source of irritation and wasted time if not properly planned and conducted. Table 10-2 provides guidelines for conducting team meetings.

TABLE 10-2   Guidelines for Productive Meetings

1. Develop and publish an agenda in advance to permit better participation by attendees.
2. List unfinished items from previous meetings on the agenda, including the names of individuals who are responsible for reporting on status.
3. Restrict attendance only to individuals who need to attend.
4. Don't waste time by discussing events that do not pertain to the purpose of the meeting.
5. Pick a meeting leader who is a leader and facilitator, not a dictator, to ensure the meeting is conducted in an informative environment.
6. Maintain strict agenda control; follow the items in order with set time limits for discussions.
7. Avoid interruptions, such as phone calls, as much as possible.

Weekly/Monthly Reports

Project management involves a never-ending process of preparing reports. To be meaningful, reports must be issued on a regular basis and should contain information that is beneficial to the receiver. There is a tendency to include everything in a report, which results in reports that are so bulky that important items may be overlooked.

In general, the project manager should prepare two routine reports, a weekly highlight report and a monthly report for each project. Much of the weekly highlight report can be obtained from the minutes of the weekly team meetings. The report should include: work completed, work in progress, work scheduled, and special problems. Generally, the weekly report is used by the project manager and his or her team to coordinate the work in progress.

The monthly report for a project should contain milestones that have been achieved, a tabulation of costs to date compared to forecast costs, and an overlay of planned and actual time schedules. Trend reports should also be included to show the anticipated project completion date and a forecast of the total cost at completion. Generally, the monthly report is used by upper management and the owner's representative and is a permanent record for the project file.

A format for both the weekly and monthly reports should be developed so all reports will be consistent, to allow comparisons of the project status, and to evaluate the progress of work and the performance of the team. In addition to communicating the status of a project, the reports serve as a means of individual accountability and recognition of good performance.

Drawing Index

The final product of design work is a set of contract documents (drawings and specifications) to guide the physical construction of the project.

As presented in Chapters 6 and 7 the project manager must develop a project work plan before starting design. A part of the work plan includes preparation of work packages by each designer. Included in each work package is a list of anticipated drawings and expected completion dates.

The project manager can develop a drawing index by assembling the list of drawings from all team members. This drawing index is extremely valuable to the project manager because it can be used as a checklist of how many drawings to expect, when to expect them, and to assist in scheduling construction. Figure 10-1 illustrates the contents of a drawing index. It is helpful to include the revision number and date for each drawing because a common problem in design coordination is keeping track of the most current issue of a drawing.

The drawing index is in a continuous state of change as design progresses. As work progresses, the number of drawings may increase or

Sheet No.		Title	Expected date	Actual date	Revision No. 1 2 3 4
Drawing Index for X Y Z Project Update as of: October 26, 2016					
C0		**Site-Work Drawings**			
C1	"	Plot Plan	09/15/2016	09/17/2016	1 - 10/05/2016
C2	"	Grading	10/01/2016	09/27/2016	
C3	"	Paving, Sh 1	10/15/2016	10/12/2016	
C4	"	Paving, Sh 2	10/20/2016	10/18/2016	3 - 10/24/2016
A0		**Architectural Drawings**			
A1	"	First Floor Plans	10/10/2016	10/09/2016	
A2	"	Second Floor Plans	10/15/2016	10/16/2016	
A3	"	Room Schedules	10/20/2016	10/20/2016	
A4	"	Door/Wall Finishes	10/25/2016	10/24/2016	
A5	"	Window Schedules	11/01/2016		
S0		**Structural Drawings**			
S1	"	Foundation	10/15/2016	10/16/2016	
S2	"	Floor Slabs	10/25/2016	10/28/2016	
S3	"	Columns & Beams, Sh 1	11/01/2016		
S4	"	Columns & Beams, Sh 2	11/05/2016		
S5	"	Roof Framing Plan	11/10/2016		
M0		**Mechanical Drawings**			
M1	"	Compressor	11/01/2016		
M2	"	HVAC Ductwork	11/05/2016		
E0		**Electrical Drawings**			
E1	"	Control Boxes	11/05/2016		
E2	"	Wiring Diagram, Sh 1	11/10/2016		
E3	"	Wiring Diagram, Sh 2	11/15/2016		
P0		**Plumbing Drawings**			
P1	"	Piping	10/10/2016	10/14/2016	
P2	"	Fixtures	10/20/2016	10/18/2016	

Figure 10-1 Illustrative drawing index.

Equipment Index for X Y Z Project Update as of: May 25, 2016					
Description	Drawing number	Vendor	Purchase order No.	Purchase order date	Expected delivery
Transformer	E14	XYZ Company	PO 73925	04/12/2016	07/15/2016
Switch Assy	E19	ABC Company	PO 83401	04/30/2016	08/01/2016
Compressor	M12	CIB Company	PO 84294	05/17/2016	07/20/2016
Pump Assy	M3	BWA Company	PO 17835	06/14/2016	08/15/2016

Figure 10-2   Illustrative equipment index.

decrease, depending upon the final design configuration that is selected. Changes in drawings are a necessary part of work and it is the duty of each designer to revise the list as necessary and to keep the project manager informed.

Equipment Index

Similar to the drawing index, an equipment index can be developed that lists major equipment to be installed in the project. The equipment index is extremely valuable to the project manager because it can serve as a reference document to track the purchase and delivery of equipment and schedule installation in the field during construction. Figure 10-2 illustrates the contents of an equipment index. Each major piece of equipment is listed along with the drawing number, vendor, purchase order date, and expected delivery date. The project manager can use this index as a reference guide and checklist to contact vendors in advance of delivery dates to ensure the equipment will arrive on schedule. A common source of delay during construction is late delivery of major equipment and material.

Distribution of Documents

The design process requires the timely distribution of documents and exchange of information. Generally, there is a sense of urgency among team members to complete work at the earliest possible date. When information is inefficiently distributed, it increases the work load of everyone and leads to delays in work, inefficient productivity, and frustration.

The project manager can develop a distribution of documents key sheet (reference Figure 10-3) that shows the routing of documents among team members and other major participants in the project. This is an effective communication tool because it allows each person to know who is receiving a particular document. Too often, a team member will receive a

Distribution of Documents Key Sheet
for X Y Z Project

Send to	Name	Title	Address	Phone	Date

Comments:

Action: See Me______ Call Me_____ Review_____

Distributed by:______________________________ Date:___________

Figure 10-3 Distribution of documents key sheet.

document, review its contents, and determine that another team member should also be aware of the document. The distribution of documents key sheet verifies the recipients, thereby eliminating the necessity of attempting to contact another person who may not be readily available.

There should be a distribution of documents key sheet that is unique to each project. It can be easily prepared on a word processor, duplicated, and bound in a tablet form for use by all team members. It differs from a traditional company routing slip because it has the project title at the top of the sheet along with the names of the team members so it easily identifies which project the document is directed to and who is receiving the document.

Authority/Responsibility Checklist

A project manager is often responsible for one or more projects at a time. Some of the projects may be in the early stage of development while others are in full progress and some are in the close-out phase. The problem that confronts a project manager who must handle multiple projects generally is not the management of any one project, but it is the difficulty in handling all projects simultaneously. This type of work

Authority/Responsibility Checklist
X Y Z Project as of February 26, 2016

Task/Work item	Authority/Responsibility	Status	Date
Soil testing specifications	John Smith, Civil	Issued	02/26/2016
Confirmation of wind loads	Tim Jones, Structural	Pending	04/22/2016
Approval of shield wire angle	Dan Banks, Owner's Rep.	Approved	03/09/2016
Final tower configuration	Tim Jones, Structural	Expected	05/26/2016
Steel supplier's bid list	Ron Mitchell, Purchasing	Will Call	04/03/2016
Right of way hearings	Joe Thomas, Legal	Pending	06/01/2016

Figure 10-4 Authority/responsibility checklist for project manager.

condition requires some systematic means of knowing the work status of numerous people.

To facilitate and keep track of team members, the project manager can develop an authority/responsibility checklist for each project (reference Figure 10-4). This list is continuously revised to show the status of completed, active, and pending work of each team member. It is especially useful in preparing agendas for team meetings. It also assists the project manager in organizing his or her own work and preparing reports to upper management and the owner's representative.

The authority/responsibility checklist can be prepared as a computer word processing file for each project. The project manager then can easily add, delete, or modify the information as it becomes known. Each update of the file can be saved on the disk under a new file name, with date, so a record can be kept of each update. This provides a thorough documentation of the history of the project, which is valuable for future reference in retracing the events in a project. The file of each project can be merged into a single file to provide a composite listing of the information that pertains to all projects for which the project manager is responsible.

A more effective means of managing the authority/responsibility checklist can be achieved by using a computer electronic spreadsheet. Each piece of information can be entered into a cell to enable the project manager to produce multiple sorts of the data. For example, a sort can be obtained for all work that is pending by a specific date, or a sort can be obtained to list all work of a particular person. The use of a spreadsheet enables the project manager to perform any desired sort and is a valuable tool for management of any single project or multiple projects.

Checklist of Duties for Design

Table 10-3 provides a comprehensive checklist of duties for the design phase of a project that is handled by the construction management (CM)

TABLE 10-3   Checklist of Duties for Design Phase

Design phase	Owner	CM	Designer*	Contractor
1. Project team	Head	Member	Member	×
2. Program information	Provide	Review	Obtain/Approve	×
3. Information required for design	Provide	Review	Determine requirements Obtain, approve	×
4. Meetings—Project team	Participate as required	Organize, conduct, record	Participate	×
5. Meetings—Design team as required	Participate as required	Participate, document	Organize, conduct	×
6. Budget	Provide	Evaluate	Evaluate	×
7. Budget contingency	Provide	Recommend	Recommend	×
8. Scheduling—Program	Participate, approve, comply	Prepare, monitor, enforce, comply	Participate, approve, comply	Comply
9. Scheduling—Milestone	Review	Provide	Review, approve	×
10. Soil bearings	Review & pay	Review	Arrange, review & recommend	×
11. Drawings—Preliminary	Review, comment, approve	Review, comment, advise	Provide	×
12. Approvals—Design	Approve	Recommend	Issue documents	×
13. Drawings—Schematic, design development, and working	Review, comment, approve	Review, comment, advise	Provide	×
14. Specifications—Conceptual	Review, comment, approve	Review, comment, advise	Provide	×
15. Specifications—Technical	Review, approve	Review, comment, advise	Provide	×
16. Design alternates	Approve as required	Recommend	Recommend, prepare	×
17. Value engineering & value management	Approve as required	Provide	Assist, review	×
18. Estimating	Approve as required	Provide	Assist, review	×
19. Agency plan reviews	Monitor	Monitor	Facilitate	×
20. Permits—Not assigned to contractors in bid documents	Pay	Arrange for & obtain	Assist, consult with agencies	×
21. Insurance—All risk-builders risk	Provide	Assist, recommend	×	×
22. Insurance—Owner's protective liability	Provide	Assist, recommend	×	×
23. Cash flow projections	Provide, utilize	Provide, update	Assist	×

(*Continued*)

TABLE 10-3   Checklist of Duties for Design Phase (*Continued*)

Design phase	Owner	CM	Designer*	Contractor
24. Specifications— Instructions to bidders	Approve	Provide	Review & advise	×
25. Specifications— Proposal section	Approve	Provide	Review & advise	×
26. Bidding alternates	Approve as required	Recommend, review	Recommend, review issue	×

Source: NSPE / PEC Reporter, Vol. VII, No. 2.
*The referenced NSPE/PEC publication cited "Architect"; however, the term "Designer" is used here to imply the architect may be the principal design professional for building projects, whereas the engineer may be the principal design professional for heavy/industrial projects.

type of contract. For projects other than the CM type of contract, the duties of the CM are distributed between the owner and designer, depending upon the contract arrangement. The architect is the principal design professional for a building type project, whereas the engineer is the principal design professional for a heavy/industrial type project.

Team Management

Effective teamwork is a key factor in the successful management of any project. Usually, the project manager is involved in three areas of responsibilities: within the project team, between the team and the client, and between the team and other management of the project manager's organization. In each of these areas, numerous situations often arise that can cause disruptions, conflicts, delays, and misunderstandings that may affect the team's performance. It is the duty of a project manager to coordinate the team in a manner that minimizes the development of these situations, to be able to identify problems immediately, and to act quickly to resolve them in a satisfactory manner. Table 10-4 is a list of typical problems that can arise related to team management.

A team is a group of individuals, each of whom is usually responsible for performing work on more than one project. As the number of projects for which a team member is involved increases, the risk of losing

TABLE 10-4   Typical Problems in Team Management

1. Differing outlooks, priorities, interests, and judgements of team members
2. Project objectives become unclear to the team members
3. Communication problems
4. Scope changes by the owner
5. Lack of coordination by team members
6. Lack of management support

priorities also increases. The project plan must include setting priorities at the start of the project. The project manager must ensure that project objectives are known and clearly understood by team members because project objectives can become unclear to the team, particularly during the design phase of a project.

Poor communication is the most common source of problems related to team management. Meetings must be held routinely to keep team members informed so information can be exchanged. Team meetings should serve as forums for getting problems out on the table so they can be resolved; however, the project manager must also be accessible to team members who may not feel comfortable airing problems or potential problems in meetings.

Scope changes by the owner can adversely affect the project budget and schedule. Team members must be cautious of expanding the scope of a project in an attempt to please the owner. The scope must be locked in at the beginning of the project and should not be changed without a written agreement between the owner's representative and the project manager. The written agreement must include an appropriate change in the budget and schedule that matches the change in scope.

The project manager is responsible for overall coordination of the entire project team. However, individual coordination between team members is also necessary. The project manager must instill an environment of cooperation that encourages the free exchange of information among team members.

The project manager must have management support from his or her own organization. This can best be achieved by keeping management informed of the status and needs of the project team. Assistance cannot be given unless the need is known. Management should be included as an integral part of project reviews.

To assist the project manager in his or her role of performing team management responsibilities, the project work plan must be kept up to date. It must be formally modified when changes are made in the scope of work, members of the project team, services and tasks to be performed, individual responsibilities, schedule, and budget.

The team must be managed in an open manner. Discussions of roles and relationships, and the rationale behind project decisions, should be done in team meetings rather than private conversations. The project manager must be consistent in making the day-to-day decisions that alter roles and relationships within the team.

The concept of "team" should be instilled in the attitude of all individuals. The project manager must emphasize commitment, clarity, and unity of the team to minimize the number of trivial conflicts. Since each person is unique, there can be differing outlooks, priorities, interests, and judgements. Because of the diversity of individuals, the project manager must deal with the human aspects in his or her role as leader of the project team. Some leadership qualities are inherent in a project manager and others require development through training and experience.

Evaluation of Design Effectiveness

Design is a complex process involving the application of technical knowledge to creative ideas in order to produce a set of specific instructions for construction of the project. It is the focal point of definition of the project and has a significant impact on cost and schedule. Thus, it is essential to have the most effective design possible.

Most of the efforts related to measurement and evaluation of productivity and performance of project work have been directed toward the construction phase. The Construction Industry Institute (CII) has sponsored research and published numerous papers on a variety of topics related to project management. *Evaluation of Design Effectiveness* is a CII document that provides a thorough description of outputs, or products, of the design process.

The following paragraphs are a summary of excerpts from the CII report *Evaluation of Design Effectiveness*. The report states this method is not intended to be, nor should it be, used as an evaluation of a designer or the design process. It is instead a measure of design effectiveness and an evaluation of design outputs.

Measurement of design productivity is perhaps more difficult than measuring productivity in the construction phase. Simplistic measurements such as cost per drawing or work-hours per drawing have obvious limitations because of variations in drawing size and content. Also there is increasing realization that the true measures of the effectiveness of the design effort are found in the construction of a project and use of the project by the owner after construction is complete. Thus, it may be more beneficial to develop a method for evaluating the effectiveness of design rather than the productivity during the design activity itself.

The ability to measure design effectiveness using the following proposed method represents an important step in a broader effort to improve the total design process. Such an effort encompasses the identification of the effect on design effectiveness of various inputs to the design process and of the systems and techniques employed by the designer.

CII researchers have suggested the use of a technique called an *objectives matrix* for productivity evaluation. The same concept can be used to develop an effectiveness measurement for design. An objectives matrix consists of four main components: criteria, weights, performance scale, and performance index.

The criteria define what is to be measured. The weights determine the relative importance of the criteria to each other and to the overall objective of the measurement. The performance scale compares the measured value of the criterion to a standard or selected benchmark value. Using these three components, the performance index is calculated, and the result is used to indicate and track performance.

Use of the objectives matrix is illustrated in Figure 10-5. The seven criteria of design effectiveness are given as column headings; the criteria weights are shown near the bottom of each column. The performance scale of 0 to 10 is given at the right of the matrix. A score of 10 represents perfection, and a score of 3 is average.

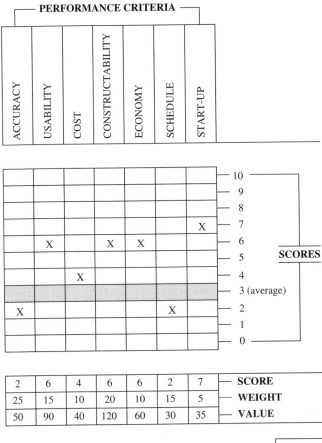

Figure 10-5 Design evaluation matrix. (*Source:* Construction Industry Institute, Publication No. 8-1.)

A score for each of the seven criteria is represented by an "X" in the appropriate box and is entered at the bottom of the matrix. Each criterion's score is multiplied by its weight to obtain its value. The sum of all values constitutes the performance index, which is shown at the bottom right of the matrix.

The score for a criterion can be obtained in at least three ways: judgmental, based upon a single quantitative measurement, or based upon a combination of several subcriteria that are represented by a matrix.

Judgmental scoring can be used for some or all criteria. The scoring in Figure 10-5 is representative of judgmental scoring. Although judgmental ratings have the limitation of subjectivity, the objectives matrix approach still allows their appropriate application with multiple criteria of different weights.

For some criteria, quantitative measures can be used instead of judgements to determine scores. For example, the accuracy of the design documents can be evaluated by measuring the amount of drawing revisions per total drawings. The score for performance against the schedule criterion can be determined by using the percent of design document release dates attained. This approach is illustrated in Figure 10-6, in which these two criteria are represented by quantitative measures.

In the example of Figure 10-6 predetermined benchmark values are entered into the boxes representing appropriate scores for each criterion. The benchmark value for a score of 3 is considered normal or average, while that for a level of 0 represents the minimum level of accomplishment realized in recent years. The benchmark value for a level of 10 represents the ultimate expected in the foreseeable future. The score of 3 is used for normal rather than 5 to allow more opportunity for improvement.

The performance value attained is entered at the top of each column. The appropriate score representing that performance is then determined and is circled in Figure 10-6. If a performance level falls between two scores, the lower score is used. A score is not given until it is attained.

A third approach is to measure each criterion through the use of several subcriteria. The subcriteria themselves can have differing weights and measurements. These can be combined into a single criterion score through the use of a submatrix for that criterion.

Given the many complexities and variables of the total design process, no measurement system can yield absolute quantitative results that are applicable without an interpretation of all the design situations and circumstances. However, the method outlined can be utilized for three purposes: to develop a common understanding between the owner, designer, and contractor concerning the criteria by which design effectiveness on a given project will be measured; to compare design effectiveness of similar projects in a systematic and reasonable quantitative

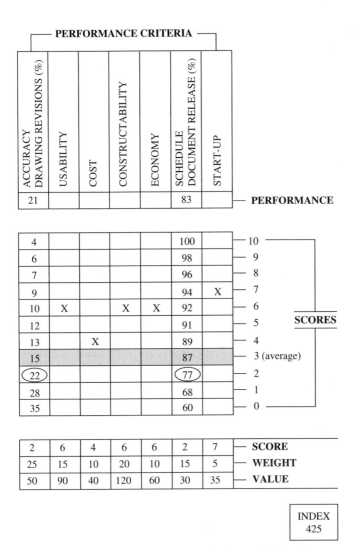

Figure 10-6 Design matrix using quantitative evaluation. (*Source:* Construction Industry Institute, Publication No. 8-1.)

manner, highlighting performance trends; and to identify opportunities to improve the effectiveness of the entire design process as well as the contributions toward the ultimate result made by all participants.

Constructability

Traditionally, engineering and construction have been separated early in the project. The adoption of new technology such as three-dimensional computer-aided drafting and design (CADD), robotics, and automation in construction has generated increased interest in the constructability of a project. With these new innovations, designs can be configured to enable efficient construction, which places more emphasis on merging engineering and construction to include constructability's input in the design effort. The desired result is to facilitate the exchange of ideas between construction and design before and during design, rather than after design.

The CII has sponsored research and published reports related to constructability. CII Publication No. 3-3, entitled *Constructability Concepts File,* provides a good description of constructability concepts related to conceptual planning, design and procurement, and field operations. The following paragraphs contain excerpts from the report to illustrate the contents of the report.

There are at least five factors that should be considered in constructability deliberations related to design configurations for efficient construction: simplicity, flexibility, sequencing, substitution, and labor skill/availability.

Simplicity is a desirable element of any constructable design. Unwarranted complexity is not in the best interest of anyone and markedly increases the probability of an unsatisfactory finished product. Special drawings and instruction may be required to improve the constructability process, particularly for retrofit or rebuild projects.

Flexibility for the field construction personnel to select alternative methods or innovative approaches is highly desirable. Designs should specify the desired results and not limit approaches to attain these results. In the fully open and competitive market, it is highly desirable to provide designs that do not limit the construction methods or approaches.

Sequencing of installation is as much a design consideration as it is a procurement and construction consideration. Many times designs have evolved that unnecessarily restrict installation sequences during construction. Design should include careful consideration of layout and spacing of facilities so more than one construction operation can occur at a time.

Substitutions or alternatives warrant attention, but too often are neglected because the attitude prevails that it has always been done a particular way. Improperly considered material applications will impact constructability, resulting in costly modifications. These impacts can be lessened and eliminated when addressed during the design phase by constructability programs.

Labor skill/availability are often not considered early enough in a project life cycle. The availability of labor and the skill level of the workers should be fully explored. The absence of either skill levels or availability of the work force can have a costly impact on a project and require consideration during the design phase.

CII research has shown that company or project size is no barrier to constructability and the implementation of a constructability program. The involvement of construction in the design phase results in better projects, lower costs, better productivity, and earlier project completion. A major obstacle to the implementation of effective constructability programs is the concept that designs are merely reviewed by construction to select the design that is easiest to build. CII Publication No. 3-2 provides guidelines for implementing a constructability program.

Post Design Review

Evaluation is a continuous process and necessary for improvement of project management. The system used by the project manager's organization to handle work must be flexible enough to respond to differences in individual projects. At the beginning of each project, the project manager must determine what modifications and improvements need to be made in the system that are appropriate for the project.

After completion of the design for each project, the project manager and his or her team should conduct a complete and candid evaluation of the design effort and the management of the design process. This evaluation should include each member of the project team as well as other key participants that were involved in design.

A checklist should be prepared to evaluate all aspects of the project, including scope growth, match of quality and scope, owner's expectations and satisfaction, conflicts within the team or other parties, excessive changes in schedules, comparison of final costs with the original budget, and a list of precautions for management of future projects.

After a thorough discussion of the design process, a brief summary report should be prepared by the project manager that should include a list of recommendations to improve the system for future projects.

References

1. Chalabi, A. F., Beaudin, B. J., and Salazar, G. F., *Input Variables Impacting Design Effectiveness,* Source Document No. 26, Construction Industry Institute, Austin, TX, April 1987.
2. *Constructability Concepts File,* Publication No. 3-3, Construction Industry Institute, Austin, TX, August 1987.
3. *Consulting Engineering: A Guide for the Engagement of Engineering Services,* Manual No. 45, American Society of Civil Engineers, Reston, VA, 1988.
4. "Effective Management of Engineering Design," *Conference Proceedings,* ASCE, Reston, VA, 1981.
5. Eldin, N. N., "Management of Engineering/Design Phase," *Journal of Construction Engineering and Management,* ASCE, Reston, VA, Vol. 117. No. 1, March 1991.

6. *Engineer, Owner, and Construction Related Documents,* Engineers' Joint Contract Documents Committee, ASCE, Reston, VA, 2013.
7. Glavinich, T. E., "Microcomputer-Based Project Management for Small Engineering Firms," *Journal of Management in Engineering,* ASCE, Reston, VA, Vol. 8, No. 1, June 1991.
8. O'Connor, J. T., Rusch, S. E., and Schulz, M. J., *Constructability Improvement During Engineering and Procurement,* Source Document No. 5, Construction Industry Institute, Austin, TX, May 1986.
9. Perkins, C., *Guidelines for Implementing a Constructability Program,* Publication No. 3-2, Construction Industry Institute, Austin, TX, July 1987.
10. "Sampling Responsibility Chart," *PEC Reporter,* Professional Engineers in Construction Practice Division, National Society of Professional Engineers, Alexandria, VA, Vol. VII, No. 2, January 1985.
11. Stull, J. O. and Tucker, R. L., *Objectives Matrix Values for Evaluation of Design Effectiveness,* Source Document No. 22, Construction Industry Institute, Austin, TX, November 1986.
12. Tatum, C. B., Vanegas, J. A., and Williams, J. M., *Constructability Improvement During Conceptual Planning,* Source Document No. 4, Construction Industry Institute, Austin, TX, March 1986.
13. Tucker, R. L. and Scarlett, B. R., *Evaluation of Design Effectiveness,* Publication No. 8-1, and Source Document No. 16, Construction Industry Institute, Austin, TX, November 1986.
14. Tucker, R. L., *Assessment of Architect / Engineer Project Management Practices and Performance,* Construction Industry Institute, Austin, TX, April 1990.
15. Vorba, L. L. and Oberlender, G. D., *A Methodology for Quality Design,* Symposium of the Project Management Institute, Newtown Square, PA, 1991.

Chapter

11

Construction Phase

Importance of Construction

The construction phase is important because the quality of the completed project is highly dependent on the workmanship and management of construction. The quality of construction depends on the completeness and quality of the contract documents that are prepared by the designer and three other factors: laborers who have the skills necessary to produce the work, field supervisors who have the ability to coordinate the numerous activities that are required to construct the project in the field, and the quality of materials that are used for construction of the project. Skilled laborers and effective management of the skilled laborers are both required to achieve a quality project.

The construction phase is also important because a majority of the total project budget and schedule is expended during construction. As presented in previous chapters, the design costs for a project generally range from 7% to 12%. Using a 10% medium value, then 90% of the cost of a project is expended during construction. Thus, a 15% variation in design costs may impact the project by only 1.5%, whereas a 15% variation in construction costs may impact the project by 13.5%.

Similar to costs, the time required to build a project is always disproportionally greater than the time required to design it. Most owners have a need for use of their projects at the earliest possible date; therefore, any delay from a planned completion date can cause significant problems for both the owner and contractor. Due to the risks that are inherent to construction, and the many tasks that must be performed, the construction contractor must carefully plan, schedule, and manage the project in the most efficient manner.

Assumptions for Construction Phase

The objective during the construction phase is to build the project in accordance with the plans and specifications, within budget and on schedule. To achieve this objective there are three assumptions as shown in Table 11-1.

Although the assumptions are reasonable, there are often variations due to the nature of construction work. A project is a single, non-repetitive enterprise. Because each project is unique, its outcome can never be predicted with absolute confidence. To construct a project the owner generally assigns a contract to a contractor who provides all labor, equipment, material, and construction services to fulfill the requirements of the plans and specifications. This requires simultaneously coordinating many tasks and operations, interpreting drawings, and contending with adverse weather conditions.

It is difficult for some individuals to acknowledge the fact that plans and specifications do have errors. The preparation of a design requires many individuals who must perform design calculations, coordinate related work, and produce many sheets of drawings that have elevations, sections, details, and dimensions. Although every designer strives to achieve a flawless set of plans and specifications, this is rarely achieved.

The owner generally accepts and approves the contract documents before commencing construction. However, the plans and specifications don't always represent what the owner wants. The interest of some owners, particularly non-profit organizations or public agencies, is represented by individuals who are members of a board of trustees, board of directors, or a commission. These individuals generally have a background in business enterprises and/or professional occupations with little or no knowledge of project work or interpretation of drawings. Thus, they may approve the selection of a material or configuration of a project without fully understanding what it looks like until it is being installed during construction.

Serious problems can arise for both the owner and contractor if the contractor submits a bid price that is lower than required to build the project, with a reasonable profit. A contractor that has underbid a

TABLE 11-1 Assumptions for Construction Phase

Scope	The design plans and specifications contain no errors and meet the owner's requirements and appropriate codes and standards.
Budget	The budget is acceptable; that is, it is what the owner can afford and what the contractor can build it for, with a reasonable profit.
Schedule	The schedule is reasonable; that is, short enough to finish when the owner needs it and long enough for the contractor to do the work.

project can also cause significant problems for the design organization. A construction company is a business enterprise that must achieve a profit to continue operations. A careful evaluation of each contractor's bid is necessary before award of a construction contract; because if a project is underbid by the construction contractor, the management of the project will be difficult regardless of the ability of the individuals that are involved.

Conditions can arise that alter the project budget and schedule, such as changes desired by the owner during construction, modifications of design, or differing site conditions. To reduce the impact of these conditions, there should be a reasonable contingency to allow for these types of variations that can adversely affect the project budget and schedule.

Sufficient time must be allowed for contractors to perform their work. If a reasonable time is not allowed, the productivity of workers and quality of the project will be adversely affected. There are always conditions that arise during construction that can disrupt the continuous flow of work, such as weather, delivery of materials, clarification of questions related to design, and inspection. The contractor must plan and anticipate the total requirements of the project and develop a schedule to allow for a reasonable variation of time that is inherent in the construction process.

The project manager must contend with problems as described above. He or she must always be alert to these situations and must continually plan, alter, and coordinate the project to handle the situations as they arise.

Contract Pricing Formats

The method that is selected to compensate the construction contractor can have a large impact on a project's cost, schedule, and the level of involvement by the owner and designer. Contract pricing may be divided into two general categories: fixed price and cost reimbursable. For fixed-price contracts the contractor may be compensated on a lump-sum or unit-price basis. Cost-reimbursable contracts may include methods of payment by any one or combination of the following: cost plus a percentage or fixed fee, guaranteed maximum price, or incentive.

Many books and articles have been written that discuss the advantages, disadvantages, and conditions that are favorable for the use of the above methods of payment for construction services. The following paragraphs are presented as a summary of what has already been written, to assist the project manager in his or her role of management.

The intent of lump-sum contracts is to fix the cost of the project by providing a complete set of plans and specifications that are prepared by the designer prior to construction. However, the contractor is entitled to extra compensation for any changes that may be necessary during construction. For lump-sum contracts, changes during construction are a major source of cost overruns. For these types of projects it is necessary to ensure a complete design that is as error free as possible and to keep any owner changes to a minimum. There should be an adequate review of the contract documents before bidding to detect any discrepancies that may exist and to confirm the constructability of the project. The project manager should work with the owner during construction to evaluate the full impact of a project change, including the effect on the project's cost and schedule, because a change in one area of the project often affects other areas of the project. There should be a rate schedule for labor and equipment for extra work related to project changes that is agreed upon before the construction contract is signed.

Unit-price contracts are awarded because the quantity of work may not be determined with a degree of accuracy to enable a contractor to submit a lump-sum bid. A major source of cost overruns for unit-price contracts is errors in the estimated pay quantities. Errors in estimated pay quantities can lead to unbalanced bids by contractors which can cause significant increases in the expected cost of the project and expensive legal disputes. There should be a thorough review of all estimated pay quantities of unit-price contracts before the request of contractor's bids. After receipt of all the bids, a careful review of each unit-cost bid item should be performed to detect any bid unbalancing. In particular there should be a review of large quantity pay items and any unusually large unit-cost bid items to detect irregularities.

For some projects it is desirable to start construction before design is complete, for example, projects that are complex in nature, or projects that must be completed due to emergency situations when it is not practical to produce a complete detailed design of the entire project before starting construction. Cost-reimbursable projects require extensive monitoring of material deliveries and measurement of work. The owner's organization must establish a field office to review and approve the costs of material, labor, equipment, and other costs associated with the project. This method of contracting can be efficient for owner organizations that need the flexibility of modifying the project as required, during construction, to produce the best end results to meet their needs. However, the owner must have extensive experience with handling projects.

Design/Bid/Build Method of Project Delivery

Design/bid/build (D/B/B) is often considered the traditional project delivery method. All design work is completed before starting the bid and construction process. This delivery method is usually selected for projects when cost is primary, schedule is secondary, and the scope is well defined.

The D/B/B project delivery method is a three-party arrangement involving the owner, designer, and contractor. The owner signs one contract with the designer for design services and signs another contract with the contractor for construction services. Both the designer and contractor work for the owner. The contractor does not work for the designer; however, the owner usually designates the designer as their representative during construction. The designer is usually paid based on a prearranged fee or on a percentage of the construction contract. The contractor is paid based on a lump-sum amount.

Since design is completed before bidding, the owner has the opportunity to know what the project will look like before proceeding into construction, when the biggest costs will be incurred. The contractor also has a clear understanding of the project requirements and is thereby able to estimate the cost of construction with a high degree of accuracy. This allows the owner to know the project cost before committing to sign construction contracts. In D/B/B the responsibility, risk, and involvement of all parties are well defined. The owner has a relatively high level of involvement and control during design, but low involvement during construction because the contract documents clearly define what the contractor must do.

The biggest disadvantage of the D/B/B project delivery method is the extended time that may be required for completing the design and bidding the project before starting actual construction. Changes made after the award of construction contracts can be costly to the owner.

Design/Build Method of Project Delivery

The design/build (D/B) project delivery method is often chosen to compress the time to complete the project. The completion time usually is reduced because construction can start before all the design is completed. The owner has considerable control and level of involvement throughout the project. This provides flexibility to the owner for revision of the design during construction. The D/B project delivery method is usually selected for projects when the schedule is primary, the cost is secondary, and the scope is not well defined.

The D/B method of project delivery is a two-party arrangement, involving the owner and the D/B company. A contract is signed between the owner and the D/B firm to perform both design and construction services. All design, including construction drawings, are done by the D/B contractor. All construction is done by the D/B contractor, although the D/B contractor may hire one or many subcontractors. The D/B firm typically has an in-house design staff as well as experienced construction personnel. This arrangement can reduce the conflicts between the designer and contractor that often occur in the D/B/B delivery method.

Sometimes a construction contractor will team with a design firm or a design company will team with a construction contractor to provide D/B work for the owner.

It is common to choose the D/B firm by a qualifications-based selection (QBS) procedure. The owner solicits proposals from a prequalified or prescreened list of firms. An evaluation process is used to assess quality, safety record, schedule, cost performance on past jobs, and other factors from each prospective D/B firm. Thus, selection is based on qualification rather than price. The cost of the D/B services is usually based on some type of cost-reimbursable arrangement, either cost plus a fixed amount or cost plus a percentage.

For projects that have a reasonably defined scope, the D/B firm is sometimes selected based on price. For incentives, the contract may be based on a guaranteed maximum amount, with bonuses when the final cost is less than the guaranteed amount and penalties when the final cost is more than the guaranteed amount.

Although the total project cost is a major consideration, the total cost is not well known at the beginning of a D/B project because the design has not been prepared. Handling inspection is an issue that must be addressed early in the project, because the designer is also the builder. If qualified individuals are available in the owner's organization, the owner may perform inspection. In some situations, an independent third party is used for inspection services.

Construction Management Method of Project Delivery

There are many variations of the construction management (CM) method of project delivery. One variation of CM is called *agency CM*, sometimes referred to as pure CM in the engineering and construction industry. In this arrangement the CM is a firm outside the owner's organization that acts as the agent for the owner. The agency CM firm performs no design or construction but assists the owner in selecting one or more design firms and one or more construction contractors to build the project. The agency CM firm assumes no risks because all contracts are signed between the owner and the designers and/or contractors. Generally the agency CM works for a fee.

Corporate CM is similar to agency CM, except the CM services are provided by personnel inside the owner's organization. The design and construction is performed by firms outside the owner's organization with the corporate CM staff managing and coordinating overall effort.

CM at risk (CM@Risk) is another variation of CM. The term CM@ Risk is used because the CM firm will actually perform some of the project work, thereby exposing themselves to risks associated with quality, cost, and schedule. The work may include some or all of the design as well as some or all of the construction. There are two sub-variations in CM@Risk, primarily construction firms that become involved in CM and primarily design firms that become involved in CM. Construction contractors that do CM@Risk work typically focus on the construction aspects of the project and hire the design work necessary to support the construction effort. Sometimes this variation of CM is referred to as *contractor CM*. Design firms that do CM@Risk are more closely affiliated with design than construction. Typically they focus on the design aspects of the project and hire the construction work. Sometimes this variation of CM is referred to as *designer CM*. The CM@Risk firm may work on a lump-sum, cost reimbursable, or a guaranteed maximum price basis.

Regardless of the variation of CM methods that is used for project delivery, CM is a single-source management of the project that allows the owner to control his or her level of involvement. To be successful the CM must be involved at the beginning of the project. Performing some of the design work without involvement of the CM prevents taking advantage of CM services. Selection of the CM early in the project increases the potential for reductions in costs by eliminating contract administration costs from designers and contractors. There can also be potential savings in time by integrating all parties from the start to the end of the project. This allows the owner to concentrate on overall aspects of the project rather than on day-to-day problems.

The CM must carefully monitor multiple construction contracts to ensure there are no gaps or overlaps in work. Tracking the costs of multiple construction contracts also requires careful attention. There also must be careful monitoring of the project to prevent delays that can occur due to interference of contractor's work at the job-site.

Bridging Project Delivery Method

Bridging is a hybrid method of project delivery for building type projects. The contract documents are prepared by the owner's designer. These documents define functional use and appearance requirements of the project. Performance specifications are used to specify the construction technology. The details of construction are developed by the construction contractor. Final design, consisting of the construction drawings, is done by an engineering/construction contractor, who performs final design and provides construction services using subcontractors.

Build-Operate-Transfer

Build-Operate-Transfer (B-O-T) involves a partnering relationship between entities in the private sector and the public sector. It is a type of arrangement where a firm in the private sector builds the project, operates it for a certain period of time, and eventually transfers ownership of the project to an entity in the public sector, typically a federal or state agency of government. This arrangement is used primarily for long-term infrastructure projects, such as toll-road highways, power plants, or water treatment facilities.

The private sector designs and builds the project, then operates it for sufficient time to recoup the cost of construction plus a reasonable return on investment. During operation of the completed project the private sector firm receives revenues from fee charges to use the project. The transfer of ownership is specified in the concession date of the B-O-T contract agreement between the private-sector firm and the public-sector agency of government.

Fast-Track Projects

Fast-track is the term commonly used for projects that must be completed in the earliest possible time. Construction work overlaps design. As soon as a portion of the design is completed, then construction work is started. Fast-track projects can be performed under the D/B or CM methods of project delivery. Fast-track applies to projects that are schedule driven at the request of the owner. For example, the owner may want to complete a process plant at the earliest possible date in order to produce and market a product before a competitor. Or a business may want to complete a building by a specific date to accommodate a special event.

Turn-Key Projects

Turn-key is the term commonly used for projects that are designed, built, and put into operation before the project is turned over to the owner. The company providing the turn-key services may secure the land for the project, perform or coordinate all aspects of the design, arrange and administer construction contracts, manage construction of the project, staff and train the personnel to operate the facility, and then turn the project over to the owner. Turn-key projects typically are manufacturing or process plant-type facilities in remote locations.

Design Development and Performance Specifications

Owner organizations that have some design capabilities, but not enough to design the complete project, will solicit bids based on preliminary design development drawings. For example, the design development drawings may only be developed to the 30% to 50% level of detail of the final drawings. Then, performance specifications are written to describe what the systems must do in the project rather than how the systems will perform. The construction drawing are completed by a D/B contractor that has special expertise related to the project. Owner organizations in the process industry sometimes handle projects in this manner.

Key Decisions for Project Delivery

The following issues can have a significant impact on the success of any project and should be considered in selecting the project delivery method:

1. Number of contracts
2. Selection criteria
3. Relationship of owner to contractor
4. Terms of payment

Number of contracts

One Contract   **▮**   Many Contracts

As illustrated above, the number of contracts can vary from one to many, depending on the chosen method of project delivery. For D/B/B projects, the owner awards contracts to two parties: one to a designer, who may contract some of the design work to other design firms, and one to the construction contractor, who may contract to numerous subcontractors who perform special construction work.

For D/B projects, the owner awards a contract to one party: the D/B firm, who may in turn subcontract to many contractors. For CM projects, the owner awards contracts to three parties: the construction manager, designer, and construction contractor. Under this scenario, there may be many subcontracts awarded under each of these three principal parties.

Selection criteria

Price   **▮**   Qualifications

As illustrated above, contractors may be selected on the basis of price or qualification. Traditionally, designers are selected based on qualification and contractors based on price. However, the trend in recent years has been to select designers based on price. The selection criteria depend on the services that are being procured. Typically price is used for selection of products that are easily defined or services that are easily evaluated. Qualifications are used for unusual products, proprietary work, or services that require special expertise, knowledge, and judgement.

Contractual relationship

Agent ——————————————————▮—————————————————————— Vendor

A contractor may be viewed as an agent or vendor. An agent represents the owner's interest, works for a fee, and usually is selected based on qualifications. A vendor delivers a specified product or service, works for a price, and generally is selected based on price. Traditionally, designers are at the agency end of the spectrum and contractors are at the vendor end of the spectrum. Some owners ask contractors to act as agents in procuring and managing construction and treat designers as vendors of plans and specifications. When owners need advice or guidance, they typically choose an agent relationship, whereas when owners know exactly what is required, they may choose a vendor relationship.

Terms of payment

Lump-Sum ————————————————▮————————————————— Cost-Plus

As illustrated above, the terms of payment can vary from fixed price to cost reimbursable. Fixed price is used when the details of the work are well understood. Cost reimbursable is used when the scope of work is unknown or not clearly defined. There are ranges of terms of payment from fixed to cost reimbursable.

Under a lump-sum arrangement for payment, the contractor is paid a fixed sum for the work based on executing the contract documents. For unit-price payments, the contractor is paid based on a predetermined amount for each unit of material that is installed.

For a cost-plus fee with a guaranteed maximum price (GMP), the contractor is paid the actual cost plus a fee. If the cost exceeds the GMP, the contractor incurs the extra cost. If the cost is under the GMP, the contractor shares in the savings.

For a cost-plus fee with a target price, the contractor is paid actual costs plus a fee. As an incentive to meet the target price, arrangements can be made for various bonuses and a penalty if the final cost deviates from the target price. For example, the contractor may incur 80% and the owner 20% of all costs over the target price. Or, the contractor may receive 70% and the owner 30% of the savings if the final cost is under the target price. Thus, the contractor shares in the savings under the target price and pays part of the overrun if the cost is over the target price. The target price can be modified by the change order process.

For cost-plus projects, the contractor is paid actual costs plus a fixed or percentage fee. The percentage fee can be a sliding scale, depending on the arrangements that are agreed to between the owner and the contractors.

The above payment terms may be combined into one contract, for example, fixed-price, lump-sum with a unit price for ordinary soil excavation and cost-plus fee for rock excavation. The terms of payment should be commensurate with the amount of risk assumed by each party.

Prospective Bidders and Bidding

The selection of the contractor is important because the successful completion of the project is highly dependent upon the contractor. The owner and/or designer must depend upon the contractor to provide labor, equipment, material, and know-how to build the project in accordance with the plans and specifications. If the contractor has problems, everyone has problems.

The owner generally requires prospective contractors to provide a bid bond before a bid is accepted. Before award of a contract, most owners require the contractor to submit a material and labor payment bond, and a performance bond. All bonds are supplied to the owner from the contractor before commencing construction in the field. Although bonds provide some degree of protection to the owner, they do not guarantee that construction will proceed in a smooth operation. In addition to bond requirements, prospective bidders should be screened by a prequalification process that evaluates their record of experience, financial capability, safety record, and general character and reputation in the industry.

For competitive-bid projects at least three bids should be received to provide a representive comparison of costs. Generally, a higher number of bidders will generate more competition, resulting in lower bids. However, the quality of the bidders is more important than the quantity of bidders. For private projects it is possible to control which companies are allowed to submit a bid. In this type of situation it is better to not allow companies to bid if their capabilities are in question or if they are simply not wanted to build the project.

Careful consideration should be given to the length of time that is allowed for contractors to submit bids. The proposed due date should be adequate for bidders to prepare a thorough bid. If there is uncertainty regarding what length of time would be adequate, a reputable contractor can be consulted to assist in developing a reasonable time for preparation of bids. If the time is too short, some bidders may decline, or worse the bid may not be properly prepared. If the bid time is too long, there is an unnecessary delay in construction.

An addendum is a change in the bid package during the bidding process to correct errors, clarify requirements of the project, or make changes before awarding a contract to the contractor. Numerous addenda may discourage reputable, prospective bidders or place the bidders in a precarious "beware" status regarding the quality of the plans and specifications, or the possibility of additional changes the owner may make during construction. These conditions can lead to costly change orders that will adversely affect the final cost of a project.

A prebid conference should be held to clarify any unique aspects of a project and assist the bidders in their preparation of a good bid. This is an opportune time to clarify scope, explain special working conditions, and answer questions of contractors. Any item that is clarified at the meeting, that is not in the bid documents, should be confirmed in writing to all parties.

For any project, the party that will administer the contract should prepare a detailed cost estimate from the same set of bid documents that the contractors are using to bid the project. This will assist in the evaluation of contractors' bids because the process of preparing an estimate requires a close scrutiny of all aspects of the project. Many problems associated with a project can be detected by thoroughly reviewing the bid documents and going through the process of preparing a detailed cost estimate. There are numerous professional estimating companies that can perform this service if the capability does not exist in the party's organization that will administer the contract.

Qualification-Based Selection (QBS)

For lump-sum, fixed-price projects, construction contractors typically are selected based on the lowest and best-qualified bidder. Determining the lowest price is relatively simple since each contractor is required to submit a bid price. To ensure qualification, often the contractors are required to submit a prequalification form prior to bidding. The form typically requests information regarding the contractor's reputation, financial stability, and capacity to perform the work. In addition most bid documents require a bid bond when the bid is submitted to the owner. Qualification for bid bonds typically assures the owner that the contractor will also qualify for the payment and performance bonds after the contract is signed.

For cost-reimbursable projects or negotiated contracts, contractors are often chosen based on QBS, because the price of the project is unknown at the time the contractor is being selected. The owner typically selects a short list of contractors, each of whom would be likely selected for the work. Meetings are held to give each prospective contractor the opportunity to ask questions and receive clarifications regarding the extent of the proposed contract and the desired project outcome by the owner. The meetings are informative in nature and may involve a visitation to the project site. Then, formal requests for proposals (RFPs) are solicited from contractors. Typically, the RFPs request information regarding overall project management and technical ability, records of relevant past jobs and performance, and the approach to be taken in handling the project, including a cost proposal. A fixed format for the RFP is provided to each contractor to allow the owner to compare proposals on a common basis.

Upon receipt of all RFPs, the owner performs an evaluation. An assessment is made of each cost proposal. Since the actual cost is not known, the owner must determine if each contractor's cost proposal is realistic. The cost proposal may involve a fee schedule for all labor, equipment, and indirect charges that would apply to the job. Typically, the owner is not looking for the least cost, but a competitive cost range. Costs are considered, but the final selection is based on the best and final offer.

To determine the best and final offer, a weighted evaluation system is used to provide a quantitative measure of each RFP. For example, the RFP is divided into categories, with a weight given to each category depending on its importance. Some categories may be given the same weight, whereas others are given different weights. Below is an illustrative example of categories for evaluation.

- Management information system
- Project schedule
- Personnel
- Contractor quality control
- Management of subcontractors
- Resource utilization
- Health and safety approach
- Financial capacity
- Experience and references

The management information system may include software for CADD, scheduling, cost estimating, accounting, submittal reporting systems, and other relevant data. The level of development of an effective management information system is a measure of a firm's ability to maintain a project budget during both design and construction. An appropriate project schedule, either a bar chart or network, that shows major milestones required to complete the project should be provided. A cost estimate should be required because price is a consideration in selecting contractors. Although the final cost is not known at the time a contractor is chosen, the cost estimate does give an indication of the final cost.

The caliber of personnel is extremely important in choosing contractors by QBS. Resumes of key personnel should demonstrate a team composition that is capable of performing the work required in the project. Formal education, project experience, and professional registration of key team personnel are indicators of their ability to perform the work. Each proposal should contain a list of personnel, resources, and major subcontractors to be utilized for the project.

Engineering and construction projects require management of multiple subcontractors. Solicitation of RFPs for choosing contractors by QBS should include a plan for acquiring material, equipment, services, and subcontractors. The owner should provide a list of their preferred vendors before requesting RFPs from prospective contractors. Any QBS process should also provide information about resource utilization, including a detailed description of the staffing plan to accommodate normal fluctuating work load to ensure an experienced work force throughout the project, including periods of work buildup and decline.

A quality assurance and quality control (QA/QC) program should be fully documented in the RFP for choosing contractors by the QBS process. The system should include quality control during development of design drawings and specifications and throughout the construction process including testing, inspection, and safety.

The factors to be considered in evaluating proposals for QBS of contractors should be tailored to each project. Generally, final selection is based on a combination of quality, technical merits, schedule, and cost. Quality may be expressed in terms of technical excellence, management capability, personnel qualifications, prior experience, past performance, and schedule compliance. Although the lowest price is appropriate for deciding on a contractor, the final selection may be based on the proposal that offers the greatest value to the owner in terms of performance.

For cost-reimbursable contracts, the cost proposal should not be the controlling factor since advance estimates of cost may not be a valid indicator of final actual costs. Typically, cost-reimbursable contracts are not awarded based on the lowest proposed cost or lowest total proposed cost plus a fee. The award of cost-reimbursable contracts primarily on the basis of estimated costs may encourage the submission of unrealistically low estimates and increase the likelihood of cost overruns. The primary consideration should be which firm has the proposal to perform the contract in a manner most advantageous to the owner, as determined by evaluation of proposals according to the established evaluation criteria.

Numerical weights are applied to each factor in consideration by the owner. Typically, the value of the weights is not disclosed to the firms submitting proposals. However, the RFP may inform prospective contractors of the minimum requirements that apply to particular evaluation factors or subfactors. During the evaluation process, the owner assigns a score to each factor, based on the weight of the factor. The selection of the contractor is based on which RFP receives the best score. The contractor is then notified and an advance agreement is made. After final arrangements are made, a contract is signed.

Checklist for Bidding

Table 11-2 provides a checklist of duties for the bidding and award phase of a project that is handled by the CM type of contract. For projects other than the CM type of contract, the duties of the CM are distributed between the owner and designer, depending upon the contract arrangement.

Keys to a Successful Project

There are several factors that are important in order to achieve a successful project during construction. A good field construction representative must be present to represent the interests of the owner and designer. He or she must know the requirements of the project and be readily available to answer questions and respond to situations as they arise. The field construction representative's authority and responsibility must be clearly defined to all parties including the owner, designer, and contractor. It should be recognized that this individual is an asset available to all the parties involved in the project: the owner, designer, and contractor.

Another important factor is a good, detailed construction schedule that is developed and used by the contractor who is performing the work, not the owner or designer. The owner should only define the start and/or end date of the project. Contractors know their capabilities, resources, and how they plan to coordinate the many activities required to build the project in the field. Thus, they are best qualified to develop a schedule to guide the numerous construction operations.

A good project control system must be developed to monitor, measure, and evaluate the cost, schedule, labor-hours, and quality of work. Chapter 9 provided a detailed discussion of project tracking and control.

The most important key factor in a successful project is good communication. Most experienced project managers readily agree that the source of most problems can be traced to poor communications. People don't intend to do poor work or make mistakes. These types of problems are a result of misunderstanding of what is to be done and when it is to be done because of poor communications. There must be open lines

TABLE 11-2   Checklist of Duties for Bidding and Award Phase

Bidding & award phase	Owner	CM	Designer*	Contractor
1. Procedures— Contracting	Approve as required	Formulate issue	Advise, comply	×
2. Bidders list	Approve	Prepare	Assist, approve	×
3. Bid documents estimate	Approve	Prepare	Review, advise	×
4. Bid division descriptions	Review, approve	Prepare	Review, approve	×
5. Proposal forms	Approve	Prepare	Review, advise	Complete
6. Specifications— Advertisement for bids	Approve	Prepare	Review, advise	Respond
7. Preparation of bid documents	Pay printing & dist. cost	Provide required forms & bidders list	Print & distribute	Receive
8. Meetings—Pre-bid	Participate as required	Organize, conduct	Participate	Participate
9. Addenda	Approve	Recommend, review	Recommend, draft & issue	Acknowledge
10. Bid Openings	Attend as required	Organize, conduct	Attend	×
11. Meetings—Post-bid	Attend as required	Organize, conduct	Attend	Attend
12. Letter of intent to award	Approve, sign	Prepare, issue	Recommend, approve	Respond
13. Bonds— Performance, labor & material	Review, file	Review & approve	Review & advise	Provide
14. Insurance—liability & property damage	Specify & approve	Advise, monitor & file certificates	Include requirements in bid documents	Provide
15. Notice to proceed	Approve, sign	Prepare, issue	Recommend, approve	Respond
16. Award of contracts	Award	Recommend	Recommend	Receive

Source: NSPE / PEC Reporter, Vol. VII, No. 2.
*The referenced NSPE/PEC publication cited "architect"; however, the term "designer" is used here to imply the architect may be the principal design professional for building projects, whereas the engineer may be the principal design professional for heavy/industrial projects.

of communications so the right people are available to respond when they are needed.

A project organizational chart normally shows vertical lines of authority. However, strictly following vertical lines of authority sometimes is not responsive to dissemination of information in a timely manner. Thus, there is a need for communications that flow horizontally between people who are actually involved in the work.

Figure 11-1 shows a project organizational chart with dotted lines that represent the horizontal communications that are necessary for a project during the construction phase. The horizontal communications between the various organizations are the most difficult to monitor and control because they are primarily verbal and often not documented, which can result in misunderstandings and possible law suits. Therefore, horizontal communications should be restricted to sharing of information, but no decision making. Although these horizontal communications are necessary, it is the responsibility of all individuals to keep their immediate supervisor informed.

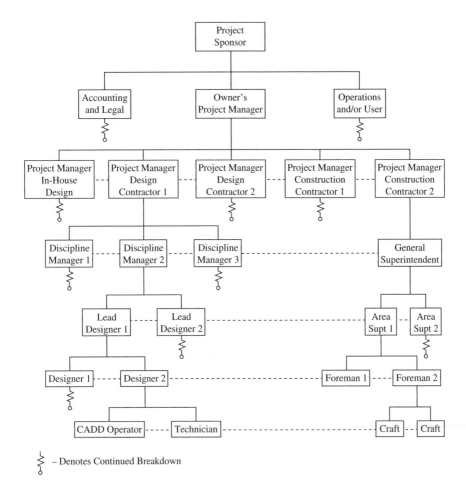

$\lessgtr$ – Denotes Continued Breakdown

Figure 11-1 Effective horizontal communication channels necessary during construction. (Note: Horizontal dashed lines denote exchange of information only and vertical solid lines denote decision-making authority.)

Construction Schedules

The purpose of scheduling a construction project is to ensure adequate planning of the work so it can be effectively and efficiently managed. A schedule must show all the activities required to complete the work, organized in a logical sequence, and generally presented in a time scale for easy interpretation. The purpose is to ensure that all activities necessary to complete the work are properly planned and coordinated. Adjustments in the schedule must be made as changes occur. The construction contractor must plan the work, and work the plan.

It is imperative for the job-site superintendent to be involved in preparing the schedule. The superintendent will be involved in the day-to-day supervision of the work, so he or she has a better idea as to how the work will be put in place. If the superintendent is involved in preparing the schedule, the schedule will be used to manage the work because the superintendent will have bought into the schedule. Schedules that are prepared without the superintendent are seldom successful. The person that estimated the job and the contractor's project manager are valuable resources for providing information and assisting the job-site superintendent in preparing schedules, but the superintendent is the person who must plan the sequence of activities the way he or she plans to work the project.

A good construction schedule is in the best interest of all parties: the owner, designer, and contractors. First and foremost, the construction schedule must be readable and easily understood by the field personnel who are involved in supervising the work. Too often, construction schedules are developed with excessive details that are of little value to the workers in the field. The schedule should only contain enough detail to enable the field supervisors to know what is expected of them. Then, area foremen in the field can develop a daily or weekly work schedule for their respective work. For example, the schedule can show an activity of "second floor painting"; then the foreman of the paint crew can develop a daily work schedule that shows the specific rooms to be painted for the week specified in the construction schedule.

Material and equipment delivery dates are extremely important during construction because late delivery is often the cause of delays. Delivery dates for critical activities must be closely monitored to ensure arrival of materials and equipment when needed at the job-site.

Problems with Construction Schedules

In some situations a contractor may avoid giving any schedule to the owner with the concept that without a schedule, the owner cannot demonstrate that the contractor failed to properly plan or coordinate the work. An ancillary theory is that an absence of a schedule on the job allows the contractor to build an as-planned versus an actual schedule at the end of the job to serve the interest of the contractor for any disputes or claims.

One method of enticing the contractor to develop an appropriate construction schedule is to include the construction schedule as a specific pay item in the contract documents. Then, the contractor would be entitled to bill for this pay item only after submittal of the baseline schedule and each monthly schedule update. For example, submission of the construction schedule can be tied to the mobilization payment. Thus, the contractor would not be paid mobilization costs until the construction schedule is submitted. This is easy to incorporate in the construction contract because most contracts provide for mobilization payments to the contractor.

Unfortunately, in recent years there has been a trend toward using construction schedules to help build claims against the owner for additional time and cost. Some contractors prepare a schedule only because it is required in the contract. It is common practice in construction contracts to tie the costs to the schedule, which impacts the monthly payment to the contractor. Some contractors prepare a schedule based on the concept that it will help their chances of recovering more money and/or time from the owner in change orders and claims. Thus, the schedule is sometimes not used solely as a planning tool to manage the project, but as a tool to obtain additional time or money.

There are risks involved in requiring, reviewing, and accepting schedules from contractors. Approved or accepted schedules can produce legal implications to the owner. For example, the owner can be placed at risk when he or she accepts a schedule, because the schedule can later be used as documentation of claims against the owner. However, there is also a risk in not requiring construction schedules. When no schedule exists, the contractor may not plan the job properly and the owner has no way of measuring progress or checking on coordination. Most experienced construction managers agree that there are fewer risks when construction schedules are required.

Whoever is assigned the responsibility of reviewing or approving construction schedules should exercise caution because a number of problems can arise with respect to construction scheduling. Some of these precautions are presented in the following paragraphs.

Precautions for Construction Submittals

Submittals are documents that are transmitted from the contractor to the owner or designer for their review and/or approval. Sometimes the contract documents specify that the engineer will review contractor submittals within a reasonable period of time so as not to delay progress on the work. Contractors usually show submittal review times in the schedule as 3 or 5 days. If the schedule is accepted, the time shown in the schedule establishes the definition of "reasonable period of time." Any review that takes more time is then considered as "unreasonable" by definition, thus establishing a case for delay or impact claims.

In other situations there may be no submittal dates shown in the schedule. Instead the submittals are considered to start on the same day, or next day, of the construction activity. Then, the contractor can raise a delay claim for any time the start of an activity is delayed while the engineer reviews the submittal for that construction activity.

To avoid problems associated with submittal review times, a provision can be written in the contract documents that states the allowable time for submittal review. For example, a paragraph can be included that states the engineer shall have a minimum of 20 calendar days, from receipt of a submittal, to review and respond to a contractor's submittal. Additionally, the contract documents can stipulate that resubmittals of the contractor's submittals shall have the same review time as the initial submittal. This clarifies entirely the issue of whether the owner and engineer should have a single time frame in which to review the submittal and all resubmittals, or whether the starting time for submittal review should start over again with each resubmittal. This is an equitable assignment of the risk of providing unacceptable submittals.

A contract that requires submittal reviews in the construction schedule can help avoid many problems related to submittal reviews and the impact of review times. A provision can be included in the contract that stipulates that the contractor prepare a construction schedule of all submittals required on the project, including when each submittal will be provided for review. This helps to ensure that the contractor has a plan for all submittals. It also allows the owner and engineer to plan their staffing and work load appropriately for the review process. A master list of all required submittals, provided by the designer, will assist in incorporating submittals in the schedule.

Delivery Dates of Owner-Furnished Equipment or Materials

It is common in the construction industry for owners to procure and furnish long lead-time equipment or materials. This is done to save time and money for the owner. The delivery dates of owner-furnished equipment or material are typically shown on the construction schedule. If the owner-furnished equipment or material delivery dates are shown on the schedule and the schedule is accepted by the owner, then this establishes an informal warranty that the items will be delivered no later than the date shown on the schedule. If the owner is unable to meet the previously scheduled and agreed-upon delivery dates, then the contractor may file a claim for extra cost due to delay of work and impact claims.

Another problem can arise when the construction schedule shows very early delivery dates for all owner-furnished items. This provides more opportunities for a delay claim against the owner if the owner-furnished items arrive late. To help avoid this type of problem, the contract documents can stipulate the earliest possible delivery dates for all owner-furnished items. These dates can be based on delivery dates established by the manufacturer, plus an appropriate contingency for time. Another approach to addressing the issue of claims related to owner-furnished items is to make provisions in the contract documents that states owner-furnished items will be delivered not earlier than a certain date but not later than another date. In this manner, the contractor is given some assurance of when to expect the items on the jobsite. Providing a window of expected delivery dates reduces risk related to owner-furnished items.

Scheduling Contractor Procured and Installed Equipment

The construction schedule that simply shows installation of equipment as one activity in the schedule does not provide an adequate description of the work necessary to procure and install the equipment. Major equipment that is procured and installed by the construction contractor should show fabrication, delivery, and installation as separate activities. This ensures adequate planning by the contractor and reduces confusion regarding fabrication and delivery times. It also clarifies the pay request from the contractor related to the equipment because the pay request for purchasing equipment will likely occur much earlier than the pay request for installing the equipment. The designer should specify in the bid documents which items of equipment are considered major.

Another method of assisting the proper planning and scheduling of contractor-furnished equipment is to include a provision in the contract that requires the contractor to provide a separate schedule that shows anticipated order and delivery dates of each piece of equipment on the project. The contract documents should specify which pieces of equipment are considered major, to prevent any misunderstandings.

Contract Schedule Constraints

During the design phase of a project the contract documents may be written to include certain schedule constraints that require construction activities to be completed before other activities can start, or sequences that must be followed to prevent disruption of other activities. This type of situation is more prevalent in renovation-type projects than construction of new facilities. The contractor's initial schedule submittal may ignore these requirements. If the owner accepts the schedule without the constraints and then later attempts to enforce the contract requirements, the contractor may argue that the owner "waived the contract requirements" by acceptance of the schedule and therefore now owes an equitable adjustment in order to reestablish these constraints.

The owner and designer should carefully review all constraints to assess their impact on the contractor's operations. This must be done during the design phase to ensure that a complete list of constraints is written into the contract documents. Constraints on the contractor usually result in additional costs and time for the contractor to complete the work.

Sequestering Float

Sequestering float is a technique of developing a construction schedule so there is little or no float for most activities. This results in a schedule with multiple critical paths or one single critical path with numerous other paths that are near critical because the non-critical paths have very small float, such as less than 5 or 10 days. Construction schedules with extensive sequestering float are sensitive to delays. Any interference with the contractor's operations by the owner increases the opportunity for the contractor to bring claims for additional costs to the owner.

The construction schedule may also contain preferential sequencing, where activities that could be performed concurrently instead are shown as sequential. Showing activities sequentially instead of concurrently reduces float in the network, thereby causing more critical activities.

A careful review of the construction schedule should be performed to evaluate artificial activity durations. For example, an activity that is shown with a 20-day duration may actually only require 10 days, which would result in reduced float in the network.

To decrease problems related to sequestering float, the contract documents can include a non-sequestering float clause. Such a clause may read as, "Pursuant to the float sharing requirement of the contract documents the use of float suppression techniques such as preferential sequencing or logic, special lead/lag logic restraints, and extended activity times are prohibited and the use of float time disclosed or implied by the use of alternative float suppression techniques shall be shared to the proportionate benefit of the owner and contractor."

Schedule Updates

Too often a good schedule at the beginning of a project becomes a schedule that is unmanageable due to poorly prepared updates. As changes occur during construction, there is a tendency to use constraints to model the actual work. The use of activity constraints such as start-to-start and finish-to-finish almost invariably cause logic that is difficult to understand and often leads to confusion and distrust of the schedule.

It can be helpful to involve the owner, designer, contractor, and all major subcontractors to review the update process. A joint discussion between all parties will enhance communications and reduce surprises when the schedule updates are provided. The discussions can often eliminate the activity constraints. The best approach is to not use activity constraints in developing the original schedule or in preparing schedule updates.

Relations with Contractors

Construction contractors have the lead role during the construction phase; however, the owner and designer have an important role as well. A cooperative environment of teamwork must be developed so all parties can work together as a unit to achieve the project.

The construction industry is unique compared to other industries. Because each construction project is different: the work force is transient, multiple crafts are involved, projects are planned and worked in short time frames, and there is a tremendous variety of material and equipment that must be installed. Also, much of the work is exposed to weather and construction workers are continually working themselves out of a job. Due to these conditions the management of construction is a challenge, and cooperation of participants is imperative.

In all situations the relations with contractors must be fair, consistent, polite, and firm. A person must conduct his or her affairs in a professional manner to gain the respect of others and to get others to do what must be done. There are times when a person must be assertive, but not obnoxious, and other times when a person must be reserved, but not a push over. The ability to work with people and to know how to react in each situation must be developed in order to work in a construction environment.

Due to the nature of construction projects, there are times when disagreements and conflicts between individuals arise. One must realize that disagreements and conflicts are not necessarily all bad because many good ideas have been developed as a result of disagreements. The attitude that should prevail is that disagreements can be changed to agreements with diplomacy. There are times when it may be desirable to take a neutral position.

Contractors are independent business organizations and are only required to produce the end product of the contract. There are times when a contractor and an owner may not agree; however, achievement of the end product should always remain a priority. Good owner/contractor relationships make the best use of the contractor's expertise, labor, and equipment.

Checklist of Duties

Table 11-3 provides a comprehensive checklist of duties for the construction phase of a project that is handled by the CM type of contract. For projects other than the CM type of contract, the duties of the CM are distributed between the owner and the designer, depending upon the contract arrangement.

Quality Control

Quality is achieved by people who take pride in their work and have the necessary skills and experience to do the work. The actual quality of construction depends largely upon the control of construction itself, which is the principal responsibility of the contractor. What is referred to today as "quality control," which is a part of a quality assurance program, is a function that has for years been recognized as the inspection and testing of materials and workmanship to see that the work meets the requirements of the drawings and specifications.

Quality is the responsibility of all participants in a project. Too often, the attitude is "what can we do to pass quality control?" or "what can we do to get past inspection?" Instead, the attitude should be, "what can we do to finish the project that we can be proud of that meets the specifications and satisfies the owner?" Without the right attitude, even the best planned quality-control program cannot be successful.

In the past, the traditional acceptance of construction work by the owner, or owner's representative, involved 100% visual inspection of the construction work. In addition, a representative sample, which often consists of a single specimen, is used to determine the true quality of the material. If the test result is within the stated tolerance, the material passes and is accepted, otherwise the material or construction fails to pass and is unacceptable. In the latter case, engineering judgement is then applied to reach a decision as to whether the material should be retested or whether it may be said to substantially comply, because the deviation will not cause a serious impairment to performance. This places the inspector in a difficult situation and can cause delays, disputes, and numerous problems for both the owner and contractor.

TABLE 11-3   Checklist of Duties during Construction Phase

Construction phase	Owner	CM	Designer*	Contractor
1. Contracts— Construction	Approve, execute, enforce as required	Prepare documents, enforce	Review, approve	Approve, execute, conform
2. Meetings— Pre-construction	Participate as required	Organize, conduct, record	Participate	Participate
3. Meetings— Monthly project	Participate	Organize, conduct, record	Participate	Participate
4. Meetings— Monthly team	Participate	Organize, conduct, record	Participate	×
5. Insurance— Workers compensation	Specify, approve	Advise, monitor, file certificates	Include requirement in bid documents	Provide
6. Submittals— Shop drawings & samples	Approve as required	Coordinate, expedite & review	Specify submittals required, review& approve	Provide
7. Scheduling— Short-term construction activities plan	Approve as required, comply	Prepare, monitor, enforce, comply	Participate, comply	Participate, comply
8. Schedule enforcement	Participate as required	Provide	Assist	Comply
9. Construction support items	Approve as required & pay	Recommend & arrange	Advise	Cooperate
10. Security—Site	Advise, review, approve, pay	Advise & arrange for	Facilitate	Facilitate
11. Field layout	Observe	Coordinate, check where practical	Design, observe	Provide & be responsible
12. Temporary— Power, water, roads, etc.	Approve, pay	Determine need, arrange & coordinate	Provide specifications as required	Use
13. Meetings— Weekly construction	Participate as required	Organize, conduct, record	Participate as necessary	Participate
14. Construction methods, procedures	Observe	Observe	Observe	Originate
15. Contractor coordination	Observe	Provide	Advise	Facilitate
16. Field reporting	Review	Provide, review	Provide review	Cooperate
17. Job-site safety	Enforce through CM	Observe, report	Observe, report	Responsible

(*Continued*)

TABLE 11-3   Checklist of Duties during Construction Phase (*Continued*)

Construction phase	Owner	CM	Designer*	Contractor
18. Payment requests	Approve, respond	Review, approve, assemble	Review, approve, endorse	Submit
19. Waivers of lien	Review & File	Coordinate, review, & assemble	Review	Provide
20. Change orders	Approve, execute	Request, review, approve, distribute	Request, prepare, approve	Execute, respond
21. Quality control	Observe	Observe, evaluate, report	Observe, evaluate	Provide & be responsible
22. Field testing	Approve, pay	Arrange, coordinate	Recommend, approve	Cooperate
23. Expediting— Construction	Participate as required	Originate, monitor, motivate	Advise, assist	Responsible
24. Equipment— Owner purchased	Specify, purchase, expedite	Coordinate scheduling, installation & start-up	Incorporate requirements into design	Hook-up
25. Expediting— Owner purchased equipment	Provide	Coordinate scheduling requirements, assist	Advise	×
26. Receiving equipment	Inspect, approve	Coordinate on-site delivery & storage requirements	×	Assist
27. Drawings— As-built	Receive & file	Coordinate, monitor	Check, draft	Provide as-built

Source: NSPE/PEC Reporter, Vol. VII, No. 2.
*The referenced NSPE/PEC publication cited "architect"; however, the term "designer" is used here to imply the architect may be the principal design professional for building projects, whereas the engineer may be the principal design professional for heavy/industrial projects.

The purpose of quality control during construction is to ensure that the work is accomplished in accordance with the requirements specified in the contract. A program of quality control can be administered by the owner, designer, contractor, or an independent consultant. In recent years the trend by many owners is to require prime contractors to take a more active role in the control of project quality by making them manage their own quality-control programs. The contractor is required to establish a quality-control plan to maintain a job surveillance system of its own, to perform tests and keep records to ensure the work conforms to contract requirements. The owner then monitors the contractor's quality-control plan and makes spot-check inspections during the construction process, which constitutes a quality-assurance program.

In modern quality-assurance programs, traditional specifications are being replaced by statistically based specifications that reflect both the variability in construction materials and the true capabilities of construction processes. Quality requirements are expressed as target values for which contractors are to accomplish, and compliance requirements are specified as plus or minus limits. Statistically based acceptance plans call for random samples, with each sample consisting of a number of specimens taken at random in order to eliminate bias. Acceptance procedures are more detailed and the risks to both the owner and contractor are made known to reduce uncertainty.

Each contractor is expected to have a process control program that will ensure he or she is meeting the acceptance requirements of the owner. Control charts, which are simple line graphs of the target quality level and of the allowable variations from this level, are used as tools in the process control. They provide early detection of trouble before rejections occur, which can result in savings to the contractor by reducing penalties and rework costs. Thus, control charts present the results in a form that enables everyone concerned to observe trends or patterns that may affect quality.

An example of a control chart for a percent passing gradation is given in Figure 11-2. Figure 11-2a is referred to as an X-chart and is used to monitor the average value of the process, that is, whether the process average has remained at a constant level or has shifted. The horizontal axis identifies the number of observations; that is, the number 1 is the first inspection and the number 30 is the last inspection. The vertical axis represents the quality characteristic under consideration, which in this case is the percent passing a given sieve size. Each small circle represents the average value of that sample, or subgroup. The two dashed outer lines are the upper control limit (UCL) and lower control limit (LCL) which are used to judge the significance of a change in the process. It is also possible to place upper and lower warning limits, (UWL) and (LWL). Warning limits are typically placed at two standard deviations from the target value. The second control chart, Figure 11-2b, is called the R-chart (range) and is used to detect changes in the process variability.

Dispute Resolutions

Due to the nature of construction projects it is almost certain that contractors, owners, and designers will be involved in disputes. The resolution of

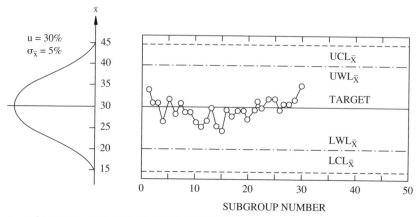

a. $\overline{x}$ CONTROL CHART FOR PERCENT PASSING

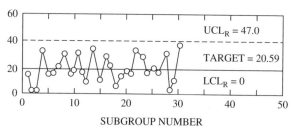

b. R CONTROL CHART FOR PERCENT PASSING

Figure 11-2   Control charts for $(a)\,X$ and $(b)\,R$.

a dispute may be by several methods: negotiation, mediation, arbitration, or litigation.

Direct negotiations between parties in the dispute can be held to openly discuss and resolve the conflict to the satisfaction of each party. Usually, no other parties are involved. Mediation allows disputing parties to use an independent, objective person to assist in resolving the dispute. The mediator has no authority to issue a final decision. Arbitration is similar to mediation except the arbitrator has the authority to issue a final binding decision that cannot be appealed by the disputing parties. For some disputes there may be a panel of arbitrators to resolve the dispute. Litigation is a resolution of disputes by lawsuits that are resolved in the legal formality of the court system. Generally this technique requires a long period of time and significantly higher legal costs than the other techniques.

The resolution of disputes by negotiations is usually the quickest and most economical technique because factual matters are discussed without the formality of legal formats. Direct negotiations are held on a voluntary informal basis by the parties involved, at a mutually agreed upon time and location. The negotiator for each party must have the authority to act for his or her company. The size of the negotiating teams varies depending upon the complexity of the issue; however, the efficiency is usually higher if there are fewer individuals involved. The success of the negotiations depends on the willingness of both parties to negotiate in good faith, the extent the contract definitively addresses the issues in the dispute, and the amount of preparation prior to negotiation. Direct negotiations can enhance good business relationships between the parties.

In the event that direct negotiations cannot result in a settlement, mediation is often the next most suitable method. Similar to negotiation, mediation is voluntary, so there must be a mutual agreement between both parties to have the mediator to serve as a catalyst to explore alternate solutions, to gather facts and clarify discrepancies, and to persuade the parties to adopt flexibility in their stances so a final settlement can be reached. During the process the mediator may have joint or private meetings with the disputing parties. The mediator is mutually agreed upon by both parties in the dispute. The American Arbitration Association (AAA) is a public-service, non-profit organization that has established guidelines and rules for mediation.

Arbitration is the submission of a dispute to a mutually agreed upon impartial third party whose decision is legally binding and enforceable. To use this method an arbitration clause is normally included in the general conditions of a contract so when the contract is signed the parties mutually agree to settle all disputes through arbitration. The Construction Industry Arbitration Rules, established by the AAA, contains 57 articles that have become the most widely recognized arbitration procedures in the construction industry. The procedures follow a five-step process: agreement to arbitrate, selection of arbitrator, preparation for hearing, hearing of the dispute, and the award. The decision by the arbitrator is usually made within 30 days from the close of the hearing. Thus, decisions are made much faster than through litigation.

Most parties strive to resolve conflicts before proceeding to litigation because it is the most costly, time-consuming, and complex method for the resolution of disputes. Legal counsels, who follow the formalities of the legal judicial system, are used to represent the interest of the parties. The final resolution of the dispute is determined by a court of law. Often, a settlement of the dispute is reached just prior to entering court proceedings.

The American Society of Civil Engineers has produced a booklet entitled *Avoiding and Resolving Disputes During Construction*, which presents a method of developing cooperative, problem-solving attitudes on projects, through a basic risk-sharing philosophy between the owner and contractor. It explains special contracting provisions and practices that have been used successfully on hundreds of projects to avoid or to resolve disputes without resorting to litigation. There are three provisions: Dispute Review Boards (DRB), Escrow Bid Documents (EBD), and Geotechnical Design Summary Reports (GDSR), which are described in detail and recommended on all projects subject to potential major disputes.

Formation of a three-member DRB at the beginning of construction assists in avoiding disputes and, if necessary, provides timely and equitable recommendations for non-binding resolutions. Escrowing of the contractor's bid documents provides a reliable database for use in negotiating price adjustments and resolving disputes. Inclusion of the Federal Differing Site Conditions clause and a GDSR in the contract acknowledges the contractor's right to compensation for materially differing site conditions and establishes a concise geotechnical baseline to clearly identify differing site conditions. These provisions are discussed in the booklet, together with guideline specifications for implementation.

Job-Site Safety

Safety is an important part of project management just as are planning, scheduling, estimating, cost control, and other project work. The concern for safety must be shown at all levels and in each phase of the project. Accidents not only affect workers, but also their families. The economic costs, liability consequences, regulatory requirements, and the image of a company all show the importance of safety related to a project.

Many of the basic principles and approaches that have been previously presented to manage a project can also be applied to safety. The element of safety is a factor that should be considered in each phase of the design and construction of a project. Safety is not included in a project by accident; it must be designed into the project and monitored just as scope, budget, and schedule.

Construction, by the nature of the work, involves many potential hazards to workers and equipment, such as heat, noise, wind, dust, vibrations, and toxic chemicals. The Occupational Safety and Health Administration (OSHA) was enacted by the federal government as a regulatory agency to ensure safety for workers. It applies to all parties involved in a project: designers, owners, workers, and contractors. Although OSHA and other legislation has been enacted for safety, the responsibility rests with management.

The project manager must work closely with his or her team members to include safety in every aspect of the project: planning, design, budget, and construction. Safety should start at the top level of an organization and, by words and actions, be infused in each level of management through to the crews and workers at the job-site.

Although the current practice places final responsibility for construction safety with the contractor, there should be a united teamwork approach to understanding and implementing a safety philosophy that improves construction job-site safety.

Accident costs include medical costs, premiums for compensation benefits, liability, and property loss. These costs have been escalating in recent years. However, there are other significant costs attributed to accidents. These include the cost of lost time of the injured employee, the cost of work stoppage of other employees due to an accident, and the lost supervisory time. Safety requires everyone's full attention.

Research has shown that safer job-site managers are also the better producers. They are better at keeping down job costs and better at keeping jobs on schedule. These facts contradict two reasons that managers sometimes give as excuses for a poor safety record: accidents are inevitable in a dangerous industry like construction, and our first priority has to be getting the job done. Such managers are misled by myths that can cost the project and the company a great deal of money. Job-site safety management should treat productivity and safety as two related parts of high job performance. A successful job-site manager has been quoted as "You don't have to sacrifice productivity for safety. The safer the crew, the quicker they work. The more safety you have, the more productivity you have."

The prequalification of contractors is usually based on bonding capacity and past experience. The selection of the construction contractor usually is based upon the lowest bid price of all prequalified contractors. Perhaps, a contractor's safety record should also be considered as a necessary part of construction bids so an unsatisfactory safety record could be one possible basis for rejection of a bid.

Management of Changes

Every project requires some changes during construction in order to complete the project. The source of changes may be the owner, designer, or contractor. An owner may wish to make a change to better achieve his or her intended use of the project after construction is complete. A designer may make a change in the original plans or specifications or the contractor may wish to make a change, because it is not always possible to accurately predict all the events that will arise during the construction process. Thus, changes during construction are almost inevitable.

The mechanism for making changes during construction is the change-order, a written document that describes the modification in the work. All approved change-orders are incorporated into the original bid documents to form the binding contract documents for the project. Although a change-order may increase or decrease the cost and/or schedule of a project, most change-orders add cost to a project and impact the schedule. Thus, every project manager must be cautious in dealing with changes during construction, because a change in the work almost always adversely affects the cost and schedule of the project.

The management of changes is greatly enhanced if the changes can be predicted in advance. There are some factors that can be determined, prior to commencing construction, that are early warning signs of future changes in a project. Research has shown that lump-sum projects that have a large difference between the low bid and next higher bid amount tend to have significant increases in project costs. In the construction industry the term *money-left-on-the-table* is commonly referred to as the difference between the bid amount of the low bidder and the next higher bidder. Thus, a project manager should expend more effort in monitoring and controlling a project if there is high money-left-on-the-table, because it is an indicator of potential cost growth.

One approach to the management of changes during construction is to request a list of anticipated change-orders on a monthly basis. The project manager, or his or her assistant, then works with the party anticipating the change, to evaluate the need and/or value of the change. Sometimes a candid discussion of the merits of a change can result in a decision that the change is not necessary afterall, when the full impact of the change is compared to the true value of the change.

There must be a thorough evaluation of all aspects of a change, because a change in any part of a project often affects other parts of a project. Sometimes the full impact on other parts of a project are not known until some later date, which can adversely affect the cost later in the project. The term *ripple-effect* is commonly used to describe a change that occurs late in the project as a result of a change made earlier in the project.

A project manager should avoid changes during construction unless they are absolutely necessary. If a change is necessary, it should be thoroughly evaluated, clearly defined, agreed on by all parties, and implemented in the most efficient and economical manner.

Resource Management

Many resources are needed to accomplish the construction phase of a project. Resources include people, equipment, materials, and subcontractors. Each resource must be managed in the most efficient manner to keep costs to a minimum during construction.

The crafts that install material and operate equipment are the most important resource on the project. These individuals gain their skills through training and experience. They have the ability to accomplish the work, provided adequate instruction, tools, and materials are available when they are needed. Too often, the skilled workers are criticized for not producing good work on a project. However, it is not a trait of people to intentionally do poor work. The cause of poor work is usually from poor instructions, late delivery of material, unavailable tools, or lack of leadership and supervision. These sources of problems are the responsibility of management on the project. Thus, there must be a well-defined work plan for the project that clearly defines the work that must be accomplished today and the work that is planned ahead. This plan must be communicated to the teams of crafts on the project.

The type and number of equipment used on a project depends on the nature of the project. For example, construction of a large earthen dam may require a large spread of scrapers, dozers, water wagons, compactors, and motor patrol graders. However, the construction of a strip shopping center may only need a small front end loader, truck, and small portable crane. The selection and utilization of equipment on a project must be an integral part of the total construction plan and schedule, just as there must be a plan for the workers on the project. It is the responsibility of the construction project manager and his or her field superintendent to develop an equipment plan for the project. Adequate consideration should be given to the down-time and maintenance of equipment, because the unavailability of equipment can have a significant impact on the schedule of a project.

A major cost of many construction projects is the acquisition and installation of materials. A materials management system includes the major functions of identifying, acquiring, storing, distributing, and disposing of materials needed in a construction project. The effective utilization of people can be greatly enhanced by ensuring that quality materials are available when and where required. A material plan will vary depending on the project size, location, cash flow requirements, and the procedure for purchasing and inspection. The timing of delivery of material to the job-site is extremely important, because the cause of delays during construction is often late delivery of material, incomplete delivery of material, or delivery of the wrong type of material. Materials requiring a long lead time must be included in the construction project schedule. It is the responsibility of the contractor to ensure that a well-defined materials management system and materials management plan are developed for the project.

For most projects, the owner assigns one contract to a prime contractor, usually called the general contractor, to accomplish the construction phase of a project. In turn, the general contractor assigns contracts to numerous subcontractors, usually called specialty contractors, to perform the construction work that requires special skills or equipment.

Thus, much of the work required on many construction projects is performed by numerous subcontractors who work for the general contractor. This multiple-contract arrangement requires careful planning, scheduling, and coordinating by the general contractor to integrate the work of all subcontractors on the job. This is necessary because the work of any subcontractor usually affects one or more other subcontractor on the project. The owner may use multiple prime contractors for a large project that extends over a long period of time.

The management of subcontractors should be accomplished using the same management principles presented throughout this book. There must be a well-defined scope of work, cost, and schedule for each subcontractor on the project. In addition, there must be a clear interfacing of the work of all subcontractors on the project. It is the responsibility of the general contractor to effectively manage his or her subcontractors.

References

1. *Avoiding and Resolving Disputes During Construction*, ASCE, New York, NY, 1991.
2. Ahmed, S. A., *Statistical Methods for QA-QC for Highway Construction*, Oklahoma State University Press, Stillwater, OK, 1990.
3. Barrie, D. S. and Paulson, B. C., *Professional Construction Management*, 3rd ed., McGraw-Hill, Inc., New York, NY, 1992.
4. Bartholomew, S. H., *Construction Contracting: Business and Legal Principles*, 2nd ed., Prentice Hall, Upper Saddle River, NJ, 2001.
5. Brockrath, J. T. and Plotnick, F. L., *Contracts and the Legal Environment for Engineers and Architects*, 7th ed., McGraw-Hill, New York, NY, 2011.
6. Collier, K., Managing Construction—The Contractual Viewpoint, Delmar Publishing, Inc., Albany, NY, 1994.
7. *Construction Industry Arbitration Rules*, American Arbitration Association, New York, NY, 1991.
8. Fisk, E. R. and Reynolds, W. D., *Construction Project Administration*, 10th ed., Pearson Prentice Hall, Upper Saddle River, NY, 2013.
9. Grant, E. I. and Leavenworth, R. S., *Statistical Quality Control*, 7th ed., McGraw-Hill, New York, NY, 1996.
10. Halpin, D. W. and Senior, B. A., *Construction Management*, 4th ed., Wiley, New York, NY, 2011.
11. Hester, W. T., Kuprenas, J. A., and Thomas, H. R., "Arbitration: A Look at Its Form and Performance," *Journal of Construction Engineering and Management*, ASCE, New York, NY, Vol. 113, No. 3, September 1987.
12. Hinze, J., *Construction Contracts*, 3rd ed., McGraw-Hill, New York, NY, 2011.
13. Jervis, B. M. and Levin, P., Construction Law: Principles and Practice, McGraw-Hill, Inc., New York, NY, 1987.
14. Knutson, K., Schexnayder, C. J., Fiori, C. M., and Mayo, R., *Construction Management Fundamentals*, McGraw-Hill, New York, NY, 2009.
15. Levitt, R. E. and Samelson, N. M., *Construction Safety Management*, 2nd ed., John Wiley & Sons, New York, NY, 1993.
16. Levy, M. L., *Project Management in Construction*, 6th ed., McGraw-Hill, New York, NY, 2012.
17. Muller, F., "Mediation: An Alternative to Litigation," *Journal of American Water Works Association*, February 1984.
18. Oglesby, C. H., Parker, H. W., and Howell, G. A., *Productivity Improvement in Construction*, McGraw-Hill, Inc., New York, NY, 1989.

19. Peurifoy, R. L. and Oberlender, G. D., *Estimating Construction Costs*, 6th ed., McGraw-Hill, Inc., New York, NY, 2014.
20. *Quality in the Constructed Project: A Guide for Owners, Designers, and Constructors*, ASCE, Reston, VA, vol. 1, Manual No. 73, 1990.
21. Ritz, G. J. and Levy, S. M., *Total Construction Project Management*, 2nd ed., McGraw-Hill, New York, NY, 2013.
22. Ryan, T. P., *Statistical Methods for Quality Improvement*, 3rd ed., Wiley, New York, NY, 2011.
23. "Sampling Responsibility Chart," *PEC Reporter*, Professional Engineers in Construction Practice Division, National Society of Professional Engineers, Alexandria, VA, Vol. VII, No. 2, January 1985.
24. Sears, S. K., Sears, R. H., and *Clough, R. H., Construction Project Management*, 5th ed., Wiley, New York, NY, 2008.
25. Stukhart, G., *Construction QA / QC Systems that Work: Case Studies*, ASCE, Reston, VA, 1985.
26. Sutliff, C. D. and Zack, J. G., Jr., "Contract Provisions that Ensure Complete Cost Disclosure," *Cost Engineering*, Vol. 28, No. 10, Morgantown, WV, October 1987.
27. Tucker, R. L., *Assessment of Construction Industry Project Management Practices and Performance*, Construction Industry Institute, Austin, TX, April 1990.
28. White, N. J., *Construction Law for Managers, Architects, and Engineers*, Cengage Learning, Scottsdale, AZ, 2007.
29. Zack, J. G. Jr., "Check Your Scheduling Practices," *Civil Engineering*, ASCE, Reston, VA, 1986.

12

Project Close Out

System Testing and Start-Up

For heavy industrial plant projects, inspection of construction is performed throughout the project; however, the owner's representative generally requires inspection of vessels before closure and the testing of major equipment upon installation. The term *mechanical completion* is often used to define the stage of project development in which these procedures are performed.

It is sometimes difficult to define mechanical completion; therefore, the project manager should develop a plant completion standard with the owner's representative to clarify what constitutes completion. This should be in the construction contract so that all involved in the project know who is to do what for each phase of the project.

For system testing, the project manager for the owner needs to coordinate the interface between the owner's representatives, principal designers, construction contractors, and equipment manufacturers. The roles and responsibilities of each must be clearly defined. There must be agreements on the systems to be tested, the types of tests required, and who will witness the tests. It is important to develop a plan and schedule for construction to support system testing in a timely manner. Planning of activities for start-up should start with the "need date" by the owner's operating personnel. Work packages should be developed to prepare for and conduct system testing. The required tests should be identified by 30% into construction duration and the start-up plan should be completed by 70% into construction duration.

A formal plan should be prepared for defining when a vessel can be closed, the lead time notice required for inspection, what is to be monitored, and a sign-off sheet for the owner's representative. This is required to eliminate unnecessary opening and closing of vessels, which can require a substantial amount of project time.

The project manager must obtain in writing from the owner's representative the procedure for turning equipment over to the owner. Care and custody are important because substantial costs are involved. It is important for each party involved to know who is responsible and when they are responsible. The project manager should notify the plant when a certain piece of equipment is complete, tested, and ready for turnover to the owner. Upon acceptance, any additional changes require a work authorization from the owner. This procedure should be handled in a formal manner with signatures required from the responsible representatives.

The project manager needs to coordinate with the contractor and designer to define the procedure for start-up. The process must be formal, but also flexible. The project manager should obtain in writing from the owner's representative what they require from the various members of the project team for support during start-up. The project manager must have the owner's input and should not make assumptions as to their needs.

Final Inspection

Inspection of construction work is performed throughout the duration of a project. Before completion of the entire project, various equipment, electrical systems, and mechanical systems may be completed and ready for testing and acceptance in accordance with the contract documents. The project manager must work closely with the owner's representative and the design professionals who are responsible for inspection, testing, and final acceptance.

A definition of mechanical completion should be developed and a formal notification should be provided to the construction contractor that allows adequate lead time for the process. This is necessary so substantial time is not lost, which may adversely affect the project completion date. There must be a clear understanding regarding what the owner wants to verify during tests, what tests they want to witness, and the types of testing required. The responsibilities of the three principal contracting parties (the owner, designer, and contractor) must be clearly defined. It is the duty of the project manager to effectively coordinate this effort.

The start of project close out begins near the end of a project, when the contractor requests a final inspection of the work. Prior to the request, a punch list is prepared listing all items still requiring completion or correction. To develop this punch list, the field inspection personnel must carefully review their daily inspector's log to note all work items which have been entered that require corrective actions. It is sometimes necessary to recycle through the punch list process several times before the work is satisfactory for acceptance. The final walk-through inspection should include representatives of the owner, contractor, and the key design professionals (the architect, as well as civil, electrical, and mechanical engineers, etc.) who have worked on the project. The project manager should schedule and conduct the final walk-through inspection.

Acceptance of the work and final payment to the contractor must be done in accordance with the specifications in the contract documents. Substantial completion of a project is the date when construction is sufficiently complete in accordance with the contract documents so the project can be used for the purposes it was intended. This means that only minor items remain to be finished and that the project is complete enough to be put in use. The contractor may issue a Certificate of Substantial Completion with an attached list of all work remaining to be done to complete the project. Approval of the Certificate of Substantial Completion, with the attached deficiency list, certifies acceptance of the work that is completed. Thus, it is important to ensure the list is complete because the contractor has no further obligations under the contract after the owner signs the certificate. Generally final payment, including the release of all retainage, is withheld for 30 to 40 days after completion of all deficiencies. Before final payment, the contractor is to submit final paperwork: warranties, lien releases, and other documents required under the contract.

Guarantee and Warranties

Generally, the contract requires the contractor to guarantee all material, equipment, and work to be of good quality and free of defects in accordance with the contract documents for a period of 1 year after completion of construction. The guarantee of the overall project can be extended beyond the normal 1-year period; however, this is not common.

Individual pieces of equipment often have warranties that extend from 1 to 5 years after installation. Operating instructions, manuals, spare parts, and warranty certificates must be supplied to the owner. The project manager must ensure that all warranties are compiled and supplied to the owner before final payment to the contractor.

Lien Releases

Material suppliers, subcontractors, or workers who have furnished materials, equipment, or performed work on a project for which payment has not been made may file a lien against the property. The unpaid party has a right to file a lien even though the owner has paid the general contractor the full contract amount. Consequently, the owner may be forced to pay for some of the contract twice if the general contractor fails to pay its subcontractors, material suppliers, or employees.

During construction, the owner can withhold payments (called retainage) from the general contractor to cover open accounts and/or liens. The general conditions of most contracts have a clause that requires the general contractor and all tiers of subcontractors to supply a lien release for all labor and material for which a lien could be filed, or a bond satisfactory to the owner that indemnifies the owner against any lien. The project manager must ensure receipt of all lien releases, or the bond, prior to approval of final payment to the contractor.

Record and As-Built Drawings

Revisions and changes to the original drawings are almost certain for any project. At least one set of the original contract documents that were issued for bidding purposes must be kept in a reproducible form. This is necessary for the resolution of claims and disputes, because inevitably the question will arise, "What did the contractor bid on?" In addition, there must be a thorough documentation of all change-orders during construction.

A common contract requirement is that the contractor must develop an as-built copy of all specifications, drawings, addenda, change-orders, and shop drawings. The as-built drawings show dimensions and details of work that were not performed exactly as they were originally shown. Examples are changes in the location of doors, the routing of electrical wires or air conditioning ducts, or the location of underground piping, utilities, and other hidden work. These documents show all the changes to the original contract bid documents and are provided to the owner upon completion of the project.

Checklist of Duties

Table 12-1 provides a brief checklist of duties for project close out for the construction management (CM) method of contracting. If the project is not handled by the CM method, the duties of the construction manager are performed by either the owner or the designer.

Disposition of Project Files

During a project, the project manager usually maintains two files: a record file and a working file. The record file contains original copies of important information related to agreements, contracts, and other legal

TABLE 12-1   Checklist of Duties for Project Close Out

Project close out	Owner	CM	Designer*	Contractor
1. Certificate of substantial completion	Approve	Review, approve, file	Review, approve	Originate
2. Clean-up	Observe & comment	Coordinate, enforce	Observe	Responsible
3. Punch list	Approve as required	Expedite & coordinate work	Prepare, evaluate work	Respond
4. Call backs (after construction)	Request	Arrange & coordinate	Review & approve work	Respond

Source: NSPE / PEC Reporter, Vol. VII, No. 2.
*The referenced NSPE/PEC publication cited "architect"; however, the term "designer" is used here to imply the architect may be the principal design professional for building projects, whereas the engineer may be the principal design professional for heavy/industrial projects.

matters. The working file is the project manager's file that is used for the day-to-day management of the project and usually contains copies of documents from the record file plus correspondence, minutes of meetings, telephone logs, reports, etc. Upon completion of a project, a large amount of information pertinent to the project, including the records and files, is accumulated.

Most organizations have a defined procedure for disposition of files. Information from the record file should be organized and indexed for easy access and retrieval for future reference. The working file often has duplicate information of the record file, some of which may have handwritten notes that should not be destroyed. Although much of the contents from the file can be discarded, sufficient information should be retained so the project manager can retrace his or her work on the project.

Post Project Critique

A post project critique should be held at the conclusion of every project, because there are lessons to be learned from every project that can be used to improve the success of future projects. Attendees should include the owner and key participants in the project, including lead designers and construction representatives. The feedback that is gained through a non-accusatory discussion of the problems and solutions encountered during a project is beneficial to all team members in the planning and execution of future projects.

For the meeting to achieve the desired results, it is important that the entire discussion be presented in a positive and professional manner. The good, as well as the bad, aspects of the project need to be discussed.

The emphasis should be on how to avoid or lessen the problems on future jobs based on problems encountered on this job; not who was at fault or caused the problem on this project. Minutes of the meeting should be distributed to others who did not attend the post project critique, so they also can benefit from the lessons learned.

It may be desirable to perform a project peer review, which is an independent evaluation of a particular project's design concepts or management procedures. Project peer reviews can address the needs of the owner, designer, or another interested party. The American Consulting Engineer's Council (ACEC) and the American Society of Civil Engineers (ASCE) have produced a document entitled *Project Peer Review: Guidelines,* which describes the process of project peer review.

Owner's Feedback

After a project is completed and in use by the owner, a formal meeting should be held with representatives from the owner's organization to obtain feedback regarding performance of the project. This is an important activity for evaluation of the quality of a completed project and satisfaction of the owner, because the true measure of the success of a project can only be determined by how well the project is utilized by the owner's organization.

References

1. Corrie, R. K., *Project Evaluation,* ASCE, Reston, VA, 1991.
2. Fisk, E. R. and Reynolds, W. D., *Construction Project Administration, 10th ed.*, Pearson/Prentice Hall, Upper Saddle River, NJ, 2006.
3. *Inspection of Building Structure During Construction,* Institution of Structural Engineers, New York, NY, 1983.
4. Levy, S. M., *Project Management in Construction, 10th ed.*, McGraw-Hill, New York, NY, 2012.
5. *Project Peer Review: Guidelines,* American Consulting Engineers Council, and the American Society of Civil Engineers, Washington, DC, 1990.
6. "Sampling Responsibility Chart," *PEC Reporter,* Professional Engineers in Construction Practice Division, National Society of Professional Engineers, Alexandria, VA, Vol. VII, No. 2, January 1985.

Personal Management Skills

Challenges and Opportunities

There is a great amount of pride and satisfaction in seeing a project completed and in operation. There are many challenges in managing engineering and construction due to the dynamic nature of projects. Solving problems as they arise instills a sense of satisfaction in the project manager and members of the project team. Most project managers agree that lifelong friends are created by working with people on projects. Years after a project is completed, conversations between people who have worked together on past projects often turn to amusing memories of the problems that occurred and the methods used to resolve the problems. Generally, the problems seem much less serious after the project is completed than they were during execution of the project.

Successful people convert challenges to opportunities. Regardless of how difficult a problem may appear, there is always a way to resolve it. After completing a project, most project managers are enthusiastically ready to start a new project. They look forward to the next project and the opportunity to apply the lessons learned from previous projects. Most project managers agree that they have fun doing their work.

Using New Innovations

The preceding chapters of this book presented the concepts, methods, tools, and techniques that have been used successfully to manage engineering and construction projects. The methods have been developed and used by engineers who are actively involved in the current practice of project management. However, to be effective the project manager must be aware of new technologies that can advance the concepts presented in this book.

New technologies sometimes are resisted by people until they are proven. Too often the technology is available, but it is not used because people sometimes resist change. The successful project manager should assess new technology and devise innovative methods to incorporate the technology to enhance the project management process.

The use of computers is a good example of the adaptation of new technologies. When first introduced, some individuals only perceived the computer as a tool to be used for technical or scientific applications. However, progressive project managers realized the potential benefits of the computer in the management of projects. Today the computer is used in virtually every aspect in the execution of projects. The computer has probably changed the way we work today more than any other technology.

The Internet is a technology that is used extensively by project managers to do their work. The advantages of managing projects via the Internet are numerous, primarily in saving time and increasing efficiency. Information can be distributed more rapidly to geographically dispersed owners, designers, contractors, and suppliers. Correspondence, drawings, and photographs can be posted on project-specific Internet sites and cataloged for easy access. Key team members can monitor data exchange, which greatly reduces turnaround time for decision making. For example, using traditional methods it may take an hour just to process a request for information (RFI) and then take days or weeks to get an answer for the RFI. Now electronic RFIs can be submitted and logged almost instantly, and responses are often made the same day or within a few days. Designers, suppliers, and contractors know they are accountable with web-enabled project management.

Today, many companies operate in a paperless environment by using the electronic media of the Internet. The increased efficiencies are valuable to project managers who are pressed to accomplish more work in less time. Projects are getting more complex, and project personnel are expected to deliver results more quickly. Furthermore, team members are spread over a wider geographic location compared to projects in the past. Today, design work done in any location can be transmitted to any other location of the world almost immediately via the Internet. Design work can progress continuously, on a twenty-four-hour basis, by transferring documents from one design office to another to countries across the world.

Websites improve both internal and external communications. It can be a significant public information tool. A company can establish public and private websites for a project. The general public can visit one website for the latest on traffic conditions and construction schedules, while team members use a different site to exchange detailed design and construction information. The private site can save a vast amount of time by providing rapid data exchange among numerous offices and subconsultants working on a project.

E-mail is the most common and often used aspect of the Internet. It allows users to send written messages with optional attachments to anyone in the world who has a computer. Attachments can include drawing files, word processing documents, spreadsheets, photos, multi-media clips, and web pages. Faxes can also be sent over the Internet instead of over regular voice lines.

Web-enabled management systems add speed and efficiency to processing project information. The systems also reduce confusion and redundancy for owners, designers, and contractors trying to accomplish work as quickly and efficiently as possible. These systems can be structured with identifying tabs to mimic traditional file folder systems. For example, the Internet can be used in recording delivery records, progress updates, and other data in the field and transferring them to a database incorporated in a company's web system.

Another example is the application of project schedules, using an Internet-enabled tool designed to allow team members to see their assignments across multiple projects using a web browser. Developing, maintaining, and communicating the project schedule is vital to the process of project management. Whether through CPM scheduling or simple calendar schedules, a website can be used to post current and near-term activities.

In place of a local area network, a web server can be used as a central repository for project documents. The documents may include word processing files, spreadsheets, photographs, and drawings. Users can add and/or check in and check out existing documents. Documents and file systems can be used during the design or construction phase.

Project administration can also be greatly enhanced via the web. Web-enabled systems can manage work flow. In addition to document management, the software can create, log, track, and index project documents. Other functions include to-do lists, notification of events for project members, or required follow-up activities. Digital technology has made it extremely easy to take pictures and video and include them with documents on the websites. Complete photo histories can be stored in a database and be available for online retrieval. This can provide documentation during execution of the project as well as resolution of disputes.

Job cost reports can also be published for review. For security, limited access to cost information can be provided without permitting access to an accounting system. Status reports can be published online with links to other information, such as schedules, cost reports, RFIs, or accident reports. This provides a way of giving users a document-based database that is easy to maintain and control.

Voice and multi-media are another application of project management via the Internet. The Internet can be used as a facilitator of voice communication. E-mail can carry attachments of audio/video clips made with a digital, still, or video camera. Inexpensive hardware and software also allow two or more users to teleconference online or allow users to get a live video feed from an online camera. This is an effective tool for resolving problems in the field by a team of experts who may be located in the home office without requiring them to travel to the job-site.

Security of information can be addressed by use of Intranets and Extranets. An Intranet is an Internet site set up for private use of a company that might include certain corporate financial data, schedules, bulletins, and human resource information or other confidential information. Intranets can save companies money by reducing the amount of hard copy notices and forms that might be printed and distributed manually or by mail. An Extranet is an Internet site set up by a company for shared use with others. It might contain a variety of information with limited or full access to its employees; trading partners, such as subcontractors; and suppliers; or the general public. Project-specific websites are included in this category.

Project management via the web can save time and increase productivity. However, precautions must be taken. Just because information is sent over the Internet does not ensure it was received. A confirmed receipt of information is necessary to ensure the information has been received.

Like computers and the Internet, advances in cell phone technology have tremendous impact on the way people communicate and share information. In some aspects the cell phone has impacted everyday life as much as computers and the Internet. Computers, cell phones, and the Internet are tools that provide almost limitless abilities to perform work and transmit the results to anywhere in the world. However, using these tools generates challenges in managing the time of the project manager and his or her team. It becomes important to separate information from "nice to know" to "need to know" to reduce time and effort of the project team. Rapidly transmitting unnecessary information leads to information overload, which leads to reduction in productivity and miscommunications. As new technologies are developed, the project manager must be responsive to potential applications and devise methods to incorporate the technology to increase the efficiency of project management.

Human Aspects

The preceding chapters defined the information that must be gathered and managed to successfully bring a project to completion. Although a system must be developed for management and control of a project, it is people who make things happen.

A project management system serves as a guide for overall coordination of the entire project; however, some refinement and modification of the system is sometimes necessary for a particular project. People are the only resource that have the capability to detect problems and make the adjustments that are necessary to successfully manage a project. Thus, a project manager should not depend solely on the system of project management and neglect the importance of the people who are associated with the project.

To summarize, successful *project management* can best be described as *effective communications* between the people who perform the work that is necessary to complete the project. Management of any project requires coordination of the work of individuals who each provide an area of specialization. A successful project manager must be a good planner, delegator, and communicator.

There is a tendency of some project managers to complain that the work is not being accomplished because of factors beyond their control. For example, a project manager may feel that team members are too inexperienced, or that it takes more time to explain what needs to be done than the time it takes to do it themselves, or fear that a mistake by a team member would be too costly. Other typical examples are the feeling that others are too busy and don't have time for additional assignments or that others avoid accepting responsibility. Although these factors are of concern, they and similar factors are a routine part of working with people and can be overcome with good management skills.

Usually project managers have many years of experience of doing the work before their assignment of managing the work. Because they are familiar with what is required, they may prefer doing the work themselves rather than delegating it to others. The result is they become overworked from attempting to both do and manage the work, spend nights and weekends devoted to their job, and then complain that the work is not accomplished because others are too inexperienced. A project manager must realize that others can only gain experience by doing the work; also that many times others can do the work just as well, if not better, than managers. The problem is acceptance of the fact that others may not do the work exactly the same way the manager would do it. The criterion should not be who can do the job best, but who can do the job satisfactorily. The project manager must balance accuracy and quality of the work for the entire project. The feeling that a person is too inexperienced can be overcome by close communications and training.

The feeling by project managers that it takes more time to explain the job than to do it themselves often results in the project manager doing the work, at nights or on weekends. Many times a person can do the work quicker than explaining what is needed by others. However, if the work must be repeated on a particular project, or on future projects, it usually is more efficient to explain it to others one time so they are experienced in doing it in the future. Project managers must realize that before telling someone what to do, they must know what needs to be done themselves. A well-defined work plan, as discussed in previous chapters, provides the plan of action to effectively communicate job assignments to team members.

The accomplishment of a project usually involves high costs over an extended period of time with many risks. The fear that a mistake by a person would be too costly is a concern by every project manager. Because of this fear, a project manager may be reluctant to assign the work to others and end up doing the work himself or herself. The problem is the lack of confidence in others and the fear that they don't have sufficient judgement to handle the situation if a problem arises. The excuse that is often given is "If you want it done right, you must do it yourself." However, a good control system will help ensure that work is done right. Fear of taking a chance can prevent finding hidden ability.

Because everyone appears busy, a manager may be reluctant to assign work because of the feeling that others don't have time for additional assignments. This situation often arises when the work to be done requires the expertise that can only be performed by a small number of individuals. A system must be developed to ensure that people who are involved in the project are using their time efficiently. This is best accomplished by developing a well-defined project schedule at the beginning of the project, with participation and input from each person who will be involved. There is always enough time to do what is required.

Another complaint of some project managers is that people avoid accepting responsibility. People will avoid accepting responsibility if they fear unjust criticism if they make a mistake or feel they will receive inadequate recognition of their work. A manager must develop a project control system that protects a project from major mistakes that are catastrophic, yet tolerates minor mistakes that are inevitable. Some people simply find it easier to ask the manager than to decide themselves because some managers want to make all the decisions. A project manager must clearly define the work required because people tend to achieve what is expected of them.

Assignment of Work

Project management involves coordinating the work rather than doing the work; therefore, the project manager delegates the work to members of the project team. Assigning work to another person delegates authority and responsibility to that person to accomplish the work and to make decisions that may be necessary. However, delegating authority, responsibility, and decision making does not necessarily cause the project manager to relinquish control. A project manager may define different levels of delegation, such as complete the work and provide the results, propose the work to be performed and inform before proceeding, or perform part of the work and submit for review and approval.

Management involves assignment of work to the right person with a clear explanation of what is expected and when it must be completed. Because miscommunication is a common problem in project management, one must be certain that the other person understands what has been assigned. A project manager must give each person on the team an opportunity to do the job the way he or she wants to do it. Simply stated, "Their way is often as good as my way of doing a task."

A project manager should be reasonable in his or her expectations. If an assignment is not reasonably achievable, the person who is assigned to do the task usually will recognize this fact and resist pressure and responsibility. The best way to evaluate the reality of achieving an assigned task is to work with the person to obtain his or her assistance in defining the work that is required to achieve the final end result.

During the course of accomplishing work, many obstacles arise. A project manager must be accessible to explain things that may not be clearly understood and to make any adjustments that may be necessary. In short, the project manager must also be available when he or she is needed. Periodic team meetings are needed to keep the work flowing in a well-identified manner so that all concerned can work as a unit.

An effective project manager must lead the project team and reinforce the confidence of individuals. Trust must be shown in their capability, intelligence, and judgement. The leader of any group must occasionally check with members to see what is going on and how they are doing. This builds confidence and respect among the team members, who in turn will strive to accomplish quality work.

A project manager must recognize and reward successful and outstanding performance. People relish this recognition and are entitled to it. Likewise, a manager should hold an employee responsible for unacceptable work, show why the work is unacceptable, where mistakes were made, determine how to improve the work, and identify ways of preventing future problems.

Many project managers are assertive with a tendency to rush in and take over. Although each person must develop their own style of management, he or she must be cautious of overreacting to situations. With the right attitude and working relationship, problems can be changed to solutions in due time.

Motivation

Experienced managers readily agree that there are all types of people: those who make things happen, those who watch things happen, those who don't know what's happening, and in some instances those who don't want to know what's happening. The project manager must devise methods to motivate each of these types of people.

Each member of a project team provides an expertise that is needed to accomplish a project. Usually individual team members are assigned to the project from various discipline departments by their respective supervisors. Although everyone works for the project, each person may report to a different supervisor. Therefore, the project manager as the leader of a project team is placed in a situation of motivating individuals who actually report to a supervisor other than the project manager. Thus, the project manager must develop effective methods of motivating team members other than the traditional methods of promotion in title or salary.

Many managers believe that money is the best motivator of individuals. Obviously few people would work if they were not being paid. Although money may be a motivator to a certain degree, there are other factors that influence the motivation of people. Unless the pay is exceptionally different, there are other factors that must be considered. If money were the only motivator, the project manager would have a problem with motivation because he or she usually is not the person who directly establishes the pay of individual team members. Because most project managers have no control over pay rates, they must motivate by providing individual recognition and, most importantly, by challenging each team member with responsibilities and a chance to grow.

Many books have been written and numerous theories have been proposed related to motivating people. Most people are motivated by needs. As presented in Chapter 2, Maslow's theory of needs identifies five levels of needs that people strive to achieve: basic survival, safety, social, ego, and self-fulfillment. The theory proposes that an individual strives to satisfy the basic survival needs of food, clothing, and shelter. Upon satisfaction of these needs, a person strives for the next level of need; safety, which may include continued employment, financial security, etc. As each level of need is satisfied, the next higher level of need is sought. A project manager must strive to identify the needs of the people who are involved in the project in order to effectively motivate them. This is difficult to do during day-to-day activities, so it is sometimes desirable to associate with individuals outside the work environment. Many times an awareness of a person's interests and recognition of their needs is a positive step in understanding why they react to situations and can lead to productive motivation. Good management recognizes the motivational needs of each member of the team and develops methods to improve the performance of people.

Professional people seek job interest, recognition, and achievement. Everyone wants to feel important and to accomplish work. People who feel good about themselves produce good results; therefore a project manager should help people reach their full potential and recognize that everyone is a potential winner. This type of attitude among team members can create motivation for everyone.

Decision Making

Numerous decisions must be made during management of a project that require a significant amount of time and effort on the part of the project manager. While many decisions are routine and can be made rapidly, others are significant and may have a major impact on the quality, cost, or schedule of a project.

Good decisions cannot be made unless the primary objectives and goals that are to be accomplished are known and understood. Decision making involves choosing a course of action from various alternatives. It is the duty of the project manager to know and clearly communicate the project objectives to all participants so their efforts can be focused on alternatives that apply to the desired end results. This is important because a significant amount of time and cost can be expended toward evaluation of alternatives that may solve a problem, but not pertain to the central objective to be achieved. The project manager must coordinate the effort of the project team to ensure a focused effort.

Decisions must be made in a timely manner to prevent delays in work that may impact the cost and/or schedule of a project. Most of the project decisions are made internally (within the project manager's organization), which can be managed relatively easy. However, some decisions are made externally (outside the project manager's organization) by owners or regulatory agencies, particularly in the review and approval process. Early in the project, the project manager must identify those activities that require external decisions so the appropriate information will be provided and the person can be identified who will be making the decisions. This must be included in the project schedule to alert the responsible parties so work is not disrupted and the project is not delayed due to lack of a decision at the proper time.

Many organizations have established policies regarding the authority for decision making that a project manager can refer to during management of a project. However, there are many times when others can be consulted who have had similar situations, and who can provide valuable opinions and assistance. Regardless of the situation, there is almost always another person who has had a similar problem.

The project manager should avoid crisis decisions, although many decisions are made under pressure. One must gather all pertinent information, forecast potential outcomes, think, and then use their best judgement to make the decision. It is not possible to anticipate all eventual outcomes, but with careful thought and review, one can eliminate the unlikely events. There is a certain amount of risk in everything we do and sometimes wrong decisions are made by good managers; however, new information may become available or changed circumstances may arise that will require a change in the previous decision.

Decisiveness is required of a project manager to gain the respect of team members. A project manager must avoid procrastination and vacillation and should encourage decision making in team members. Indecision creates tension in most people, which causes stress and further indecisiveness. Failure to be decisive can cause many things to go wrong; no one knows what to do, work is not done because of lack of direction, all of which causes a waste in talent, resources, and time.

It is the responsibility of the project manager to ensure that appropriate decisions are made by the right people, at the right time and based on correct information. Once a decision is made it should be communicated to all participants involved in the project so all concerned will know what is to be done. This can be easily accomplished by distributing the minutes of the meeting (or record of the conversation) where the decision was made, with a highlight or flag to denote the decision.

Time Management

Time is irreplaceable and vital to the personal and business life of everyone. A project manager spends a large amount of time communicating and interacting with others who are involved in the project. Therefore, it is important that time is spent in a productive and effective manner. A project manager must be cautious, because there are always more interesting and worthwhile things to do than time allowed to do them.

An analysis of how time is spent is necessary in order to determine how effective time is used. Occasionally a time log of how major portions of one's time is spent should be maintained. A daily log should be compiled over 2 or 3 weeks that shows how much time is spent doing each activity, who was involved, and what was accomplished. Activities can be grouped by categories, such as telephone, meetings, unscheduled

TABLE 13-1   Common Time Wasters

1. Unproductive telephone calls
2. Unproductive meetings
3. Unscheduled visitors
4. Special requests
5. Attempting too much at once
6. Lack of goals and objectives
7. Procrastination on decisions
8. Involvement in routine items that others can handle
9. Inability to set and keep priorities
10. Inability to say no

visitors, and special requests. An analysis of the distribution of time by categories will enable the project manager to determine where his or her time is spent most, so improvements can be made. It is usually easier to reduce a category of high expenditure of time by a small amount than it is to reduce a category of low expenditure of time.

Common time wasters of project managers are unproductive telephone calls and meetings. Although the telephone is necessary for a manager to perform his or her work, it can be quite disruptive. There are times when calls should not be received so other tasks can be done. A secretary, assistant, or answering machine can intercept calls to assist in the management of telephone usage. Meetings are mandatory for project management. The most effective way to conduct a productive meeting is to prepare an agenda that is distributed to all attendees prior to the meeting. An agenda serves as a means of focusing discussions and following an organized coverage of information that should be disseminated. Table 13-1 provides a partial listing of common time wasters.

The project manager must set priorities and develop a system to manage his or her time. Tasks that are the least interesting can be scheduled at the peak of one's energy. There should be a thorough review to evaluate job activities that can be assigned to others and an analysis of work to determine how and what can be combined or eliminated. Emphasis should be placed on long-term items, rather than short-term items that can often be delegated to others. Most people are more motivated by work that is planned than work that "just happens" at the moment. Priorities should be set and kept to effectively manage time.

Communications

One of the most frequent sources of errors and misunderstandings in the management of a project and working with people is miscommunications. Too often, the "other person" does not hear or interpret the information the way it was intended. Communications may be oral (both speaking and listening) or written (both writing and reading). In each instance it is important that clear, coherent, and efficient communication skills exist to ensure successful work by all participants in a project. The project manager must realize that all people do not interpret the same thing in the same way and that a communication is of no value unless it is both received and understood.

The role of the project manager is analogous to the central server in the local area network of a computer system. He or she is responsible for the continuous and comprehensive flow of information to and from team members, with special attention to communicating information and decisions that may influence the project team's work. These communications include conversations, meetings, minutes, correspondence, reports, and presentations.

Much of the day-to-day work on a project is accomplished by informal exchanges of information among team members. Examples are telephone calls and informal meetings between two or more individuals. Although most of these exchanges are routine, some may impact the work of others or may impact project decisions related to scope, budget, or schedule. Informal exchanges of information that affect the scope, budget, or schedule must be documented into the written record at the next regularly scheduled team meeting.

The project manager should maintain a record of telephone conversations, including the names of parties in the conversation, date, time, place, items discussed, and any pertinent information resulting from the exchange (reference Figure 13-1). Copies of telephone records should be maintained and filed with each project. It is sometimes helpful to maintain a master phone log that records each call for all projects that the project manager is responsible for. Each call can be recorded on one line which contains the date, time, number, person's name, and brief remarks that provide an outline of the conversation (see Figure 13-2). This master log can be cross-referenced to the individual telephone record, which contains the details of each conversation. A master log is helpful in keeping track of the overall work of the project manager.

The project manager must develop and practice good speaking skills. Conversations must be clear, coherent, and to the point without rambling. Thoughts and ideas should be organized in a systematic manner before communicating. This can be accomplished by knowing the objectives of the communication, for example, to give information, to get information, to make decisions, or to persuade someone. Consideration must also be given to timing and location to ensure the other person has your attention, because listening is an important part of communicating. It is often necessary to follow up a conversation to be certain the communication is received and understood. This can be done by obtaining feedback.

Project Title:_____

Name:_____   Date:_____

Title:_____   Time:_____

Firm:_____   Initiated by:_____

Items Discussed:_____
_____
_____
_____
_____
_____
_____
_____
_____
_____
_____
_____
_____
_____

Conclusions:_____
_____
_____
_____

Future Actions:_____
_____
_____
_____

Figure 13-1   Individual telephone log.

Presentations

As the prime contact person for the project, the project manager is the spokesperson who often makes presentations to the owner, agencies, boards, and other interested parties. For an effective presentation, it is important to know the audience and to convey information that is of value and interest to the audience. A presentation is given for the audience, not the individual giving the presentation. It should be prepared from the audience's point of view and organized in a logical pattern so each part of the presentation will relate to other parts. Examples are problem to solution, unknown to known, cause to effect, or a chronological order.

A flaw of many presentations is an attempt to tell the audience too much, giving a step-by-step description of everything on the subject. A presentation normally has a limited time and the audience frequently

Date	Time	Number	Name	Company	Project	Remarks

Figure 13-2 Project manager's master telephone log.

is a busy group. Therefore, the presentation should be more of a summary of important factors of direct interest, leaving detailed information in a report that can be read later. Only a limited number of graphs, tables, or computer printouts can be presented so they must be carefully chosen.

A presentation should begin with a title, which is a simple statement of the subject, followed by a brief overview of the material that is to be presented. During a presentation, the individual presenting the material must realize that the audience cannot remember every word that is said. To increase clarity and to emphasize key points, the key points can be repeated by selecting alternate words and phrases to bring out the same major ideas. This is necessary for effective speaking, but is not, and should not, be done in writing because a reader can reread material to clarify or understand what is written.

The speaker should define words and acronyms that he or she thinks the audience may not know or understand. This should be done at the time the words are used, not at the beginning or end of the presentation. Defining or clarifying words ensures that the listener hears and understands what is being said. It also ensures that the listener is thinking about the presentation and focusing on the speaker's key points.

Visual aids greatly enhance any presentation, particularly tables of numbers, equations, and technical data. The value of visual aids is that the audience both hears and sees the presentation, which greatly increases their understanding and the amount of information they retain from the presentation. Visual aids also help the speaker keep the presentation flowing in a continuous manner. Computers that are used throughout the industry are capable of producing slides for presentation purposes from the data generated in graphic form. Current copy machines with enlarging and reducing capabilities can also be used to develop overhead transparencies of printed material, including computer-generated reports from laser printers.

Few people are impressed with complicated sentences or an attempt by the speaker to impress the audience. The level of detail that should be presented depends upon how much the audience knows about the subject. Therefore, it is best to know the audience. Simple and direct language should be used that presents the material so it is easily understood. To gain the attention of the audience, the speaker should not make the audience insecure. Apologies and negative comments should be avoided. A positive attitude should prevail, even when controversial subjects are being discussed.

The presentation should be summarized at the end, just as the audience was told the purpose of the presentation at the beginning. Also, adequate time should be allowed for questions and answers at the conclusion of the presentation.

Meetings

Numerous meetings are held throughout the duration of a project to exchange information and make decisions. The schedule for the regular team meetings should be defined at the beginning of the project as a part of the project work plan. The project manager chairs team meetings, which should be held weekly, preferably on the same day and time. Minutes of the meeting should record items discussed, decisions made, and actions to be taken (including the responsible person and the due date). Occasionally, special meetings may be held to discuss special problems or unforeseen situations. Minutes of these meetings should also be kept in the project files.

Other meetings are held with the owner to report progress or obtain approvals. Special meetings may be held with other interested parties, such as regulatory agencies or the general public. The project manager may not chair these meetings and often is accompanied by lead project team members to assist in discussing the issues related to the project.

Meetings should be conducted in an efficient manner because those in attendance are usually busy people who have other work that must be accomplished. An agenda provides an effective means of organizing a meeting to define and sequence the items that are to be discussed, to prevent discussions from wandering. The time required to conduct a meeting is also significantly reduced when an agenda is used to guide the discussions. An agenda can also prevent an individual from dominating the discussions and give each person an opportunity to participate.

Meetings should be started and finished on time. Starting late penalizes those who arrive on time and rewards those who arrive late. It is preferable to define the start and end time on the agenda. Meetings with constrained times will often cover as much material, if not more, than meetings that have an open-ended time.

Minutes should be prepared for every formal meeting. As previously noted, minutes of the meeting should record items discussed, decisions made, and actions to be taken (including the responsible person and the due date). Copies of the minutes should be distributed to all attendees and a record copy placed in the project file. Minutes provide each attendee with the opportunity to verify the items discussed and decisions made. Minutes also assist the project manager in preparing the agenda for the next meeting.

Reports and Letters

Written communications document most of the activities of a project and often have major impacts related to decisions, costs, schedules, and legal matters. Because of this, all written material should be dated and written in a clear, concise, coherent, and legible manner.

To be effective, one must consider who is going to read the material and the purpose of the writing, such as to obtain information, to give information, to clarify an item, or to submit a proposal for approval. As the leader of the project team the project manager must prepare status reports that describe the progress of a project. Many of these reports contain a significant amount of computer-generated graphs and tables with short written narratives. Other typical correspondence is in the form of letters or inter-office memorandums and minutes of meetings.

Although computers with graphic capabilities, word processing software, and laser printers have automated and enhanced written communication capabilities, the project manager must still rely on his or her own writing skills to produce a coherent written document. Cutting,

TABLE 13-2   Report Format

1. **Title page**
   A. Report title, author, location, date
2. **Table of contents**
   A. List of subject matter in chronological order with page numbers
3. **Front matter**
   A. List of figures
   B. List of tables
   C. List of abbreviations
4. **Introduction**
   A. Purpose of report
   B. Scope of report, including what is contained in the report and what is not contained in the report
5. **Main text of report**
   A. Test equipment used
   B. Procedures used
   C. Data obtained
   D. Analysis of data
6. **Summary/conclusions/recommendations**
7. **Appendix, for bulk data**
8. **List of references**

pasting, merging, and spell checkers do not necessarily ensure a coherent document. For example, a spelling check of a document does not verify the correctness of the use of the words of to, too, or two. Likewise deleting or adding lines or paragraphs in one portion of a document can have a major impact on other portions of the document. Even small errors can lead to big problems and misunderstandings. Thus, a project manager must use special precautions to ensure a coherent document.

The arrangement and format of reports vary, depending on the material to be produced. Table 13-2 is a suggested format for a report. Letters should be written with at least three paragraphs: an introductory paragraph, the main body of the letter (one or more paragraphs), and the final closing paragraph, (see Table 13-3).

TABLE 13-3   Letter Format

1. **Introductory paragraph**
   A. State purpose of letter
   B. State how the letter came about
2. **Main body of letter**
   A. One or more paragraphs covering the subject matter
   B. Bulk data should be placed in a separate attachment or enclosure
3. **Final paragraph**
   A. Summary and conclusions
   B. Leave open invitation for follow-up from reader
   C. Close with final good will sentences, such as, Thank you for your consideration in this matter.

References

1. Blanchard, K. and Johnson, S., *The One Minute Manager,* William Morrow and Company, Inc., New York, NY, 1982.
2. Culp, G. and Smith, A., *Managing People (Including Yourself) for Project Success,* Van Nostrand Reinhold, NY, 1999.
3. Dinsmore, P. C., Martin, M. D., and Huettel, G. T., *The Project Manager's Work Environment: Coping with Stress,* The Project Management Institute, Newtown Square, PA, 1981.
4. Herzberg, F., "One More Time: How Do You Motivate Employees?," *Harvard Business Review,* 46, No. 1, January–February 1968.
5. Hopper, J. R., *Human Factors of Project Organization,* Source Document No. 58, Construction Industry Institute, Austin, TX, September 1990.
6. Kirchof, N. and Adams, J. R., *Conflict Management for Project Managers,* Project Management Institute, Newtown Square, PA, 1982.
7. Luthens, F., *Organizational Behavior,* 2nd ed., McGraw-Hill, Inc., New York, NY, 1977.
8. Maslow, A. H., *Motivation and Personality,* Harper & Row Publishing Co., New York, NY, 1954.
9. McGregor, D., *The Human Side of Enterprise,* McGraw-Hill, Inc., New York, NY, 1960.
10. Mitchell, T., R., *People in Organizations,* McGraw-Hill, Inc., New York, NY, 1982.
11. Peters, T. J., *In Search of Excellence,* Harper & Row Publishing Co., New York, NY, 1982.
12. Stuckenbruck, L. C. and Marchall, D., *Team Building for Project Managers,* Project Management Institute, Newtown Square, PA, 1985.

14

Risk Management

Introduction

Management of a project involves decisions and actions that are intended to move a project toward completion. However, sometimes decision making and actions can result in consequences that adversely affect the forward momentum of completing projects. Therefore, risks are inherent in the process of project management.

Risk should not be neglected, but should be recognized and managed so any adverse effects can be predicted, reduced, or entirely avoided. It is important to realize that risk is a part of doing business, especially in the engineering and construction professions. Risks should be recognized, analyzed, and managed just like other parts of a project, such as managing costs, schedules, and quality.

There are many risks in the design and construction of any project, regardless of size, type, or complexity of the project. Many things can go wrong in all phases of a project, including design, procurement, and construction. Risks are any external or internal events that could cause a deviation from the plan for project delivery. Table 14-1 shows examples of risks during design, procurement, and construction of a project.

The construction industry is one of the few industries that quote prices, promise schedules, and make commitments to perform high-risk work before work is started. Few contracts allow repricing of the work in the middle of construction. The designer and contractor are bound to perform their work in accordance with the contract price, scope, and schedule that was agreed upon at the time the contract was signed. Additional time or cost to complete a project beyond what is specified in the contract documents is typically permitted only by changed conditions from the scope, time, and quality requirements.

TABLE 14-1   Risks during Design, Procurement, and Construction

Design	Procurement	Construction
Miscommunication of design criteria with the client	Issuing timely orders for purchase of material and equipment	Reworking unacceptable quality of work at the job-site
Miscommunication of design criteria within the design team	Vendor/Supplier delays in delivery of material/ equipment	Misinterpretation, conflicts, or errors in drawings and specs
Incomplete description of the scope of work	Misinterpretation of purchase contract terms and conditions	Misinterpretation of construction contract terms and conditions
Errors or omissions in developing the design drawings and specs	Inadequate methods of handling materials and equipment	Conflicts in schedules, schedule slippage, or cost overruns
Conflicting interests between client and permitting agencies, or regulatory agencies	Availability of adequate inventory to fulfill orders	Inadequate attention to safe work practices
Availability of experienced labor resources or access to technology	Storage and handling of delivered materials or equipment	Problems in start-up and testing of equipment and processes

Risk Management Process

The owner, designer, and contractor are all exposed to risk from the start of a project through its completion, and sometimes beyond. The risks may involve cost, schedule, quality, environmental, health, safety, or customer satisfaction. Each party must identify the risks, evaluate probability of occurrence, analyze the severity and impact, and then develop a plan to prevent each risk and mitigations to rectify each risk if it occurs. Risks in the project need to be recognized and managed. Risk management consists of three steps:

Step 1—Risk Assessment
> Identify project risks (cost, schedule, scope, quality)
> Identify human risks (environmental, health, safety)

Step 2—Risk Analysis
> Determine probability of occurrence (likelihood) of the risk
> Evaluate severity (impact) of the risk

Step 3—Mitigation
> Develop plans to prevent or reduce the risk
> Develop strategy to rectify the risk if it occurs

TABLE 14-2   Activities and Deliverables of Risk Management Process

Process	Activities	Deliverables
Risk Assessment	Identify and document risks Document causes and effects	Risk Register
Risk Analysis	Perform quantitative analysis Perform qualitative analysis	Risk Assessment Report
Risk Mitigation	Select risk response strategies Develop risk mitigation plans	Risk Mitigation Plan
Lessons Learned	Conduct post project review Document lessons learned	Lessons Learned Log

Assessment of risk involves identifying the risk (what can go wrong) into categories, such as health, safety, environmental, quality, cost, or schedule. It should be noted that some risks may fall into several categories. After each risk is identified, an analysis can be made to determine the probability of occurrence (likelihood it may occur) and the severity (impact of risk) if the risk occurs. After the probability and severity of the risk are determined, a mitigation strategy can be established to prevent the risk (risk prevention) and/or develop a plan to rectify the risk (how to fix or reduce the problem) if it occurs. Thus, risk management involves risk assessment, analysis, and mitigation.

It is important to identify and manage risks early in a project, rather than later. Attempting to deal with risks late in a project puts pressure on people, which can lead to frustration and poor decision making.

Table 14-2 shows activities and deliverables in a risk management process. Methods of identifying, analyzing, and mitigating risks are obvious parts of the risk management process. However, it is also important to capture and document lessons learned from executing a risk management process. In particular, it is important to document issues related to risk mitigation strategies that were effective and therefore should be used on future projects. Also, methods that were not effective should be documented with recommendations to modify or discard. Lessons learned from previous projects are extremely useful for effectively managing future project.

Guidelines for Risk Management Process

There are at least seven guiding principles that should be followed by everyone involved in the risk management process.

1. After decisions are made, they should not be revisited unless substantively new facts become available.

2. A single person should be assigned responsibility for each risk, even if there are several people working to mitigate that risk.

3. The most severe risks should receive high priority and be tracked closely.

4. Realistic due dates should be established and the team should then work to meet the dates.

5. Risks should be mitigated at the appropriate level, such as owner, project team, subcontractor, and vendor.

6. The project team should determine and agree on the level of severity for each risk.

7. There should be documentation of the planned and actual mitigation of each risk to serve as input into lessons learned for future projects.

Risks of Owner, Designer, and Contractor

The principal parties in a project are the owner, designer, and contractor. Each party plays an important role in a successful project. When problems arise in a project, the fault often can be attributed to a combination of the owner, designer, and contractor. When something goes wrong for one party, it usually affects other parties in the project. Therefore, it is important for each party to work together to share risks, rather than trying to assign risk to another party. It should also be noted that communications play an important role in risk management. Communicating information that is misunderstood, inadequate, or received too late can cause many things to go wrong in a project.

The first consideration in risk management is the type of contract, its terms and conditions. Negotiations between parties should involve identifying potential risks. One of the primary purposes of the contract, regardless of the type, is to bind the parties together and allocate known risks. It is best to share risks equitably by apportioning and assigning risks to the party that is best capable of controlling the risk. Whether a project is defined by one contract or many, the owner, designer, and contractor must recognize that teamwork is essential to risk management.

The principal components of a project are cost, schedule, and quality of work. Jointly and individually the owner, designer, and contractor have important roles that expose them to risks related to cost, schedule, and quality. Owners are responsible for risks associated to setting performance expectations of the completed work, which subsequently establishes the cost, schedule, and quality of a project. Designers and contractors have risks in meeting the owner's expectations. Designers are in the best position to control risks related to quality, although owners and contractors have quality-related risks. Typically, contractors are in the best position to control risks related to cost and schedule on a project, although owners and designers are also exposed to risks related to costs and schedules.

All parties are exposed to risks related to costs. Owners in the private sector develop cost budgets based on economic feasibility of the project and consideration of available funds or sources of financing. Owners in the public sector develop cost budgets for their projects based on benefit/cost ratios to the public and consideration of available funds. Any deviation of design or construction costs from the owner's budget presents a risk to the owner. Contractors prepare cost estimates for bid purposes. Any underestimate of costs presents a risk to the contractor. Designers have responsibilities to meet the expectations of the owner and protect the health and safety of the public. Overruns in the design budget typically pose less significant risk than overruns of construction costs. However, overruns in design budgets impact the funds available to build the project.

The owner, designer, and contractor are exposed to risks related to schedule. The time to complete a project is a major concern of all parties. The length of time to finish a project should be long enough for the designer and contractor to do their respective work in accordance with the contract documents, but short enough for the owner to have possession of the project so it can be used for its intended purpose. The time allowed for design and construction will dramatically affect costs and quality of the completed work. For private sector owners the intended use is typically to make a profit on use of the project, whereas for public sector owners the intended use is to make the project available for use by the general public. The expected return on investment or benefit are significant drivers on risk planning.

The owner defines the level of quality that is expected in the project. It is the responsibility of the designer to prepare drawings and specifications in conformance to appropriate standards, codes, and regulations. The end products of the designer are design documents that will achieve the performance requirements of the owner. The contractor is responsible for implementing the quality in accordance with the contract documents during construction. Inspection of the work during construction is necessary to ensure quality. The inspection may be done by the owner, but more often is performed by the designer's organization. Thus, all parties have a responsibility and are exposed to risk to ensure quality in the finished project.

Precautions for owners

The owner plays an important role in the success of any project. The owner is the only participant that is involved from the start to the end of the project. There are at least four issues when the owner needs to exercise caution:

1. Project changes
2. Funding of the project
3. Owner furnished material and equipment
4. Public and regulatory environment

The owner is ultimately responsible for defining the scope of a project and the level of quality. During the conceptual phase and early design, the scope of a project will likely change many times. The changes may be due to the owner's desire to modify the project based on available funds or constraints of time, results from design reviews, safety and constructability issues, or concerns from the owner's operations and maintenance staff that will use the project after it is completed. Frequent changes during early design of a project generally have the least negative impact on cost and schedule. However, changes that occur after detailed design, and especially during construction, can have a significant impact on the cost and schedule of the project, see Figure 3-3. Therefore, the owner plays a key role in managing change.

Essentially, the owner pays for everything in a project. Therefore, owners have the burden of securing financing for their projects. Failure to have adequate funding available throughout the project places all parties at risk, especially the contractor because the contractor's cost is typically much larger than the designer's cost. The owner needs adequate funding to make progress payments to the contractor for completed work because a common cause of failure of construction companies is their inability to meet short-term obligations, such as payments to material suppliers, laborers, and rental equipment. Thus, failure of the owner to make timely progress payments places the contractor at risk as well as timely completion of the project.

For some projects the owner procures and furnishes bulk material or special equipment for installation by the contractor at the job-site. For example, an electric utility company may procure steel poles and conductor wire for an electric transmission line construction project. Or, an oil and gas company may procure special equipment, such as a pump, compressor, or heat exchanger for delivery to the job-site for installation by the contractor. Owner supplied material and equipment can be a significant risk factor for the contractor. Extensive coordination is necessary between the owner, designer, and contractor to ensure that owner procured items are delivered in a timely manner to the job-site. Owner furnished material and equipment must be on the job-site when the contractor needs it. Most contractors have an installation schedule for storage and installation of equipment at the job-site. Late delivery of owner furnished equipment is a common cause of delays of construction work and late completion dates of projects.

Public perceptions of a project, or extensive regulatory requirements, can create an adverse and complex environment for projects, which can pose risks for owners. Sometimes there is resistance to construction of a project because of the type and/or location of the project.

For example, there may be resistance from an environmental group to build a nuclear power plant, or people in a residential neighborhood may resist construction of a commercial business that is located adjacent to their houses. Other factors that can pose risks to the owner include: site security, traffic control for safety, and dust suppression at the job-site. These competing interests pose risks that may also affect the owner's ability to attract the best designers and contractors to work on the project and require owners to take precautions to manage these risks.

Precautions for designers

The designer plays a key role between the owner who needs the project and the contractor who is responsible for building the project. The designer must convert the needs of the owner into a set of plans and specifications that are constructible by the contractor and within the cost and completion date that is acceptable by the owner.

Design work involves collecting site information, performing many calculations, using computer applications, developing design drawings with details, and writing numerous pages of technical specifications. The chance of missing critical information and creating errors is enormous. There are at least six issues when designers need to exercise caution:

1. Control of scope growth
2. Quality and completeness of plans and specifications
3. Approval of shop drawings and materials
4. Response to change orders
5. Inspection during construction
6. Managing the budget and schedule for design

Designers have to continually monitor and control scope growth during the design process. The designer should only produce drawings and specifications that pertain to the needs and requirements of the owner. The designer should carefully document all criteria that define appearance, function, and performance requirements that are identified by the owner. It is especially important to also document any items that deviate or enhance the project performance. There needs to be separation between "needs" and "wants" because changes in the project's performance criteria and functionality frequently lead to scope growth. Overdesigning, overdrafting, and overwriting specifications also lead to scope growth, which increases costs and schedule. These risks not only affect the budget and schedule of the designer, but can also impact construction costs and schedule.

The quality and completeness of plans and specifications have a tremendous impact on the cost, schedule, and quality of a project. Plans and specifications should be thoroughly reviewed and checked for errors and omissions before they are released for construction. Defects in the design documents, whether perceived or real, are a common cause of construction claims and disputes. It is prudent to have persons with extensive construction experience to check drawings to ensure constructability and review specifications for clarity.

Early in a project the construction contractor issues submittals for review of shop drawings and approval of materials. The shop drawings are produced from the design documents. The designer is responsible for review and approval of the shop drawings before the contractor can begin work. Any delay in the time allowed for review and approval of shop drawings can cause a delay in the contractor's operations, which may cause a delay claim from the contractor. Delays in the contractor's schedule caused by the designer, or owner, usually lead to contractor claims for additional time and associated cost. Thus, it is important for the designer to respond to review of shop drawings in a timely manner.

Change orders during construction may be issued by the owner, designer, or contractor. The owner may want to add to the project, the designer may want to correct an error in the drawings or clarify a specification. The contractor may request a change order for additional time and money to handle a differing site condition or cleanup after a severe weather event. Change orders typically increase the time and cost of construction. The owner usually asks the designer to review and evaluate the scope, cost, and schedule of change orders prior to approval or denial by the owner. Failure to respond to change orders in a timely manner is a common cause of claims from the contractor for additional time and money.

Inspection of construction work is required to ensure quality of materials and workmanship. Usually, the designer is responsible for inspection at the request of the owner. The contractor is obligated to only perform the work in accordance with the contract documents, no more and no less. There are risks associated with performing inspection. The inspector also needs to be responsive to the contractor's request for inspection of their work. Lack of timely responsiveness of inspection can lead to delay claims from the contractor.

The cost of design work is often measured in work-hours and bar charts are commonly used for design schedules. Managing design budgets usually involves comparing actual hours worked to budgeted hours for the project. For small projects many designers work on multiple projects at the same time, which makes the process of managing design budgets more difficult. A cost control system that accurately charges design hours to each project is necessary, otherwise there is a risk of cost overruns. Most designers are more interested in designing the work than keeping track of the cost or schedule. To prevent schedule slippage, periodic design meetings need to be conducted to compare accomplished work to the scheduled work for key milestone dates in the design effort.

Precautions for contractors

During construction the contractor is in control of the project, which exposes the contractor to risks. The amount of control is defined in the contract documents. Contractors estimate costs, prepare schedules, purchase materials, and manage labor and equipment. There are at least four issues when contractors need to exercise caution:

1. Underestimated costs
2. Unrealistic schedules
3. Ineffective control systems
4. Management of field labor

Most contractors do a relatively good job of preparing cost estimates. However, there is always the possibility of underestimating a job. Low or inaccurate estimates lead to low bids, which lets the contractor get a contract that will not sustain its business. An underfunded job places the contractor in the risky situation of continually having to find ways to recover money through change orders, claims, or temptations to consider questionable business practices.

Prior to construction the only scheduling information usually given to contractors is the approved start date and the required end date for a project. For some projects the contractor may be given some critical intermediate milestones dates. Contractors are responsible for preparing a comprehensive master schedule to direct and control all aspects of their operations. The schedule should be detailed enough to serve as a guide to workers at the job-site and to ensure timely delivery of materials. The contractor's schedule must be periodically updated based on events that occur on the job and to show actual work progress compared to planned progress.

Project control systems are crucial in construction. Control systems include costs, schedules, quality, safety, and materials management. To be effective, a project control system must be simple to administer and easily understood by all participants in a project. An efficient controls system must be developed so information can be routinely collected, verified, evaluated, and communicated to the persons responsible for managing the work. It should be recognized that project control systems are most valuable when there is a good plan and schedule for the job. A good control system provides a mechanism to identify and prevent risks.

Construction workers at the job-site are the most important resource of the contractor. They operate equipment, install materials, detect problems, solve problems, and get the work accomplished despite the many obstacles that may arise. The contractor needs to make a special effort to give proper instructions for work assignments, provide adequate training, ensure a safe working environment, and create a sense of teamwork. When the contractor takes care of the workers, the workers will take care of the contractor. Well-managed workers are more productive and have better safety records, which reduces other risks of the contractor.

Risk Assessment

Risk assessment involves identifying potential risks. In simple terms, a list of risks is the answer to the question of "What can go wrong?" Members in disciplines from all areas of the project team should assess the work for which they are responsible and develop a list of risks. It is especially important to identify the most critical risks, which are the risks that are most likely to occur and the risks that could have the most impact on the project, and then document those risks.

Any member of the project team should identify potential risks in a project. Early in the life of a project a risk workshop should be conducted to identify risks based on lessons learned and experience from previous projects. Typically participants include the project manager, risk specialists, and people from project control, scheduling, contracts, and other subject matter experts. Participants in the workshop may also include risk management consultants from outside the company. A variety of techniques can be used for risk identification, including brainstorming, interviews, checklists, and review of project risk registers from previous projects.

Table 14-3 is a sample list of common risk factors that may occur during the design, procurement, and construction phases of a project. Table 14-4 is an example of a checklist of risks for construction of an electrical transmission line project. A risk analysis of the items in this checklist is presented in the next section.

Most companies establish policies and develop standards that are used to guide the risk management team in risk assessment. Typically, risk assessment is made using qualitative methods in accordance with the risk standard of a company. After a risk is defined, a subjective evaluation of the risk is made based on the judgment and collective thoughts of the risk management team. For example, a company standard for evaluating the level of occurrence of a risk may be classified by a scale, such as very low, low, medium, high, or very high. A low level of occurrence designates a small chance the risk will occur, while a high level of occurrence designates a high chance the identified risk will occur.

TABLE 14-3   Risk Factors during Design, Procurement, and Construction

Design

Changes (owner design changes, scope growth, cost growth, schedule slippage)
Design errors or omissions (calculations, spreadsheets, software)
Compliance (owner's standards, regulatory requirements)
Drawing errors (conflicts in dimensions, plans, elevations, sections, details)
Specifications (conflicts in writings of related specs and/or omissions)
Conflicts in documents (between drawings and specs)
Underestimating construction costs of design alternatives
Working relationships with owner and contractor

Procurement

Prequalification of suppliers and vendors
Late issue of requisitions for long lead-time equipment
Partial shipment of materials and/or equipment to job-site
Damaged shipment of special materials/equipment

Construction

Physical risks (site access/storage, subsurface conditions, weather)
Contractual (approval process, subcontractor failure, supplier defaults)
Costs (underestimated costs, cash flows, cost overruns, liquidated damages)
Schedule (unrealistic schedules, delays in work, weather, subcontractor schedule conflicts)
Labor (availability, productivity, strikes, safety, accidents, defective work, disputes)
Delivery of material/equipment (late/partial/damaged delivery to job-site)
Drawings and specifications (errors in drawings, conflicts in specs)
Working relationships with owner, designer, vendors, and other contractors
Coordination with owner's staff and staff of other contractors working on project
Construction traffic, proximity to public thoroughfares
Construction equipment availability and maintenance

TABLE 14-4   Checklist of Risks for Construction of an Electrical Transmission Line Project

Material delivery delays
Steel poles
Conductor wire
Shield wire

Foundation problems
Encountering rock instead of clay in drilling foundations
Moving underground utilities

Right-of-ways
Delays in securing right-of-way
Condemnation hearings to secure right-of-ways

Energizing line
Schedule slippage due to adverse weather conditions
Loss of tax advantage if line is not energized by end of calendar year

TABLE 14-5   Converting Qualitative to Quantitative Values

Scale	Schedule impact	Cost impact
Very High	> 40%	> 20%
High	20%–40%	10%–20%
Medium	10%–20%	5%–10%
Low	1%–10%	1%–5%
Very Low	> 1%	> 1%

A similar initial evaluation of the impact or severity of a risk is often made on a qualitative scale from very low to very high. Typically, a company risk management standard defines guidelines for converting qualitative evaluations to quantitative numerical values as illustrated in Table 14-5.

Risk Analysis

Risk analysis involves analyzing the probability of occurrence and the degree of severity of each risk. Although statistical models, simulations, and other numerical methods can be used in risk analysis, the judgment and wisdom of competent people who have extensive experience in project work should be included. Numerical analysis of risk is not a stand-alone substitute for common sense and judgment of experienced people.

Table 14-6 shows a risk analysis of the checklist of items for the electrical transmission line project shown in Table 14-4. A probability of occurrence and a level of severity are given to each item in the checklist. The risk score is calculated by multiplying the probability times the severity factor. A legend of the criteria for the analysis of severity is shown at the bottom of the table. The risk score for the project is calculated as the sum of the risk score for each item in the checklist. Results of the analysis show the highest risk is delivery of material, in particular delivery of conductor wire to the job-site. Therefore, special attention should be devoted to ensure delivery of conductor wire. Actions that can be taken to ensure timely delivery of conductor wire include verifying availability from the supplier, issuing an early purchase order, securing a confirmation delivery date in the purchase order, and follow up by contacting the vendor on shipping dates.

Risk Analysis of Costs

There are many risks in preparing a cost estimate for a project. There are risks in calculating the quantities of materials, extension of prices, and variations in quotes from subcontractors and suppliers of materials and equipment. There are also risks in assumed production rates of laborers and construction equipment.

TABLE 14-6   Risk Analysis of Electrical Transmission Line Project

Risk Item	Probability (Likelihood)	Severity (Impact)	Risk Score
Delivery of materials			
Steel poles	70%	Low (3)	2.1
Conductor wire	50%	High (9)	4.5
Shield wire	25%	Medium (6)	1.5
		Subtotal = 8.1 material delivery	
Foundation problems			
Encountered rock	40%	High (8)	3.2
Underground utilities	10%	Low (2)	0.2
		Subtotal = 3.4 foundation problems	
Right-of-way easements			
Delays in securing	50%	Medium (7)	3.5
Condemnation	15%	Medium (6)	4.4
		Subtotal = 7.9 easements	
Energizing Line			
Schedule slippage	30%	Low (3)	0.9
Lost tax advantage	30%	High (9)	2.7
		Subtotal = 3.6 energizing line	
		Total Risk Score = 23 for Project	

Company established criteria for severity:

Severity	*Cost*	*Schedule*
Low (1-4)	*reduced profit*	*no penalties*
Medium (5-7)	*lost profit*	*minimal penalties*
High (8-10)	*catastrophic loss*	*severe penalties*

Typically the basis of the cost estimate consists of the estimated cost plus a contingency amount that is applied to each bid item. After an estimate is prepared a thorough review of bid items in the estimate should be conducted to identify potential risks. Then each bid item in the list can be evaluated for gaps or overlaps in scope, budget, quality, or other potential risks that might occur. The first step in risk analysis involves determining the maximum possible risk for each bid item, recognizing that it is unlikely that all of the bid items will incur risk. The next step involves determining the probability of occurrence of each risk. The maximum possible risk multiplied by the probability of occurrence percentage gives the expected net risk for each bid item, see Table 14-7.

The risk analysis data in Table 14-7 shows a base estimate of $2,108,000 with a maximum anticipated cost of $2,582,000, a difference of $474,000. However, it is unlikely that all the risk will occur for each item in the estimate. Therefore, the expected net risk for each bid item is calculated by multiplying the maximum risk by the probability percentage of each item. The total expected net risk for the project is $206,500, which is calculated as the sum of the net risk for each item in

TABLE 14-7   Risk Analysis of Cost

Estimate item	Base estimate	Maximum cost	Maximum risk	Probability percentage	Expected net risk
1	$40,000	$50,000	$10,000	20%	$2,000
2	100,000	150,000	50,000	30%	15,000
3	250,000	320,000	70,000	50%	35,000
4	237,000	320,000	83,000	60%	49,800
5	94,000	135,000	41,000	30%	12,300
6	730,000	870,000	140,000	40%	56,000
7	572,000	640,000	68,000	50%	34,000
8	85,000	97,000	12,000	20%	2,400
Totals	$2,108,000	$2,582,000	$474,000		$206,500

the estimate. The $206,500 represents a contingency markup of 9.8% on the base cost of $2,108,000, which is calculated as ($206,500/$2,108,000) × 100 = 9.8%. For this project the final estimate can be calculated as the base estimate plus the total expected net risk, $2,108,000 + $206,500 = $2,314,500.

Risk Analysis of Schedule Using PERT

Most contract bid documents require the contractor to submit a schedule of construction activities that shows the sequencing of activities and the total time to complete the project. The time required to complete the project is generally established from a schedule that is developed from a bar chart or CPM network analysis. As presented in Chapter 8, the bar chart and CPM methods of scheduling are based on assigning a specific duration for each activity. However, for some projects it is difficult to estimate a single duration for one or more activities in the schedule. Typically there is a range of time that may apply to a particular activity. For example, the time to drive piles on a project may vary from 14 days to 27 days, with a most likely expected time of 19 days. In a CPM schedule the 19 days likely would be used to develop the project schedule.

In Chapter 8 the PERT method of scheduling is presented and discussed. The PERT method is seldom used by the contractor to develop a project schedule for submittal to the owner. The lack of use is generally attributed to the need to use multiple durations for activities and the application of statistical methods. However, in early stages of project development the PERT method is a valuable tool for analyzing risk in a project schedule.

Network analysis systems, both CPM and PERT, are based on arranging activities in sequential manner, which shows the interrelationship of activities. A single duration is assigned for each activity using the CPM method, whereas three durations are assigned to each activity using the

PERT method. There are many paths of activities from the start to end of the network diagram. The longest path in the network establishes the total project duration, which identifies the critical path and critical activities. Typically, the critical activities consist of 10% to 20% of the total number of activities in the project schedule. Example 14-1 illustrates the use of PERT for risk analysis of a project schedule.

Example 14-1   A project team has developed a PERT schedule for the project. Below are the activities on the critical path of the PERT diagram. These six activities are used in the risk analysis of the project schedule because the planned project duration is governed by critical path activities. The optimistic time (a), most likely time (m), and pessimistic time (b) are shown at the bottom of each activity. All times are measured in months. Perform a risk analysis of the project schedule.

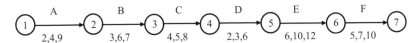

A. **Calculate the most likely time (t_e) for each activity in the critical path.**
The expected time for each activity can be calculated by the following equation:

$$t_e = (a + 4m + b)/6 \text{ where } t_e \text{ is the expected time of an activity}$$

Activity	$t_e = (a + 4m + b)/6$	Expected time (t_e)
A	[2 + 4(4) + 9]/6	4.500 months
B	[3 + 4(6) + 7]/6	5.667 months
C	[4 + 4(5) + 8]/6	5.333 months
D	[2 + 4(3) + 6]/6	3.333 months
E	[6 + 4(10) + 12]/6	9.666 months
F	[5 + 4(7) + 10]/6	7.167 months

B. **What is the optimistic, pessimistic, and expected time for completing the project?**
Optimistic duration = 2 + 3 + 4 + 2 + 6 + 5 = 22 months
Pessimistic duration = 9 + 7 + 8 + 6 + 12 + 10 = 52 months
Expected duration, T_E = summation of the expected time of each activity
= 4.50 + 5.667 + 5.333 + 3.333 + 9.666 + 7.167
= 35.7 months

C. **What is the probability that this project can be completed in 37 months?**
The σ_{TE} for the final event, number 7, is the square root of the variances of these events, which can be calculated as follows:

$$\sigma_{TE} = \sqrt{v_{1-2} + v_{2-3} + v_{3-4} + v_{4-5} + v_{5-6} + v_{6-7}}$$
$$= \sqrt{[(9-2)/6]^2 + [7-3]/6]^2 + [(8-4)/6]^2 + [(6-2)/6]^2 + [(12-6)/6]^2 + [(10-5)/6]^2}$$
$$= \sqrt{1.3611 + 0.4444 + 0.4444 + 0.4444 + 1.0000 + 0.6944}$$
$$= \sqrt{4.3887}$$
$$= 2.095$$

As calculated on the previous page, $T_E = 35.7$. The probability of completing the project by the 37th month can be calculated as follows:

$$z = (T_S - T_E)/\sigma_{TE}$$
$$= (37 - 35.7)/2.095$$
$$= 0.621$$

From Table 8-6, based on a deviation z of 0.621, the probability is 0.74. Thus, there is a 74% probability that the project will be completed by the 37th month.

D. What is the probability that this project will be completed by the 32nd month?
As calculated above, $\sigma_{TE} = 2.095$ and $T_E = 35.7$

$$z = (T_S - T_E)/\sigma_{TE}$$
$$= (32 - 35.7)/2.095$$
$$= -1.77$$

From Table 8-6, based on a deviation z of −1.77, the probability is 0.04. Thus, there is a 4.0% probability that the project will be completed by the 32th month.

E. For a 99% probability, what is expected number of months for completion?
As calculated in C. above, $\sigma_{TE} = 2.095$ and $T_E = 35.7$
For a 99% probability, $z = + 2.5$

$$z = (T_S - T_E)/\sigma_{TE}$$
$$+2.5 = (T_S - 35.7)/2.095$$
$$T_S = 40.9 \text{ months}$$

F. For a 1% probability, what is the expected number of months for completion?
As calculated in C. above, $\sigma_{TE} = 2.095$ and $T_E = 35.7$
For a 1% probability, $z = -2.5$

$$z = (T_S - T_E)/\sigma_{TE}$$
$$-2.5 = (T_S - 35.7)/2.095$$
$$T_S = 30.5 \text{ months}$$

G. Develop a graphical plot of the probability versus the project duration for the full range of probabilities for assessing the risks of finishing the project ahead or behind the expected duration of 35.667 months.

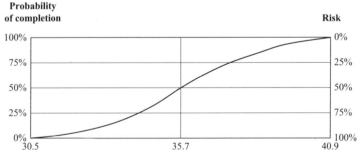

This risk versus project duration graph can be used by upper management in their process of risk management. Based on the analyses it is unlikely the project will be completed in 30 months, whereas it is highly likely the project will be completed in 41 months. There is a 50/50 chance the project will be completed in 35.7 months. Providing this information is better than simply giving a specific project duration without regards to the probability of completion. PERT analysis of the project schedule is best suited for high-risk projects, where there are many uncertainties in the duration that are assigned to activities.

Risk Analysis Using Simulation

Monte Carlo simulation is a statistical method for interactively evaluating a deterministic model using sets of random numbers as inputs. It can be used as a tool for risk analysis of schedules and/or cost estimates. Monte Carlo analysis begins with a deterministic model, such as a baseline schedule or base cost estimate.

Computer software programs are available to mathematically compute and track many different possible risk scenarios and the probabilities of occurrence associated with each risk. Analysis of risks can be made for cost estimates, project schedules, labor productivity, or resource allocations. Software packages can analyze spreadsheets based on variable inputs in terms of realistic ranges of possible values and probable outcomes. The analysis assists in making decisions on which risks to take and which risks that should be avoided.

Monte Carlo simulations are typically applied and most helpful in the process of front-end planning by owners for high-risk projects, prior to actual work. In general, contractors find limited use of Monte Carlo simulations in active projects due to limited time and budget constraints in ongoing projects. The larger the project, the more need for sophisticated risk analysis tools like Monte Carlo simulation to analyze a range of risk probabilities.

It is important to note that the credibility of Monte Carlo simulations is only as effective as the cumulative experience of those who develop the variables that are used in the simulation. The reliability of a simulation of a project schedule is highly dependent on the quality of the baseline schedule for the project. For example, the project schedule must have appropriate sequencing of activities and realistic durations before attempting to run a simulation analysis. Likewise, the reliability of a simulation analysis of a cost estimate is highly dependent on the quality of the baseline estimate. Attempting to run a simulation analysis on a baseline cost estimate that is full of errors gives unreliable results.

Risk Analysis of Schedule Using Simulation

Simulation allows much greater flexibility than PERT for risk analysis of schedules. A CPM schedule is a deterministic model that consists of relationship of activities and their durations. Simulation is a technique that uses a computer to evaluate the model numerically. Monte Carlo is a simulation technique that requires a computer to perform thousands of calculations to evaluate activities in a CPM network. Many of the computer scheduling software packages have the capability to perform a Monte Carlo simulation.

Monte Carlo simulation of a project schedule involves developing multiple schedules based on durations that are randomly selected for each activity. First, a range of durations is defined and a duration distribution is established for each activity. The distributions may be linear or nonlinear, such as uniform, triangular, or exponential. Figure 14-1 shows examples of duration distributions that are typically used for simulations of costs and schedules for engineering and construction projects. The minimum duration is (*a*) and the maximum is (*b*) and the most likely is (*m*).

After the range and duration distribution is defined for each activity, a schedule for the project is developed by assigning a duration for each activity that is randomly selected from the duration distribution of each activity. For example, the first schedule may calculate a project duration of 380 days. Then, a second schedule is developed by assigning a duration for each activity that is again randomly selected from the duration distribution of each activity. The second schedule may calculate a project duration of 420 days. This process is repeated thousands of times by the computer to develop a profile of the project duration. The end result may be a range of durations from 360 days to 450 days. Then the mean and standard deviation of the project durations can be calculated and a probability curve can be developed.

Figure 14-2 is an example of using simulation for risk analysis of a project schedule. The left-hand ordinate shows the number of calculations in the simulation process and the right-hand ordinate shows the cumulative frequency and range of working days. The abscissa shows the range of working days from a minimum of 162 to a maximum of 237 days. The simulation analysis shows there is a 10% probability the project will

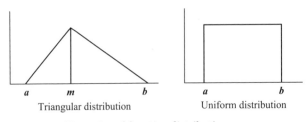

Triangular distribution   Uniform distribution

Figure 14-1   Examples of duration distribution.

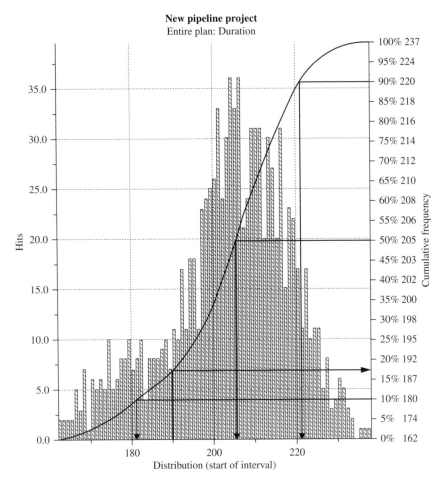

New pipeline project
Entire plan: Duration

Figure 14-2 Risk analysis of project schedule using simulation.

be completed in 182 days and a 90% probability it will finish in 221 days. There is a 50% probability the project will be completed in 206 days. The probability of completing the project in 190 days is 17%. Risk analysis by simulation allows the project team and upper management to make decisions using the probability of achieving different durations, rather than reporting a single duration for the project.

Risk Analysis of Costs Using Simulation

Monte Carlo simulation of costs begins with the deterministic cost estimate, which is commonly called the baseline cost estimate. The baseline cost estimate is calculated using the most likely cost of items in the estimate.

A range of costs for each item in the estimate varies from an optimistic minimum value (**a**) to a pessimistic maximum value (**b**). A cost distribution is established for each cost item in the estimate. Using a computational algorithm, the Monte Carlo simulation program runs thousands of simulations, each time using a different set of values that are randomly selected from the cost distributions. The repeated thousands of simulations by the computer are used to develop a profile of the cost estimate of the project. The end result is a range of estimates with a probability percentage. Then the mean and standard deviation of the project cost estimates can be calculated and a probability curve can be developed.

Figure 14-3 is an example of using simulation for risk analysis of a cost estimate. The left-hand ordinate shows the number of calculations in the simulation process and the right-hand ordinate shows the

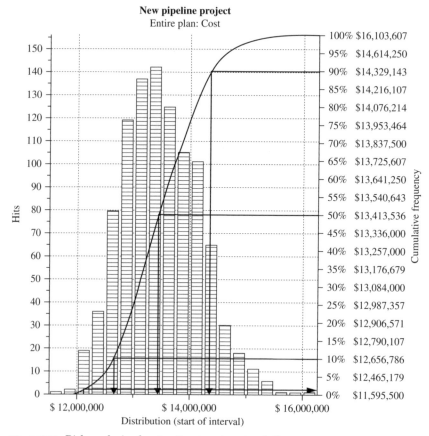

Figure 14-3 Risk analysis of cost estimate using simulation.

cumulative frequency and range of costs. The abscissa shows the range of cost estimates from a minimum of $11,595,500 to a maximum of $16,103,607. The simulation analysis shows there is a 10% probability for a cost estimate of $12,656,786 and a 90% probability for a cost estimate of $14,329,143. There is a 50% probability for a cost estimate of $13,413,536. Risk analysis of a cost estimate by simulation allows upper management to make decisions based on a range of costs with associated probabilities, rather than reporting a single number for the cost estimate.

Mitigation of Risk

A key attribute of experienced and successful project managers is their ability to identify and manage risk. Successful project managers are always looking for what can go wrong and they have a plan to rectify the problem. A project manager should be proactive in managing risk, rather than responding to risks after they occur.

It is always much easier to prevent a risk from happening than it is to fix problems that occur after the risk is realized. In simple analogous terms, an ounce of prevention is worth a pound of cure. Many risks can be easily prevented by effective communications and instructions. For example, a simple call to a supplier to confirm delivery dates, performing certified welds in the shop instead of field welds at the job-site, etc. Fixing problems from the occurrence of a risk can be challenging and costly with respect to time, money, and quality.

Experienced project managers generally have good ideas of what can go wrong in projects. Their knowledge and wisdom is gained from "lessons learned" while working on previous projects. Capturing the experience gained in handling risk issues on previous projects and bringing those "lessons learned" to risk management of new projects is a tremendous advantage in mitigating future risks.

The contract for a project plays a major role in risk management. Contracts govern the process of design and construction. The contract is a legal document that defines the roles and responsibilities of the parties and the requirements for performance of the work. Thus, the contract delegates risk to participants in the project. There is detailed information that is written in the contract, but too often the contract is not read. Understanding the terms and conditions of the contract that is signed by both parties is important to identifying, assessing, and mitigating risks.

There are multiple risks for any project. Therefore, the risks in a project should be considered collectively, rather than individually. To effectively manage and mitigate risks a project risk register should be developed prior to the start of a project. A project risk register is a tool for collecting information about each risk and tracking the risk management process throughout the life cycle of a project. The project risk register should be regularly evaluated at team meetings and updated when new risks are identified as a project moves through design, construction execution, and final completion of a project.

Project Risk Register

A project risk register is a document that contains information about each risk, see Figure 14-4. The information should include a description of the risk, level of risk assessment, probability of occurrence, impact of the risk (in schedule and costs), and a risk management action plan. Following is the type of information that should be contained in a project risk register:

- Description of Identified Risk—a description that uniquely describes the risk and its origin

- Risk Manager—(sometimes called risk coordinator or risk owner) the person responsible for maintaining and implementing the risk mitigation plan

- Risk Area—primary area of the project to which the risk pertains; risks are generally organized in categories of scope, budget, schedule, and quality

- Status—indicates the risk's standing in the risk management process; status may be defined as new, open, canceled, mitigating, and completed

- Risk Probability—indicates the probability of the risk occurring on a percentage basis

- Estimated Value of Risk —estimated value of remediating the realized risk based on an estimated, or known, project cost impact and remedial measure

- Potential Risk Impact—calculated value by multiplying the risk probability times the estimated risk value

- Allocation—identifies that element of the project team that is best able to control the risk mitigation

- Risk Mitigation Plan—summary narrative of the actions, schedule, and assigned resources required to mitigate the risk.

Any member of the project team can, and should, identify potential risks in a project and each risk should be entered in a project risk register. Each risk is identified in a risk area, such as scope, budget, schedule,

Description of Identified Risk	Area of Risk	Probability of Occurrence	Estimated Value of Risk	Potential Impact of Risk	Status of Risk	Risk Mitigation Action Item
Design completion schedule (4-week slippage due to changed process)	Schedule	60%	$250,000	$150,000	Open	Dependent on extent of client comments
New equipment for revised process is manufactured overseas	Quality	33%	$48,000	$16,000	Open	Arrange for factory acceptance test at origin by expert witness
Schedule reduction (25 mo. vs. 24 mo.) to reduce insurance	Schedule	55%	$20,000	$11,000	Open	Start construction as soon as possible (< 5 months from notice to proceed)
Estimated labor rates are $1/hr above local average (consider using local labor?)	Cost	40%	$170,000	$68,000	Closed	Accounted for in estimate
What is the owner's time frame to provide approval of submittals?	Contract	40%	$39,000	$15,600	New	Need 8 days, not 28, approval of shop drawings (collaborate with submittor and reviewer)
Expedite procurement of long lead-time material/equipment (piping and compressors)	Cost and Schedule	10%	$750,000	$75,000	Open	Obtain commitment letter from owner and check cancelation costs
Significant dewatering requirements to install piping along river alignment	Cost and Schedule	60%	$500,000	$300,000	Open	Designate hydrogeologist to create dewater plan with dewatering subcontractor
Power lines cross the work area where our cranes will be working	Safety	100%	$58,000	$58,000	Closed	Coordinate with power company to relocate line (buy and install line-guards)
Weather-construction will be in adverse winter conditions	Schedule	25%	$120,000	$30,000	Mitigating	Check local area history for typical severe weather days and prepare accordingly
Obtaining right-of-way easements are taking longer than expected	Schedule	40%	$90,000	$36,000	Open	Employ additional land agents and/or consider alternate routes
Original land acquisition plan was based on known 45.3 mile route, but now find that state does not allow eminent domain	Scope	90%	$1,894,000	$1,704,600	New	Rerouting of pipeline is expected, which may result in 48.8 miles of pipe rather than planned 45.3 miles
Client expects first system in service not later than 180 days after award of contract	Schedule	80%	$350,000	$280,000	Open	Coordinate sufficient resources for multiple shifts or extended work-hours

Figure 14-4 Illustration of project risk register.

health, safety, or quality. Persons that identify the risk are responsible for an initial assessment of the risk based on their judgment, or the collective thoughts of the project team. Subsequent analysis of each risk is performed that includes the probability of occurrence, level of severity, and impact of the risk. The person that defines the risk is sometimes called the risk originator.

The risk is then assigned to a risk manager (most often doubling as the project manager) who is empowered to ensure the risk is mitigated. The assignment is based on the type of risk and expertise of the risk manager responsible for mitigating the risk. The risk manager is responsible for successfully mitigating the risk to a target (lower) risk severity level. The target severity level is determined by the risk manager, and it represents the expected result of implementing the plan to mitigate the risk. The risk manager is also responsible for providing the progress of mitigation to the project team.

It should be noted that often the effect of mitigating a risk is not to eliminate the possibility that a risk could occur, but instead to reduce the likelihood and/or impact of the risk to an acceptable level.

In the process of identifying risk in a project, it is helpful to ask "What If" questions. For example, in Figure 14-4 the last risk would be identified by the question *"What if the owner wants the first system to be in service not later than 180 days after award of the contract."* For this risk the 80% probability of occurrence is based on the current late finish date and the estimated schedule delays included in risk register to-date. The estimated value of the risk is calculated as (liquidated damages cost) × (potential days past the service date.). The potential impact of the risk is calculated as (probability of occurrence) × (estimated value of the risk), (80% × \$350,000 = \$280,000). A unique analysis is made of each item in the project risk register.

Development of the Initial Project Risk Register

Generally the initial development of the project risk register is performed in a brainstorming workshop. Those who participate in the identification of project risks will include staff from all parties who are involved in the project. For example, for a design/build project the participants should include the project management from each organization, lead design engineers, construction manager, lead cost estimator and scheduler, procurement manager, start-up and commissioning people, and the owner's operations and maintenance personnel. These key staff people fulfill a broader role on the project as members of the risk management team.

Team members selected to participate in the risk identification process are provided background information in the form of proposal documents, the designer's scope of work, predesign studies, design development deliverables, geotechnical reports, and cost estimates. This information is used by each participant to become familiar with the project and its requirements as well as the owner's goals and objectives.

After the brainstorming session, the identified risks are placed in the project risk register. Thus, the project risk register becomes a project management tool that should be regularly evaluated at project team meetings. Additionally, as the project delivery cycle moves forward, the project risk register is updated with newly identified risks as the project progresses through design milestones into construction execution and the start-up and commissioning phases of the project.

Strategies to Mitigate a Risk

The first step in developing a risk mitigation plan is selection of a strategy to mitigate the risk. The strategy selected depends on the level of impact of the risk and the potential to successfully apply the strategy to mitigate the risk. Strategies for risk mitigation include:

- Accept the risk
- Avoid the risk
- Control the risk
- Investigate the risk
- Reduce the risk
- Transfer the risk

Acceptance of a risk is normally used for risks that have low probabilities of occurrence and minimal impacts. This technique may also apply to risks that are difficult to control. It is a passive technique that focuses on allowing the outcome of the risk to occur without trying to prevent that outcome.

Avoidance is a technique that prevents the occurrence of the risk. The risk is circumvented by changing the contract parameters that were established between the owner and contractor. Examples include reduction in scope of work, changing the requirements in the specifications, changing the statement of work, and submitting waivers and deviations.

Controlling the risk involves taking actions to reduce the likelihood of occurrence or impact of the risk. The control-based actions can occur at all phases of the project's lifecycle. The actions selected to control the risk are monitored and reported as part of the regular performance analysis of the project.

Investigation is a technique that defers all actions until more work is done and/or more facts become known. This technique does not define any mitigation for reducing an individual risk. It is applied to risks where no clear solution is identified and further research is required.

Reduction is a technique to actively lower the risk by a planned series of activities. The method of reduction is unique for each risk.

Transferring a risk is a process of moving the risk to another party in the project. Typically, the risk is transferred to subcontractors or suppliers who are more able to reduce the overall risk exposure. This technique is mostly used during the proposal process of a project. The transfer of risk can also include the use of third-party guarantees, such as insurance or performance bonds.

Methods to Prevent or Mitigate Risks

Methods of risk prevention or mitigation depend on the type of project and type of risk. Engineering and construction projects are generally categorized into three sectors: buildings, infrastructure, or process industry. The building sector may be subdivided into subdivisions, such as residential, commercial, or industrial buildings. Examples of infrastructure industry projects include highways and bridges, water and sewer systems, electrical transmission and distribution systems, etc. The types of projects in the process industry include oil and gas, chemical plants, manufacturing, pharmaceutical, etc. Figure 14-5 gives examples of methods of preventing or mitigating risks for a project in the infrastructure sector.

Development of a Risk Mitigation Plan

After a risk mitigation strategy is selected, a risk mitigation plan can be developed by the risk manager. The risk mitigation plan is essentially a response to the risk. It contains a series of action items in a sequential order that will successfully mitigate the risk to the target severity level and by the due date for mitigation of the risk. Each action item should have a defined start and finish date.

Each action item should also have a target severity level, which is the severity level that will be obtained when the action item is completed. It should also identify persons who will be responsible for performing the action items. The risk manager may contact the project manager or any member of the project team for assistance in developing and implementing the action plan. The project team should be contacted immediately if the risk manager is not able to mitigate the risk. There may be situations when it becomes necessary to reassign the risk to another risk manager who is capable of resolving the risk.

Risk Description	Method to Prevention/Mitigation
Potential safety risks constructing underground intake shaft	Manage the project safety plan closely, ensure all personnel are well-trained and adhere to regulations.
Public interest group opposition to project	Develop public relations plan and establish effective communications to address issues
Distraction of client representatives from core project roles	Transparent analysis of alternatives; effort to hear and address concerns
Schedule delay	Weekly short-interval schedules; design and construction milestones; early start on construction before full design completion; stacking trades; multiple shifts; overtime
Accidental damage to environment	Conduct environmental compliance training; set up a full-time environmental monitoring system
Access shaft proximity to pump station construction and trucks crossing the site	Employ a conveyor system to transport spoils to common loading point adjacent to site entrance
Evaluation and classification of geotechnical conditions	Contingency plan developed to shift crews to an alternate location
Misalignment impacts performance and schedule of tunneling operation	Bore the intake shaft first and then bring the tunnel to it
Risks to diving personnel due to depth and visibility conditions at underwater intake assembly and installation	Maintain the allowable bottom time per dive plus decompression time. Preassemble as many components as possible. Use remotely operated vehicles (ROVs) for inspection task (when visibility allows) to reduce diving requirement. Formulate means and methods (templates, stabbing guides, remotely activated equipment, etc.) to minimize diving requirements
Water source for testing tunnel pipelines	Develop flushing and testing plan; coordinate the connection of pipelines with large basin construction as a potential reservoir to flush
Risks to ongoing plant operations	Develop a plan to maintain plant operations during construction; construct bypass systems; install temporary systems
Average of 60 in. of snowfall per year; the average winter temperature is below 40°F	Site-specific safety procedures; training requirements, administrative controls, engineering controls, personal protective equipment; protection of the ongoing and completed work from extreme weather

Figure 14-5 Methods to prevent/mitigate risks.

After the risk mitigation plan is developed, it should be presented to the project team for their review and final approval. Upon approval of the risk management plan the project risk register is updated and the risk manager can commence the risk mitigation. During implementation of the risk mitigation plan, the risk manager provides status updates as they occur through the risk register and prior to each project team meeting.

When the risk is mitigated, the risk manager updates the action items and provides adequate detail that describes the final status in the project risk register. If the project team agrees that mitigation is completed, the risk status is shown as "completed" or "closed" in the risk register.

References

1. Cooper, D., Stephen G., Raymond, G., and Walker, P., *Project Risk Management Guideline-Managing Risk in Large Projects and Complex Procurement*, John Wiley & Sons, The Atrium, Southern Gate, Chichester, West Sussex, England, 2005.
2. Crouhy, M. C., Galai, D., and Mark, R., *The Essentials of Risk Management*, McGraw-Hill Inc., New York, NY, 2006.
3. Fenton, N. and Neil, M., *Risk Assessment and Decision Analysis with Bayesian Networks*, CRC Press, Taylor and Francis Group, Boca Raton, FL, 2013.
4. Gupta, A., *Risk Management and Simulation*, CRC Press, Taylor and Francis Group, New York, NY, 2014.
5. Hassett, M. J. and Steward, D. G., *Probability for Risks Management*, 2nd ed., ACTEX Publications, Inc., Winsted, CT, 2009.
6. *Practical Standard for Project Risk Management*, Project Management Institute, Newton Square, PA, 2009.
7. Raydugin, Y., Project *Risk Management-Essential Methods for Project Teams and Decision Makers*, John Wiley & Sons, Hoboken, NJ, 2013.

Example Project

Statement of Work

This example project is an actual project that has been reduced in size and simplified for illustrative purposes. The management of this type of project is discussed in the previous chapters of this book. Chapter 3 presented a discussion of the owner's study, including the needs assessment and project objectives. A discussion of project budgeting based on square foot estimating is presented in Chapter 5. The work breakdown structure and development of the project work plan (Chapter 6), a discussion of project scheduling (Chapter 8), and project tracking (Chapter 9) presented pertinent information for this project.

The project is a service facility that consists of an industrial building (Building A) for servicing equipment and vehicles and an employee's administrative office building (Building B) for management of the operation of the service facility. A warehouse building (Building C) is proposed as a future addition to the project, but is not initially included in the project scope due to constraints in the owner's budget. Site-work consists of grading, drainage, and all on-site utilities that are required for operation of the service facility. The project is located on 100 acres of land. Figure A-1 is a site plan that shows the layout of buildings, driveways, and parking areas.

The scope of work includes engineering, procurement, and construction for the site-work and two buildings. A soil investigation, legal boundary survey, and contour map is provided by the owner. Figure A-2 shows the contour map. Special procurement includes the overhead crane for Building A and the elevator conveying system for Building B. The maintenance building has a 40-foot clear ceiling height, with a 20-ton overhead crane. The owner anticipates 45 employees will be using the building during a day shift and plans expansion of work to include a night shift at a future date. A small office area and a shop with machining equipment is to be located in the maintenance building. A wash-down area is to be provided for washing and servicing of truck vehicles.

The employee's office building is a two-story structure that will be used by 70 employees who will be involved in clerical work. A conference room, training facilities, computer work stations, and small cafeteria are to be included in the building.

All sealed surface paving and parking areas are to be constructed with portland cement concrete. The layout for the crushed aggregate paving area, for storage of major equipment and materials, is shown on the site plan.

The owner's feasibility study involved a needs assessment to produce the project budget that is shown in Figure A-3. Acquisition of land and permits are completed by the owner. Anticipated design fees are included in the total project budget. The owner's required time of completion of the project is 17 months, including engineering, procurement, and construction.

Development of Project Work Plan

The responsibility of the project manager is to develop a comprehensive work plan to guide all aspects of the project. The first step in the process is a detailed review of the two sets of project data: the owner's study that developed the statement of work, budget, and schedule for the project; and the proposal that was approved to issue the contract to the project manager's organization. Ideally, the project manager and key members of his or her project team would have been involved in the development of both of these sets of project data; however, many project managers are assigned the responsibility of managing a project in which they have had no prior involvement.

Chapter 6 presents a discussion of the items to be reviewed by the project manager at the start of a project. A thorough review must be performed so the project manager has a clear understanding of the work that must be accomplished. In particular, the project budget must be evaluated to ensure it is realistic, and the project schedule must be evaluated to ensure it is reasonably attainable.

After the project manager's initial review, he or she develops a work breakdown structure (WBS) in sufficient detail to identify major areas of work to be performed. The purpose of this initial development of the WBS is to define the required disciplines for selection of project team members. The project manager must work with the managers of the different disciplines to staff the project based upon available resources from the organizational breakdown structure (OBS). For this example project, the design work for Building B is to be assigned to an outside design organization because the building requires architectural expertise that is not available in the project manager's organization. Sometimes work is assigned to an outside design organization due to time constraints that would be placed on in-house personnel to accomplish the work.

Figure A-4 shows the WBS for the project, which indicates the tasks to be performed, the grouping of the tasks, and the person responsible for each part of the project. Each project team member is responsible for development of a work package for the work that he or she is to perform. The format for the contents of a work package is shown in Figure A-5. The total cost of the project is derived from the sum of the costs of the work packages from all team members. The coding system for the example project is shown in Figure A-6.

As discussed in Chapter 8, a project schedule (CPM) is developed by the project manager in cooperation with the team members based upon an integration of the schedule of the work packages (reference Figure A-7). The project manager works with the team members to develop a coding system for the project to sort and report project reports.

A complete listing of all computer input data for the project is shown in Figure A-8. The information that is contained in Figure A-8 is the only project data that is used to generate all the reports (Figures A-9 through A-14) that are shown in the remainder of this Appendix. These reports are illustrative examples of typical project reports. The title at the top of each report identifies the contents of the report.

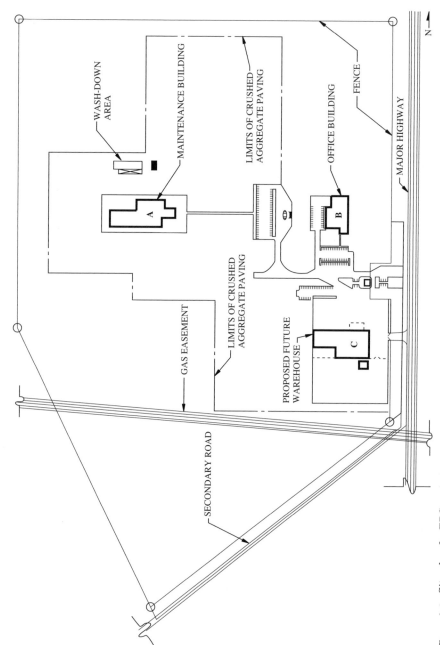

Figure A-1 Site plan for EPC project.

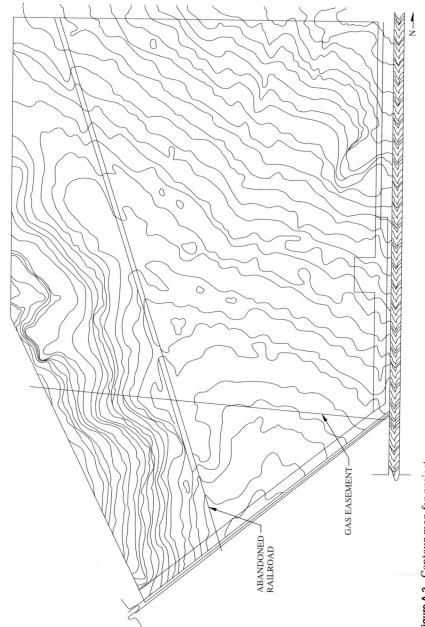

N

ABANDONED
RAILROAD

GAS EASEMENT

Figure A-2 Contour map for project.

Facility	Budget

Site-Work.................................... $522,400
Grading
Paving
Landscape

On-Site Utilities.............................. $173,700
Stormwater
Sanitary Sewer
Domestic Water
Underground Electrical
Natural Gas

Maintenance Building A.................... $1,083,600
Architectural
Structural
Electrical
Plumbing
Heat & Air

Maintenance Building B.................... $1,097,100
Architectural
Structural
Electrical
Plumbing
Heat & Air

Project Management.......................... $19,350
Clerical

Procurement................................. $157,000
Crane
Elevators
Construction Bids

Total Directs = $3,053,150

Contingency................................. $150,000

Total Approved Budget = $3,203,150

Figure A-3   Approved EPC project budget.

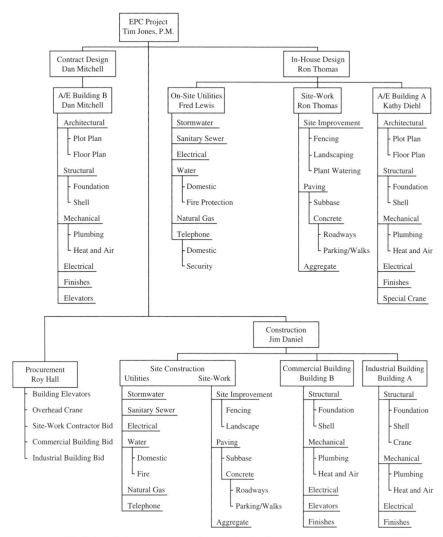

Figure A-4 Work breakdown structure for service facility project.

WORK PACKAGE

Title: _____

WBS Code: _____

1. Scope

Required Scope of Work: _____
_____
_____

Services to Be Provided: _____
_____
_____

Services not included in this Work Package, but included in another work package: _____
_____

Services not included in this Work Package, but will be performed by: _____
_____

2. Budget

Personnel Assigned to Job	Work-Hours	$-Cost	CBS Code Acct	Computer Services Type	Hours	$-Cost
_____	____	____	____	____	____	____
_____	____	____	____	____	____	____
_____	____	____	____	____	____	____

Total Work-Hours = _____   Personnel Costs = $ _____

Computer Hours = _____   Computer Costs = $ _____

Travel Expenses		Reproduction Expenses		Other Expenses	
_____	+	_____	+	_____	= $ _____

Total Budget = $-Labor + $-Computer + $-Travel + $-Other = $ _____

3. Schedule

OBS Code	Work Task	Responsible Person	Start Date	End Date
___	_____	_____	_____	_____
___	_____	_____	_____	_____
___	_____	_____	_____	_____

Work Package: Start Date: _____ End Date: _____

Additional Comments: _____
_____
_____

Prepared by: _____ Date: _____

Approved by: _____ Date: _____

Figure A-5   Team member's work package.

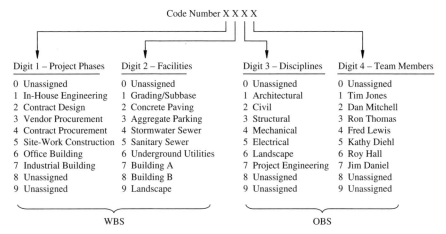

Code Number X X X X

Digit 1 – Project Phases	Digit 2 – Facilities	Digit 3 – Disciplines	Digit 4 – Team Members
0 Unassigned	0 Unassigned	0 Unassigned	0 Unassigned
1 In-House Engineering	1 Grading/Subbase	1 Architectural	1 Tim Jones
2 Contract Design	2 Concrete Paving	2 Civil	2 Dan Mitchell
3 Vendor Procurement	3 Aggregate Parking	3 Structural	3 Ron Thomas
4 Contract Procurement	4 Stormwater Sewer	4 Mechanical	4 Fred Lewis
5 Site-Work Construction	5 Sanitary Sewer	5 Electrical	5 Kathy Diehl
6 Office Building	6 Underground Utilities	6 Landscape	6 Roy Hall
7 Industrial Building	7 Building A	7 Project Engineering	7 Jim Daniel
8 Unassigned	8 Building B	8 Unassigned	8 Unassigned
9 Unassigned	9 Landscape	9 Unassigned	9 Unassigned

WBS   OBS

Figure A-6   Coding system for EPC service facility project.

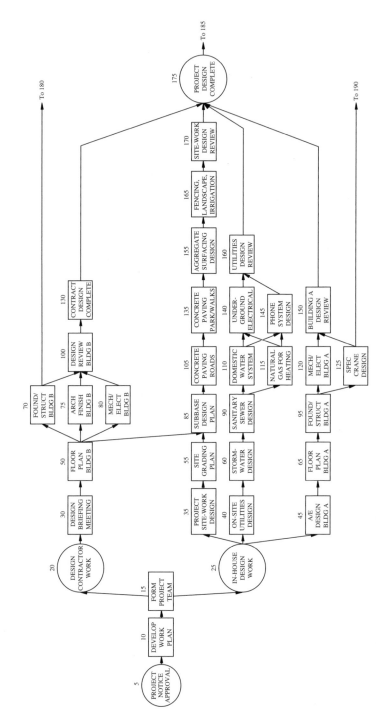

Figure A-7a CPM diagram for EPC service facility project.

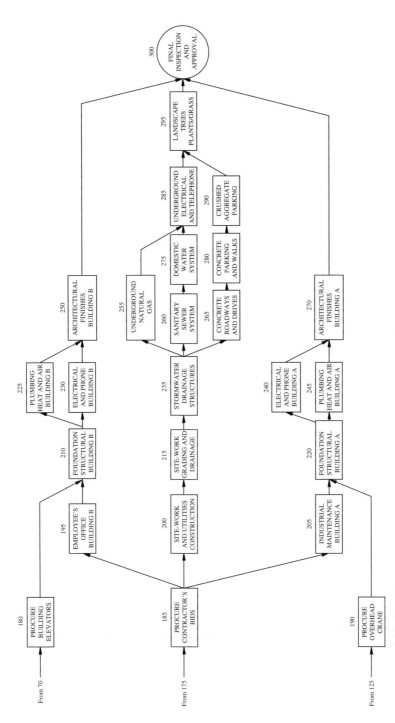

Figure A-7b (*Continued*)

```
                  ****************************
                  **    Input Details    **
                  ****************************

              PROJECT NAME:  SERVICE MAINTENANCE FACILITY
ACTIVITY                                                           ASSIGNED
NUMBER   CODE    DESCRIPTION                     DURATION    COST    START
---------------------------------------------------------------------------

    5      71    PROJECT NOTICE APPROVAL             3     $    500.
   10      71    DEVELOP WORK PLAN                   7     $  12000.
   15      71    FORM PROJECT TEAM                   5     $    850.
   20    2872    DESIGN CONTRACTOR'S WORK            2     $   3000.
   25    1073    IN-HOUSE DESIGN WORK                3     $   1500.
   30    2872    DESIGN BRIEFING MEETING             1     $   1200.
   35    1073    PROJECT SITE-WORK DESIGN            1     $   1400.
   40    1624    ON-SITE UTILITIES DESIGN            1     $   1200.
   45    1715    A/E DESIGN BUILDING A               1     $   1500.
   50    2812    FLOOR PLAN BUILDING B              10     $   9900.
   55    1123    SITE GRADING PLAN                  12     $  14000.
   60    1424    STORM-WATER DESIGN                 10     $   2000.
   65    1715    FLOOR PLAN BUILDING A              15     $  26000.
   70    2832    FOUND/STRUCT BUILDING B            45     $  31200.
   75    2812    ARCH FINISHES BUILDING B           30     $  49500.
   80    2842    MECH/ELECT BUILDING B              45     $  37300.
   85    1123    SUBBASE DESIGN PLAN                 5     $   4000.
   90    1524    SANITARY SEWER DESIGN              10     $  12000.
   95    1735    FOUND/STRUCT BUILDING A            30     $  92700.
  100    2871    DESIGN REVIEW BUILDING B           10     $   8000.
  105    1223    CONCRETE PAVING ROADS              20     $  12000.
  110    1624    DOMESTIC WATER SYSTEM               7     $   9000.
  115    1624    NATURAL GAS SYSTEM                  8     $   6000.
  120    1745    MECH/ELECT BUILDING A              30     $  22200.
  125    1735    SPECIAL OVERHEAD CRANE             11     $  10800.
  130    2872    CONTRACT DESIGN COMPLETE            1     $   1000.
  135    1223    CONTRACT PAVING PARKING/WALKS      10     $   7000.
  140    1654    UNDERGROUND ELECTRICAL             14     $  12000.
  145    1654    UNDERGROUND TELEPHONE SYSTEM        4     $   3000.
  150    1771    BUILDING A DESIGN REVIEW            3     $   5000.
  155    1323    AGGREGATE SURFACING DESIGN          8     $   6000.
  160    1677    UTILITIES DESIGN REVIEW             1     $   1100.
  165    1963    FENCING/LANDSCAPE/IRRIGATION       14     $  28000.
  170    1071    SITE-WORK DESIGN REVIEW             5     $   7000.
  175      71    PROJECT DESIGN COMPLETE             1     $   1000.
  180    3876    PROCURE BUILDING ELEVATORS         25     $  95000.
  185    4076    PROCURE CONTRACTOR'S BIDS          20     $   7000.
  190    3776    PROCURE OVERHEAD CRANE             40     $  55000.
  195    6887    EMPLOYEES' OFFICE BUILDING B        3     $   1000.
  200    5087    SITE-WORK/UTILITIES CONSTRUCTION    4     $   1500.
  205    7787    INDUSTRIAL/MAINTENANCE BLDG A       2     $   1400.
  210    6882    FOUND/STRUCT BUILDING B            45     $ 195000.
  215    5083    SITE-WORK/GRADING/DRAINAGE         18     $  85000.
  220    7785    FOUND/STRUCT BUILDING A           110     $ 390000.
  225    6882    PLUMBING, HEAT, & AIR BLDG B       75     $ 285000.
  230    6882    ELECTRICAL/PHONE BUILDING B        60     $ 215000.
  235    5484    STORMWATER/DRAINAGE STRUCTURES     15     $  22000.
  240    7785    ELECTRICAL/PHONE BUILDING A        65     $ 167000.
  245    7785    PLUMBING, HEAT & AIR BLDG A        85     $ 192000.
  250    6882    ARCH FINISHES BUILDING B           50     $ 260000.
  255    5684    UNDERGROUND NATURAL GAS             5     $  10500.
  260    5584    SANITARY SEWER SYSTEM              21     $  33200.
  265    5283    CONCRETE PAVING ROADS & DRIVES     60     $ 185000.
  270    7785    ARCH FINISHES BUILDING A           30     $ 175000.
  275    5684    DOMESTIC WATER SYSTEM               7     $  13200.
  280    5683    CONCRETE PARKING & WALKWAYS        15     $  35000.
  285    5684    UNDERGROUND ELECT & PHONE          14     $  47000.
  290    5383    CRUSHED AGGREGATE PARKING          40     $  76000.
  295    5983    LANDSCAPE TREES/PLANTS/GRASS       20     $  62000.
  300    9977    FINAL INSPECTION & APPROVAL         3     $   3500.
```

Figure A-8a Computer input data for generation of project reports that are shown on the following pages in this Appendix.

```
           SEQUENCE ORDER
         ********************
         I NODE      JNODE
         --------   --------
             5         10
            10         15
            15         20
            15         25
            20         30
            25         35
            25         40
            25         45
            30         50
            35         55
            40         60
            45         65
            50         70
            50         75
            50         80
            50         85
            55         85
            60         90
            65         95
            70        100
            70        180
            75        100
            80        100
            85        105
            90        110
            90        115
            95        120
            95        125
           100        130
           105        135
           110        140
           110        145
           115        140
           115        145
           120        150
           125        150
           125        190
           130        175
           135        155
           140        160
           145        160
           150        175
           155        165
           160        175
           165        170
           170        175
           175        185
           180        210
           185        195
           185        200
           185        205
           190        220
           195        210
           200        215
           205        220
           210        225
           210        230
           215        235
           220        240
           220        245
           225        250
           230        250
           235        255
           235        260
           235        265
           240        270
           245        270
           250        300
           255        285
           260        275
           265        280
           270        300
           275        285
           280        290
           285        295
           290        295
           295        300
```

```
     PROJECT START DATE :  5/26/2016
 **  NO HOLIDAYS REPORTED FOR THE PROJECT  **
```

Figure A-8b *(Continued)*

```
                        ******************************
                        **   ACTIVITY  SCHEDULE   **
                        ******************************

PROJECT: SERVICE MAINTENANCE FACILITY
SCHEDULE FOR ALL ACTIVITIES - ISSUED TO TIM JONES ON 4/20/2016                    ** PAGE  1 **
                                                                                  ACTIVITY SCHEDULE
```

C	ACTIVITY NUMBER	ACTIVITY DESCRIPTION	DURA-TION	EARLY START	EARLY FINISH	LATE START	LATE FINISH	TOTAL FLOAT	FREE FLOAT
C	5	PROJECT NOTICE APPROVAL	3	26MAY2016 / 1	30MAY2016 / 3	26MAY2016 / 1	30MAY2016 / 3	0	0
C	10	DEVELOP WORK PLAN	7	31MAY2016 / 4	8JUN2016 / 10	31MAY2016 / 4	8JUN2016 / 10	0	0
C	15	FORM PROJECT TEAM	5	9JUN2016 / 11	15JUN2016 / 15	9JUN2016 / 11	15JUN2016 / 15	0	0
	20	DESIGN CONTRACTOR'S WORK	2	16JUN2016 / 16	17JUN2016 / 17	27JUN2016 / 23	28JUN2016 / 24	7	0
C	25	IN-HOUSE DESIGN WORK	3	16JUN2016 / 16	20APR2016 / 18	16JUN2016 / 16	20JUN2016 / 18	0	0
	30	DESIGN BRIEFING MEETING	1	20JUN2016 / 18	20JUN2016 / 18	29JUN2016 / 25	29JUN2016 / 25	7	0
	35	PROJECT SITE-WORK DESIGN	1	21JUN2016 / 19	21JUN2016 / 19	27JUN2016 / 23	27JUN2016 / 23	4	0
	40	ON-SITE UTILITIES DESIGN	1	21JUN2016 / 19	21JUN2016 / 19	9AUG2016 / 54	9AUG2016 / 54	35	0
C	45	A/E DESIGN BUILDING A	1	21JUN2016 / 19	21JUN2016 / 19	21JUN2016 / 19	21JUN2016 / 19	0	0
	50	FLOOR PLAN BUILDING B	10	21JUN2016 / 19	4JUL2016 / 28	30JUN2016 / 26	13JUL2016 / 35	7	3
	55	SITE GRADING PLAN	12	22JUN2016 / 20	7JUL2016 / 31	28JUN2016 / 24	13JUL2016 / 35	4	0
	60	STORM-WATER DESIGN	10	22JUN2016 / 20	5JUL2016 / 29	10AUG2016 / 55	23AUG2016 / 64	35	0
C	65	FLOOR PLAN BUILDING A	15	22JUN2016 / 20	12JUL2016 / 34	22JUN2016 / 20	12JUL2016 / 34	0	0
	70	FOUND/STRUCT BUILDING B	45	5JUL2016 / 29	5SEP2016 / 73	22JUL2016 / 42	22SEP2016 / 86	13	0
	75	ARCH FINISHES BUILDING B	30	5JUL2016 / 29	15AUG2016 / 58	12AUG2016 / 57	22SEP2016 / 86	28	15
	80	MECH/ELECT BUILDING B	45	5JUL2016 / 29	5SEP2016 / 73	22JUL2016 / 42	22SEP2016 / 86	13	0
	90	SANITARY SEWER DESIGN	10	6JUL2016 / 30	19JUL2016 / 39	24AUG2016 / 65	6SEP2016 / 74	35	0
	85	SUBBASE DESIGN PLAN	5	8JUL2016 / 32	14JUL2016 / 36	14JUL2016 / 36	20JUL2016 / 40	4	0
C	95	FOUND/STRUCT BUILDING A	30	13JUL2016 / 35	23AUG2016 / 64	13JUL2016 / 35	23AUG2016 / 64	0	0
	105	CONCRETE PAVING ROADS	20	15JUL2016 / 37	11JUL2016 / 56	21JUL2016 / 41	17AUG2016 / 60	4	0
	115	NATURAL GAS SYSTEM	8	20JUL2016 / 40	29JUL2016 / 47	7SEP2016 / 75	16SEP2016 / 82	35	0
	110	DOMESTIC WATER SYSTEM	7	20JUL2016 / 40	28JUL2016 / 46	8SEP2016 / 76	16SEP2016 / 82	36	1
	145	UNDERGROUND TELEPHONE SYSTEM	4	1AUG2016 / 48	4AUG2016 / 51	3OCT2016 / 93	6OCT2016 / 96	45	10
	140	UNDERGROUND ELECTRICAL	14	1AUG2016 / 48	18AUG2016 / 61	19SEP2016 / 83	6OCT2016 / 96	35	0
	135	CONCRETE PAVING PARKING/WALKS	10	12AUG2016 / 57	25AUG2016 / 66	18AUG2016 / 61	31AUG2016 / 70	4	0
	160	UTILITIES DESIGN REVIEW	1	19AUG2016 / 62	19AUG2016 / 62	7OCT2016 / 97	7OCT2016 / 97	35	35
C	120	MECH/ELECT BUILDING A	30	24AUG2016 / 65	4OCT2016 / 94	24AUG2016 / 65	4OCT2016 / 94	0	0
	125	SPECIAL OVERHEAD CRANE	11	24AUG2016 / 65	7SEP2016 / 75	31AUG2016 / 70	14SEP2016 / 80	5	19
	155	AGGREGATE SURFACING DESIGN	8	26AUG2016 / 67	6SEP2016 / 74	1SEP2016 / 71	12SEP2016 / 78	4	0
	100	DESIGN REVIEW BUILDING B	10	6SEP2016 / 74	19SEP2016 / 83	23SEP2016 / 87	6OCT2016 / 96	13	0

Figure A-9 Schedule of all activities for EPC of service facility project.

```
            *****************************
            **   ACTIVITY  SCHEDULE   **
            *****************************
```

PROJECT: SERVICE MAINTENANCE FACILITY
SCHEDULE FOR ALL ACTIVITIES - ISSUED TO TIM JONES ON 4/20/2016

** PAGE 2 **
ACTIVITY SCHEDULE

	ACTIVITY NUMBER	ACTIVITY DESCRIPTION	DURA-TION	EARLY START	EARLY FINISH	LATE START	LATE FINISH	TOTAL FLOAT	FREE FLOAT
	180	PROCURE BUILDING ELEVATORS	25	6SEP2016 74	10SEP2016 98	22DEC2016 151	25JAN2017 175	77	23
	165	FENCING/LANDSCAPE/IRRIGATION	14	7SEP2016 75	26SEP2016 88	13SEP2016 79	30SEP2016 92	4	0
	190	PROCURE OVERHEAD CRANE	40	8SEP2016 76	2NOV2016 115	15SEP2016 81	9NOV2016 120	5	5
	130	CONTRACT DESIGN COMPLETE	1	20SEP2016 84	20SEP2016 84	7OCT2016 97	7OCT2016 97	13	13
	170	SITE-WORK DESIGN REVIEW	5	27SEP2016 89	30CT2016 93	30CT2016 93	7OCT2016 97	4	4
C	150	BUILDING A DESIGN REVIEW	3	5OCT2016 95	7OCT2016 97	5OCT2016 95	7OCT2016 97	0	0
C	175	PROJECT DESIGN COMPLETE	1	10OCT2016 98	10OCT2016 98	10OCT2016 98	10OCT2016 98	0	0
C	185	PROCURE CONTRACTORS' BIDS	20	11OCT2016 99	7NOV2016 118	11OCT2016 99	7NOV2016 118	0	0
	195	EMPLOYEE'S OFFICE BUILDING B	3	8NOV2016 119	10NOV2016 121	23JAN2017 173	25JAN2017 175	54	0
	200	SITE-WORK/UTILITIES CONSTRUCTION	4	8NOV2016 119	11NOV2016 122	24JAN2017 174	27JAN2017 177	55	0
C	205	INDUSTRIAL/MAINTENANCE BLDG A	2	8NOV2016 119	9NOV2016 120	8NOV2016 119	9NOV2016 120	0	0
C	220	FOUND/STRUCT BUILDING A	100	10NOV2016 121	12APR2017 230	10NOV2016 121	12APR2017 230	0	0
	210	FOUND/STRUCT BUILDING B	45	11NOV2016 122	12JAN2017 166	26JAN2017 176	29MAR2017 220	54	0
	215	SITE-WORK/GRADING/DRAINAGE	18	14NOV2016 123	7DEC2016 140	30JAN2017 178	22FEB2017 195	55	0
	235	STORMWATER/DRAINAGE STRUCTURES	15	8DEC2016 141	28DEC2016 155	23FEB2017 196	15MAR2017 210	55	0
	255	UNDERGROUND NATURAL GAS	5	29DEC2016 156	4JAN2017 160	28JUL2017 307	3AUG2017 311	151	23
	260	SANITARY SEWER SYSTEM	21	29DEC2016 156	26JAN2017 176	27JUN2017 284	25JUL2017 304	128	0
	265	CONCRETE PAVING ROADS & DRIVES	60	29DEC2016 156	22MAR2017 215	16MAR2017 211	7JUN2017 270	55	0
	225	PLUMBING, HEAT, & AIR BLDG B	75	13JAN2017 167	27APR2017 241	30MAR2017 221	12JUL2017 295	54	0
	230	ELECTRICAL/PHONE BUILDING B	60	13JAN2017 167	6APR2017 226	20APR2017 236	12JUL2017 295	69	15
	275	DOMESTIC WATER SYSTEM	7	27JAN2017 177	6FEB2017 183	26JUL2017 305	3AUG2017 311	128	0
	285	UNDERGROUND ELECT & PHONE	14	7FEB2017 184	24FEB2017 197	4AUG2017 312	23AUG2017 325	128	73
	280	CONCRETE PARKING AND WALKWAYS	15	23MAR2017 216	12APR2017 230	8JUN2017 271	28JUN2017 285	55	0
	240	ELECTRICAL/PHONE BUILDING A	65	13APR2017 231	12JUL2017 295	11MAY2017 251	9AUG2017 315	20	20
C	245	PLUMBING, HEAT, & AIR BLDG A	85	13APR2017 231	9AUG2017 315	13APR2017 231	9AUG2017 315	0	0
	290	CRUSHED AGGREGATE PARKING	40	13APR2017 231	7JUN2017 270	29JUN2017 286	23AUG2017 325	55	0
	250	ARCH FINISHES BUILDING B	50	28APR2017 242	6JUL2017 291	13JUL2017 296	20SEP2017 345	54	54
	295	LANDSCAPE TREES/PLANTS/GRASS	20	8JUN2017 271	2JUL2017 290	24AUG2017 326	20SEP2017 345	55	55
C	270	ARCH FINISHES BUILDING A	30	10AUG2017 316	20SEP2017 345	10AUG2017 316	20SEP2017 345	0	0
C	300	FINAL INSPECTION & APPROVAL	3	21SEP2017 346	25SEP2017 348	21SEP2017 346	25SEP2017 348	0	0

```
************************************** END OF SCHEDULE ***************************************
```

Figure A-9 *(Continued)*

```
                              *******************************
                              **  MONTHLY COST SCHEDULE  **
                              *******************************
```

```
PROJECT: SERVICE MAINTENANCE FACILITY
SCHEDULE FOR ALL ACTIVITIES - ISSUED TO TIM JONES ON 4/15/2016                    MONTHLY COST SCHEDULE
START DATE: 26 MAY 2003   FINISH DATE: 22 SEP 2004                                 - For all activities -
```

NO.	MONTH YEAR	EARLY START COST/MON	EARLY START CUMULATIVE COST	LATE START COST/MON	LATE START CUMULATIVE COST	TARGET SCHEDULE COST/MON	TARGET SCHEDULE CUMULATIVE COST	%TIME	%COST
1	MAY 2016	$ 3929.	$ 3929.	$ 3929.	$ 3929.	$ 3929.	$ 3929.	1.4%	.1%
2	JUN 2016	$ 48841.	$ 52770.	$ 34645.	$ 38573.	$ 41743.	$ 45672.	7.5%	1.5%
3	JUL 2016	$ 177261.	$ 230031.	$ 101204.	$ 139778.	$ 139233.	$ 184904.	14.1%	6.1%
4	AUG 2016	$ 130583.	$ 360614.	$ 130838.	$ 270616.	$ 130711.	$ 315615.	20.1%	10.3%
5	SEP 2016	$ 166931.	$ 527545.	$ 160525.	$ 431141.	$ 163728.	$ 479343.	26.4%	15.7%
6	OCT 2016	$ 69255.	$ 596800.	$ 63784.	$ 494925.	$ 66519.	$ 545863.	33.0%	17.9%
7	NOV 2016	$ 180187.	$ 776987.	$ 62507.	$ 557432.	$ 121347.	$ 667210.	38.8%	21.9%
8	DEC 2016	$ 247116.	$ 1024103.	$ 111945.	$ 669377.	$ 179531.	$ 846740.	45.4%	27.7%
9	JAN 2017	$ 324067.	$ 1348170.	$ 180933.	$ 850311.	$ 252500.	$ 1099240.	51.7%	36.0%
10	FEB 2017	$ 332900.	$ 1681070.	$ 235742.	$ 1086053.	$ 284321.	$ 1383561.	57.5%	45.3%
11	MAR 2017	$ 316279.	$ 1997348.	$ 234362.	$ 1320415.	$ 275320.	$ 1658882.	64.1%	54.3%
12	APR 2017	$ 242022.	$ 2239371.	$ 245967.	$ 1566382.	$ 243995.	$ 1902877.	70.4%	62.3%
13	MAY 2017	$ 250489.	$ 2489860.	$ 308343.	$ 1874725.	$ 279416.	$ 2182293.	76.4%	71.5%
14	JUN 2017	$ 284017.	$ 2773877.	$ 329589.	$ 2204314.	$ 306803.	$ 2489096.	82.8%	81.5%
15	JUL 2017	$ 89479.	$ 2863356.	$ 322710.	$ 2527024.	$ 206094.	$ 2695190.	89.1%	88.3%
16	AUG 2017	$ 110461.	$ 2973817.	$ 338893.	$ 2865917.	$ 224677.	$ 2919867.	95.4%	95.6%
17	SEP 2017	$ 79333.	$ 3053150.	$ 187233.	$ 3053150.	$ 133283.	$ 3053150.	100.0%	100.0%

```
**************************************************  END OF MONTHLY COST SCHEDULE  **************************************************
```

Figure A-10 Monthly distribution of costs for all activities of EPC service facility project.

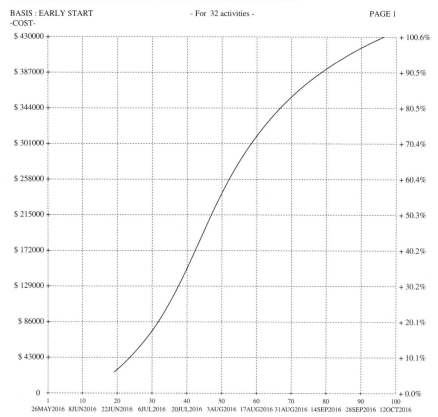

Figure A-11 S-curve for design activities only (sort by code digit 1 greater than 0 and less than 3).

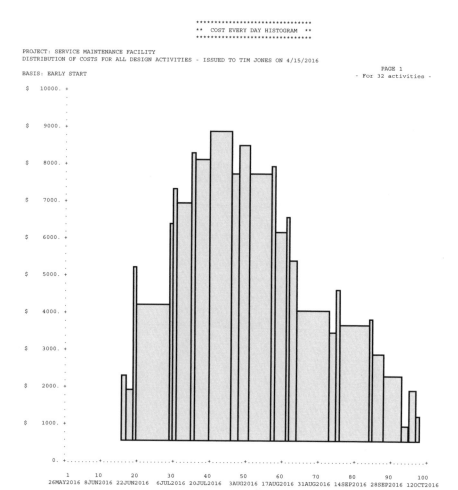

Figure A-12 Daily distribution of costs for design work only (sort by code digit 1 greater than 0 and less than 3).

```
                                       ********************
                                       **   BAR  CHART   **
                                       ********************
                                                                                          ** PAGE 1 **
PROJECT: SERVICE MAINTENANCE FACILITY                                                        BAR CHART
SCHEDULE FOR ALL WORK OF TEAM MEMBER RON THOMAS - ISSUED 4/20/2016

                                                1        10        20         30        40         50        60        70        80        90
ACTIVITY  DESCRIPTION               DURATION  26MAY2016 8JUN2016 22JUN2016  6JUL2016 20JUL2016  3AUG2016 17AUG2016 31Aug2016 14SEP2016 28SEP2016
                                                +.........+.........+.........+.........+.........+.........+.........+.........+.........+
C 25    IN-HOUSE DESIGN WORK           3                         XXX
                                                .    .    .    .    .    .    .    .    .    .    .    .    .    .    .    .    .    .    .
  35    PROJECT SITE-WORK DESIGN       1                            X
                                                .    .    .    .    .    .    .    .    .    .    .    .    .    .    .    .    .    .    .
  55    SITE GRADING PLAN             12                             XXXXXXXXXXXX
                                                .    .    .    .    .    .    .    .    .    .    .    .    .    .    .    .    .    .    .
  85    SUBBASE DESIGN PLAN            5                                         XXXXX
                                                .    .    .    .    .    .    .    .    .    .    .    .    .    .    .    .    .    .    .
 105    CONCRETE PAVING ROADS         20                                              XXXXXXXXXXXXXXXXXXXX
                                                .    .    .    .    .    .    .    .    .    .    .    .    .    .    .    .    .    .    .
 135    CONCRETE PAVING PARKING/WALKS 10                                                                  XXXXXXXXXX
                                                .    .    .    .    .    .    .    .    .    .    .    .    .    .    .    .    .    .    .
 155    AGGREGATE SURFACING DESIGN     8                                                                            XXXXXXXX
                                                .    .    .    .    .    .    .    .    .    .    .    .    .    .    .    .    .    .    .
 165    FENCING/LANDSCAPE/IRRIGATION  14                                                                                    XXXXXXXXXXXXXX
                                                .    .    .    .    .    .    .    .    .    .    .    .    .    .    .    .    .    .    .
                                                +.........+.........+.........+.........+.........+.........+.........+.........+.........+
                              WORK DAYS         1        10        20         30        40         50        60        70        80        90
                         CALENDAR DATES   26MAY2016 8JUN2016 22JUN2016  6JUL2016 20JUL2016  3AUG2016 17AUG2016 31AUG2016 14SEP2016 28SEP2016
```

Figure A-13 Bar chart for all work of team member Ron Thomas (sort by code digit 4 equal to 3).

```
*****************************
**   ACTIVITY  SCHEDULE   **
*****************************
```

PROJECT: SERVICE MAINTENANCE FACILITY
SCHEDULE FOR ALL ACTIVITIES RELATED TO BUILDING A - ISSUED TO KATHY DIEHL ON 4/20/2016   ** PAGE 1 **
                                                                                                ACTIVITY SCHEDULE
```
*************************************************************************************************************
```

	ACTIVITY NUMBER	ACTIVITY DESCRIPTION	DURA-TION	EARLY START	EARLY FINISH	LATE START	LATE FINISH	TOTAL FLOAT	FREE FLOAT
C	45	A/E DESIGN BUILDING A	1	21JUN2016 19	21JUN2016 19	21JUN2016 19	21JUN2016 19	0	0
C	65	FLOOR PLAN BUILDING A	15	22JUN2016 20	12JUL2016 34	22JUN2016 20	12JUL2016 34	0	0
C	95	FOUND/STRUCT BUILDING A	30	13JUL2016 35	23AUG2016 64	13JUL2016 35	23AUG2016 64	0	0
C	120	MECH/ELECT BUILDING A	30	24AUG2016 65	4OCT2016 94	24AUG2016 65	4OCT2016 94	0	0
	125	SPECIAL OVERHEAD CRANE	11	24AUG2016 65	7SEP2016 75	31AUG2016 70	14SEP2016 80	5	19
	190	PROCURE OVERHEAD CRANE	40	8SEP2016 76	2NOV2016 115	15SEP2016 81	9NOV2016 120	5	5
C	150	BUILDING A DESIGN REVIEW	3	5OCT2016 95	7OCT2016 97	5OCT2016 95	7OCT2016 97	0	0
C	205	INDUSTRIAL/MAINTENANCE BLDG A	2	8NOV2016 119	9NOV2016 120	8NOV2016 119	9NOV2016 120	0	0
C	220	FOUN/STRUCT BUILDING A	100	10NOV2016 121	12APR2017 230	10NOV2016 121	12APR2017 230	0	0
	240	ELECTRICAL/PHONE BUILDING A	65	13APR2017 231	12JUL2017 295	11MAY2017 251	9AUG2017 315	20	20
C	245	PLUMBING, HEAT & AIR BLDG A	85	13APR2017 231	9AUG2017 315	13APR2017 231	9AUG2017 315	0	0
	270	ARCH FINISHES BUILDING A	30	10AUG2017 316	20SEP2017 345	10AUG2017 316	20SEP2017 345	0	0

```
*********************************************   END OF SCHEDULE   *****************************************
```

Figure A-14   Schedule for all activities related to Building A (sort by code digit 2 equal to 7).

Index